Green Materials from Plant Oils

RSC Green Chemistry

Editor-in-Chief:
Professor James Clark, *Department of Chemistry, University of York, UK*

Series Editors:
Professor George A Kraus, *Department of Chemistry, Iowa State University, Ames, Iowa, USA*
Professor dr. ir. Andrzej Stankiewicz, *Delft University of Technology, The Netherlands*
Professor Peter Siedl, *Federal University of Rio de Janeiro, Brazil*
Professor Yuan Kou, *Peking University, China*

How to obtain future titles on publication:
A standing order plan is available for this series. A standing order will bring delivery of each new volume immediately on publication.

For further information please contact:
Book Sales Department, Royal Society of Chemistry, Thomas Graham House, Science Park, Milton Road, Cambridge, CB4 0WF, UK
Telephone: +44 (0)1223 420066, Fax: +44 (0)1223 420247
Email: booksales@rsc.org
Visit our website at www.rsc.org/books

Green Materials from Plant Oils

Edited by

Zengshe Liu
NCAUR, ARS/USDA, Peoria, Illinois, USA
Email: kevin.liu@ars.usda.gov

George Kraus
Iowa State University, Ames, Iowa, USA
Email: gakraus@iastate.edu

THE QUEEN'S AWARDS
FOR ENTERPRISE:
INTERNATIONAL TRADE
2013

RSC Green Chemistry No. 29

Print ISBN: 978-1-84973-901-6
PDF eISBN: 978-1-78262-185-0
ISSN: 1757-7039

A catalogue record for this book is available from the British Library

Published by The Royal Society of Chemistry,
Thomas Graham House, Science Park, Milton Road,
Cambridge, CB4 0WF, UK

Registered Charity Number 207890

For further information see our web site at www.rsc.org

Printed and bound by CPI Group (UK) Ltd, Croydon, CR0 4YY

Preface

The central theme of this book is the functionalization and utilization of plant oils to prepare sustainable materials. This book brings together for the first time researchers from one of the premier agricultural research laboratories in the USA with scientists from renowned university research centers around the world. This book was conceived and assembled by Dr Zengshe Liu, a research chemist in the Bio-Oils Research Unit of the National Center for Agricultural Utilization Research (NCAUR) located in Peoria, Illinois. Dr Liu has had a longstanding research interest in sustainable chemistry. He contributes a chapter on the polymerization of oils in carbon dioxide media.

This book contains contributions from several of his colleagues. A chapter from Dr Brent Tisserat describes the preparation of co-polymers *via* microwave heating. A chapter from Dr Grigor Bantchev outlines the development and use of thiol-based radical chemistry. Dr Rogers Harry-O'kuru describes the lubricity characteristics of modified plant seed oils. A chapter by Dr Kenneth Doll focuses on the production of cyclic carbonate materials from natural oils.

Several notable academic researchers have also contributed chapters. Drs Ruijun Gu and Mohini Sain from the University of Toronto, Canada outline the preparation of natural polyurethanes from plant oil polyols. Drs Brajendra Sharma and Sevim Erhan from the University of Illinois, with Dr Liu from NCAUR, have contributed a chapter summarizing the screening of modified seed oils as feedstocks for lubricants. A chapter from Drs Xuesong Jiang and Yanchang Gan from Shanghai Jiao Tong University, China reports new photo-cured monomers based on vegetable oils. Drs Fei Liu and Jin Zhu from the Ningbo Institute of Material Technology and Engineering, China describe plant-oil-based polymers. Drs Rongpeng Wang and Thomas Schuman from Missouri University of Science and Technology, USA have contributed a chapter on the polymers generated from epoxidized

RSC Green Chemistry No. 29
Green Materials from Plant Oils
Edited by Zengshe Liu and George Kraus
© The Royal Society of Chemistry 2015
Published by the Royal Society of Chemistry, www.rsc.org

vegetable oils. Drs Shailesh Shah and Shelby Thames from the University of Southern Mississippi, USA describe renewable materials for coating applications. A chapter by Drs Chengguo Liu and Yonghong Zhou from the Chinese Academy of Forestry covers green polymers from triglyceride oils.

George A. Kraus
Iowa State University, USA

Contents

RSC Green Chemistry No. 29
Green Materials from Plant Oils
Edited by Zengshe Liu and George Kraus
© The Royal Society of Chemistry 2015
Published by the Royal Society of Chemistry, www.rsc.org

CHAPTER 1

Photo-cured Materials from Vegetable Oils

YANCHANG GAN AND XUESONG JIANG*

School of Chemistry & Chemical Engineering, Shanghai Jiao Tong University, Shanghai 200240, People's Republic of China
*Email: ponygle@sjtu.edu.cn

1.1 Introduction

Photo-curing is one of the most effective processes for the rapid transformation of liquid multifunctional monomer resins to cross-linked polymer networks.[1–3] Photo-curing technology has found an increasing number of industrial applications over the past decade due to its unique benefits, for example solvent-free formulations, and high-speed and room-temperature processing.[4,5] Photo-curing technology has become attractive, especially in the paint, ink, adhesive, optical disk, photo-lithography and coating industries due to its very low consumption of energy and its low emissions of volatile organic compounds.[6,7]

The raw materials for photo-curing technology such as monomer resins, photo-initiators and functional additives, are usually produced from fossil oil. However, due to the growing demand for petroleum-based products and the resulting negative impact on the environment, there has been a growing interest in the utilization of renewable resources as an alternative to petroleum-based polymers.[8] The replacement of petroleum-based raw materials with renewable resources constitutes a major contemporary challenge in terms of both economical and environmental aspects.[9]

RSC Green Chemistry No. 29
Green Materials from Plant Oils
Edited by Zengshe Liu and George Kraus
© The Royal Society of Chemistry 2015
Published by the Royal Society of Chemistry, www.rsc.org

Vegetable oils are inexpensive, environmentally friendly, renewable, naturally raw materials with low toxicity and functional groups such as hydroxy, epoxy, carboxyl and C=C. Vegetable oils are extracted primarily from the seeds of oilseed plants. Their competitive cost, worldwide availability, and built-in functionality make them attractive. In recent years, there has been a growing trend in using vegetable oils as renewable resources, especially in oleochemical products. Several derivatives of vegetable oils are used as poly-merizable monomers in radiation-curable systems due to their environ-mentally friendly character and low cost.[10,11] For example, the long fatty acid chains of vegetable oils provide some brittle resin systems such as epoxy, urethane and polyester resins, which have good flexibility and toughness.[12] Vegetable oils exhibit many excellent properties which can be utilized in preparing valuable polymeric materials such as polyester amide, epoxy, polyurethane, alkyd polymers and have many applications in other areas.[13–15] Epoxidized vegetable oils, such as soybean oil and epoxidized palm oil have been used in UV-curable coating systems.[16,17] Vernonia oil contains epoxide groups, which means it can be utilized as a polymerizable monomer directly in cationic UV-cured coatings.[18] In this chapter, we summarize the photo-cured materials that can be obtained from vegetable oils including photo-curable monomers, photo-initiator systems and the photo-curing approach.

1.2   Photo-curable Monomers Derived from Vegetable Oils

Vegetable oils consist of mainly triglycerides formed between glycerol and various fatty acids, which have a three-armed star structure (Figure 1.1). Triglycerides are comprised of three fatty acids joined at a glycerol junction. Most of the common oils contain fatty acids that vary from 14 to 22 carbons in length, with 0 to 3 double bonds per fatty acid.[19] Table 1.1 summarizes the most common fatty acids present in vegetable oils.[9,19,20] As shown in Table 1.1, most fatty acids are long straight-chain compounds with an even number of carbons, and the double bonds in most of these unsaturated fatty acids possess a *cis* configuration.

Because the internal double bonds in the triglyceride structure are not sufficiently reactive for various polymerization processes, the vegetable oils

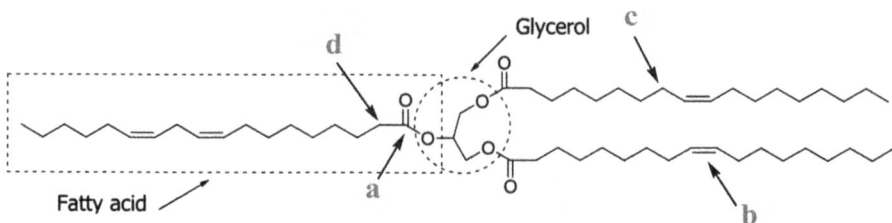

Figure 1.1   The triglyceride molecular structure of vegetable oils.[19]

Table 1.1 Formulae and structures of widely used fatty acids.[9,19,20]

Fatty acid	Formula	Structure
Caprylic	$C_8H_{16}O_2$	
Capric	$C_{10}H_{20}O_2$	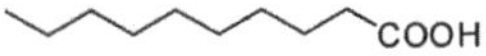
Lauric	$C_{12}H_{24}O_2$	
Myristic	$C_{14}H_{28}O_2$	
Palmitic	$C_{16}H_{32}O_2$	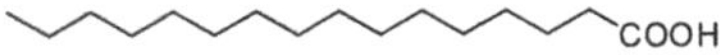
Palmitoleic	$C_{16}H_{30}O_2$	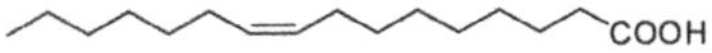
Stearic	$C_{18}H_{36}O_2$	
Oleic	$C_{18}H_{34}O_2$	
Linoleic	$C_{18}H_{32}O_2$	
Linolenic	$C_{18}H_{30}O_2$	
α-Eleostearic	$C_{18}H_{30}O_2$	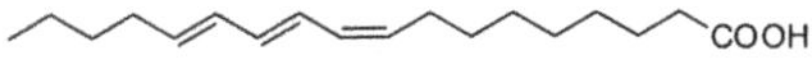
Ricinoleic	$C_{18}H_{34}O_3$	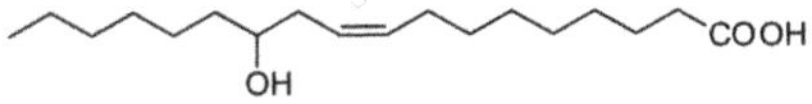
Vernolic	$C_{18}H_{32}O_3$	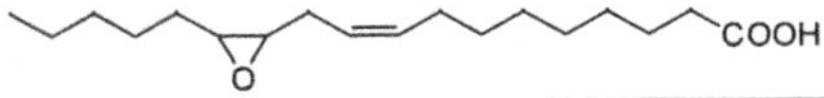

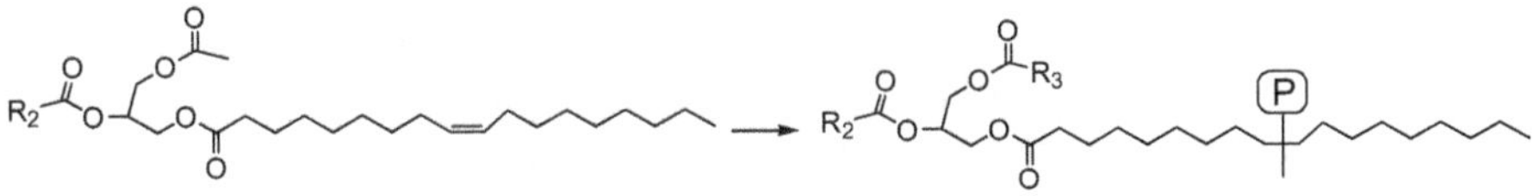

Figure 1.2 General modification pathway for the introduction of photo-polymerizable groups into the triglyceride molecule.[19] Here, 'P' means the photo-polymerizable groups.

must be modified with efficient photo-polymerizable groups such as acrylate or epoxy when being used as photo-curing monomers.

The triglycerides contain several reactive positions, as shown in Figure 1.1: ester groups (a); C=C double bonds (b); acrylic positions (c); and the α-position of ester groups (d) can act as the starting points in different reactions. The C=C double bond reactive positions are usually used as the starting points for introducing highly efficient reactive groups. The general modification pathway is shown in Figure 1.2.

1.2.1 Epoxy Monomers based on Vegetable Oils

Epoxidation is one of the most important functionalization reactions involving C=C double bonds, and epoxidized vegetable oils show excellent

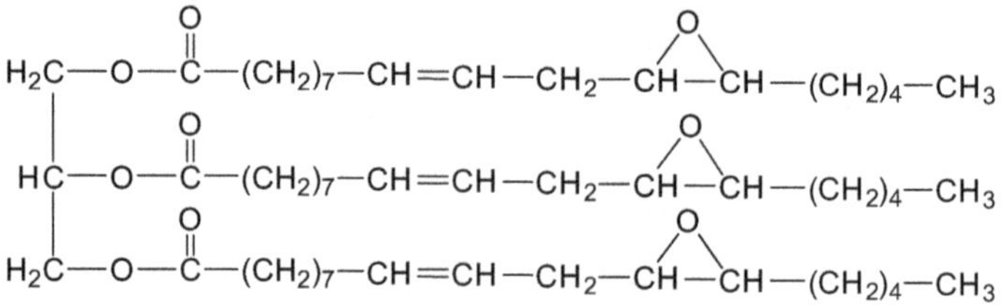

Figure 1.3   The structure of the triglyceride of vernolic acid.[23]

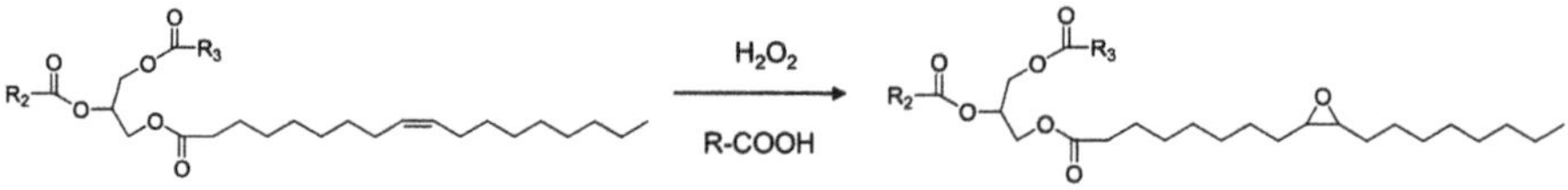

Figure 1.4   General conventional method for the introduction of epoxy groups into a triglyceride.[19]

promise as inexpensive, renewable monomers for photo-curing industrial applications.[21,22] Some raw vegetable oil such as vernonia oil already contain large numbers of epoxide groups.[23] The vernonia oil is extracted from the seeds of *Vernonia galamensis* with petroleum ether or hexane after the seeds are lipase-deactivated and coarse ground, and yields of up to 42% have been reported.[24,25] One unique characteristic of vernonia oil is that about 80% of the oil is the triglyceride of vernolic acid. The structure is shown in Figure 1.3. Vernonia oil can be utilized as a polymerizable monomer directly in cationic UV-curing coating due to its high content of epoxy groups.[23]

Most of the multifunctional epoxy monomers for photo-curing based on vegetable oils are prepared from the epoxidation of unsaturated fatty acids or triglycerides. The epoxidation of triglycerides or unsaturated fatty acids can be achieved in a straightforward fashion by reaction with molecular oxygen, hydrogen peroxide, or by chemo-enzymatic reactions.[26] The chemistry of the Prileshajev epoxidation of unsaturated fatty compounds is well known.[27] A short-chain peroxy acid, usually peracetic acid, is prepared from hydrogen peroxide (H_2O_2) and the corresponding acid either in a separate step or *in situ* (Figure 1.4).

This process is performed industrially on large scale, and more research is currently focusing on how to improve the conversion rate.[28] An epoxidation reaction of mahua oil using hydrogen peroxide was done by Goud *et al.* They used H_2O_2 as the oxygen donor and glacial acetic acid as the oxygen carrier in the presence of sulfuric acid (H_2SO_4) and nitric acid (HNO_3), and found that H_2SO_4 is the best inorganic catalyst for this system, producing a high conversion of double bonds to epoxide groups.[29] Dinda *et al.* worked on the epoxidation kinetics of cottonseed oil using H_2O_2 and liquid inorganic acids *i.e.* hydrochloric (HCl) and phosphoric (H_3PO_4) acids, and HNO_3 and H_2SO_4 as catalysts. They used carboxylic acid *i.e.* CH_3COOH and HCOOH as oxygen carriers, but they found that acetic acid is a more effective oxygen carrier than formic acid. Among all the liquid inorganic acid catalysts, H_2SO_4 was found to be most efficient and effective.[30]

Cai *et al.* also studied the kinetics of the *in situ* epoxidation of soybean oil, sunflower oil and corn oil by peroxyacetic acid using H_2SO_4 as the catalyst. They found that soybean oil had the highest conversion rate and the lowest activation energy for epoxidation using peroxyacetic acid,[31] and an 87.4% conversion rate for the epoxidation of jatropha oil.[32] Moreover, the epoxidation of soybean oil and the extent of side-reactions were studied using an ion-exchange resin as the catalyst. The results revealed that the reactions were first-order with respect to the double bond concentration and that side-reactions did not occur on a large scale.[8,33]

The catalytic epoxidation of methyl linoleate with different transition metal complexes as catalysts was studied by Woo's group. Complete epoxidation with aqueous H_2O_2 (30%) can be obtained within four hours for methyltrioxorhenium (4 mol%) and pyridine.[34] Gerbase's group proved the same catalyst could be successfully applied for the direct epoxidation of soybean oil in a bi-phasic system showing complete double bond conversion within two hours.[35] Moreover, enzymes have been widely studied for the epoxidation of plant oils and their derivatives. The reaction proceeds *via* the enzymatic *in situ* formation of the peracids required for the chemical epoxidation of the double bonds. The general advantage of this kind of epoxidation is that there are no undesired ring-opening reactions of the epoxides obtained.[36–41]

1.2.2 Acrylated Monomers based on Vegetable Oils

The double bonds present in vegetable oils can be transformed into acrylate groups through two steps, then the acrylated vegetable oils can be used as a binder in fast UV-curable coating mixtures. In a first modification step, the double bonds are converted to epoxide groups, then the epoxide groups are further converted to acrylate groups (Figure 1.5).[42,43] Vegetable oils like soybean, castor, lesquerella, palm and vernonia have been successfully converted to acrylates and methacrylates respectively.[42,44–46] As reported by Bajpai and co-workers, acrylated epoxidized soybean oil (AESO) has been prepared, using triethylamine and hydroquinone as a catalyst and gelling inhibitor, respectively.[10] Similar strategies have been applied for the synthesis of AESO, which, along with maleinized soy oil monoglyceride and maleinized hydroxylated oil, were used to prepare composite materials with glass fibers and natural flax and hemp fibers.[8,47]

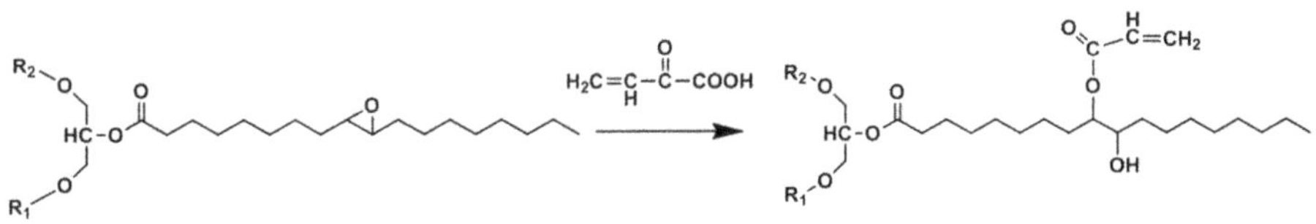

Figure 1.5   General process for the synthesis of acrylated oil from epoxidized oil.

In the report of Wuzella *et al.*,[48] an acrylated epoxidized linseed oil (AELO) was synthesized from epoxidized linseed oil (ELO) through the ring opening of the oxirane group using acrylic acid as the ring-opening agent (Figure 1.6).

Esterification of hydroxylated vegetable oils using acrylic acid or acryloyl chloride is another efficient method for the preparation of acrylated vegetable oils for UV-curing. The reaction is active and usually happens at low temperatures, which may minimize side-reactions such as the homopolymerization of acrylate monomers. The naturally occurring hydroxyl groups in castor oil are usually used to attach polymerizable acrylic moieties, by reacting castor oil with acryloyl chloride. The Applewhite[49] and Pelletier[50] groups reacted the hydroxyl groups of castor oil with acryloyl chloride to prepared acrylated castor oil (ACO). We also obtained ACO from castor oil using the same strategy (Figure 1.7).[51]

Epoxidized vegetable oils can be reacted with polyhydric alcohols to prepare vegetable-oil-based polyols. Cheong's group used a vegetable-oil-based polyol to prepare an acrylated polyol ester pre-polymer *via* the polycondensation esterification between polyol and acrylic acid, and thereafter produced a radiation-curable formulation from the pre-polymer (Figure 1.8).[52]

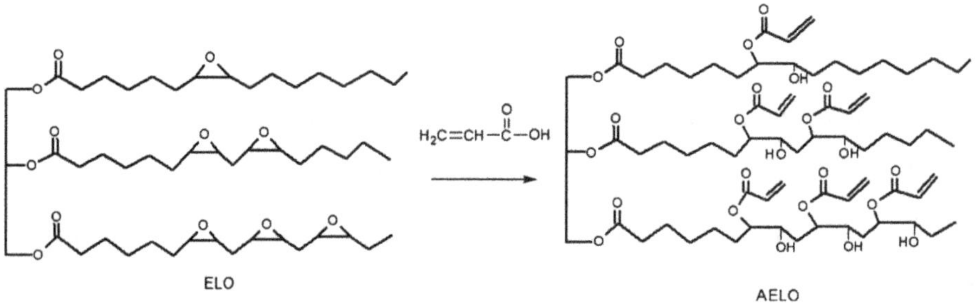

Figure 1.6 Synthesis of acrylated epoxidized linseed oil (AELO) using acrylic acid monomer from epoxidized linseed oil (ELO).[48]

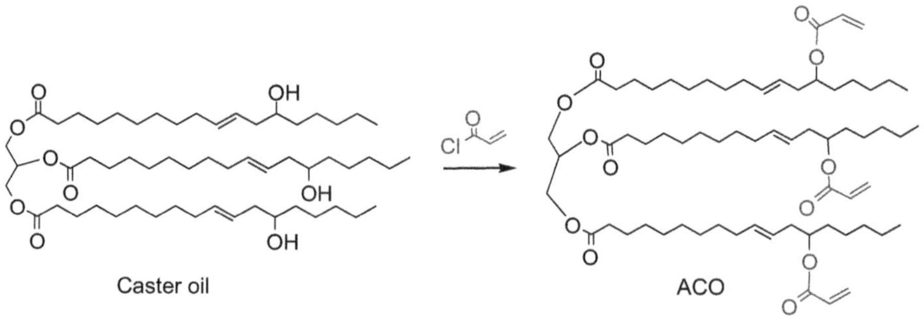

Figure 1.7 Synthesis of acrylated caster oil (ACO).[51]

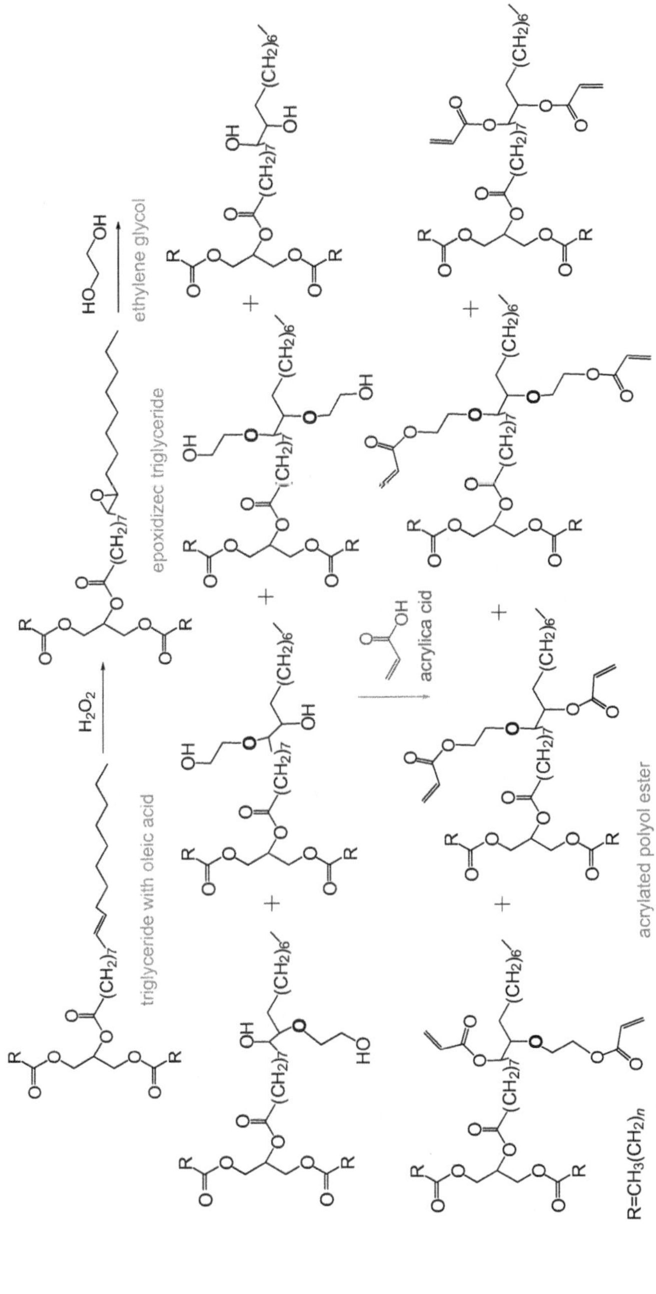

Figure 1.8 Reaction pathway of an acrylated polyol ester pre-polymer synthesis.[52]

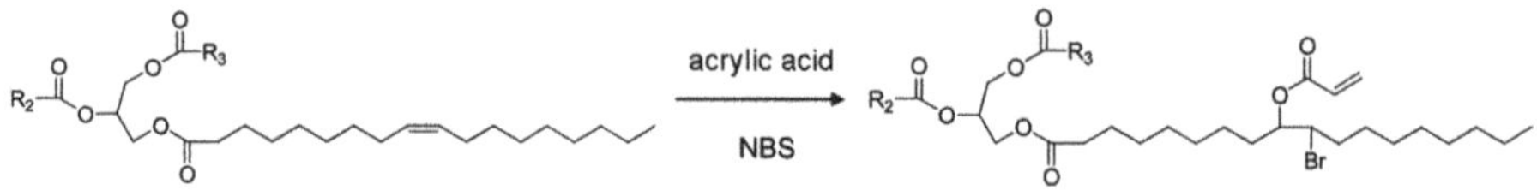

Figure 1.9 Preparation of acrylated vegetable oils in one step.[19,53]

Acrylate moieties have also been attached to triglyceride structures by the one-step addition of bromide and acrylate groups to a C=C bond. Soybean and sunflower oils have been modified in the presence of acrylic acid and *N*-bromosuccinimide (NBS, Figure 1.9).[19,53]

The Patel group[54] has synthesized a series of UV-curable polyurethane acrylate pre-polymer monomers by reacting polyols from sesame oil (edible) and using different ratio of polyols, aromatic isocyanate, and aliphatic isocyanate, 2-hydroxy ethyl methacrylate (HEMA) and dibutyltin dilaurate (DBTDL) as the catalyst. Polyols were prepared *via* the alcoholysis of trigly-ceride oil using a proprietary method, which was further reacted with toluene diisocyanate and isophorone diisocyanate in different ratios to develop a series of polyurethanes (Figure 1.10).

Homan *et al.* also employed an acrylate-bearing isocyanate group to produce acrylate castor oil.[55] Patel *et al.* modified monoacylglycerol (MAG) with the diisocyanate reagents methylene bis(4-phenylisocyanate) (MDI) and toluene diisocyanate (TDI). The free terminal isocyanate groups of MAG reacted with the acrylate monomer, which bears a free –OH group.[56] A final route for urethane-acrylated vegetable oils is the reaction of an acrylate bearing an isocyanate group and fatty chains, with a hyperbranched hydroxyl-terminated polyester.[57]

1.3 Photo-cured Materials from Vegetable Oil Monomers

Photo-curing formulations are usually comprised of multifunctional monomers and oligomers, with small amounts of a photo-initiator which generates reactive species (free radicals or ions) upon UV exposure. The overall process can be represented schematically as shown in Figure 1.11.[7] There are two major classes of UV-curable resins, and they differ in their polymerization mechanism of monomers *i.e.* acrylates or unsaturated polyesters (free radical polymerization) *vs.* photo-initiated cat-ionic polymerization of multifunctional epoxides and vinyl ethers (cationic polymerization).

Many researchers pay attention to the chemistry behind the photo-curing process, and especially the photo-curing kinetics of ultrafast reactions for both cationic-type and radical-type polymerization of the multifunctional monomers from vegetable oils. Polymer networks based on vegetable oils with different structures and tailor-made properties have been obtained by photo-curing formulations containing one or more type(s) of monomer.[58–61]

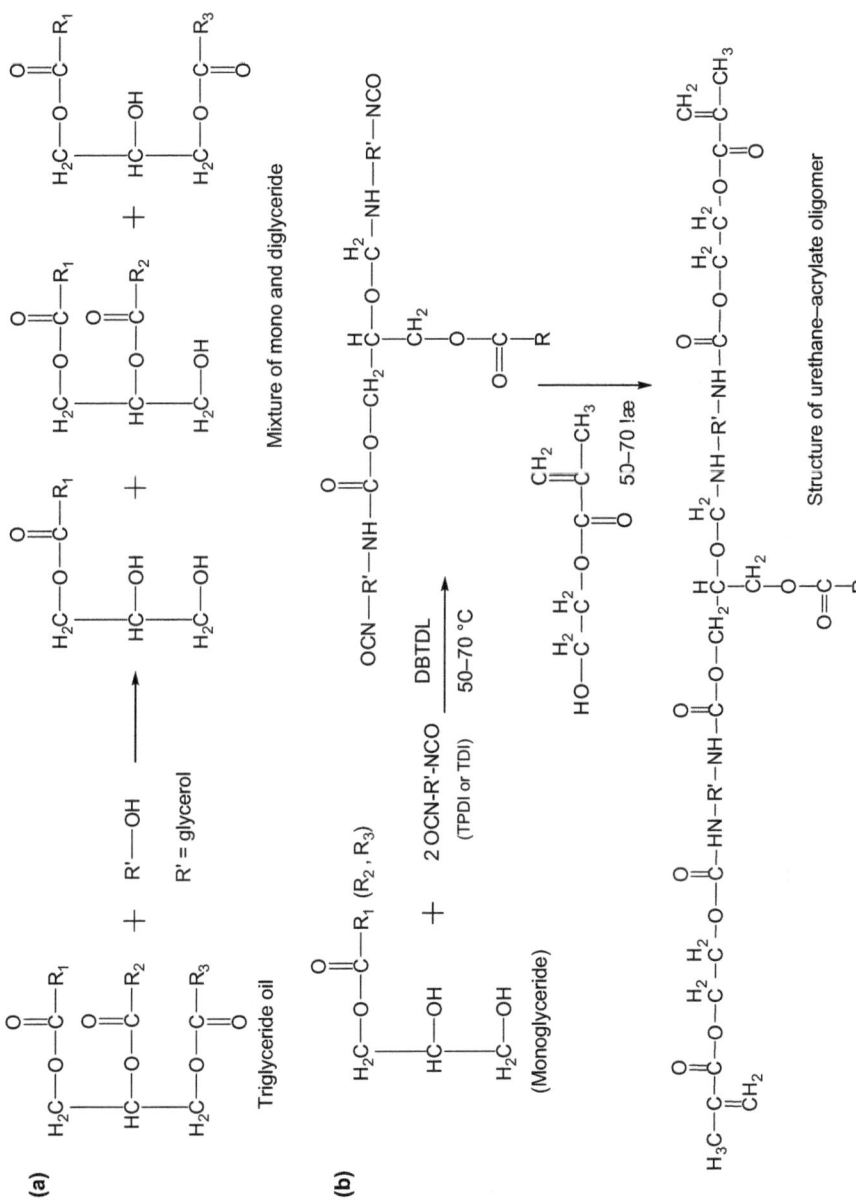

Figure 1.10 Preparation of a urethane-acrylate oligomer.[54]

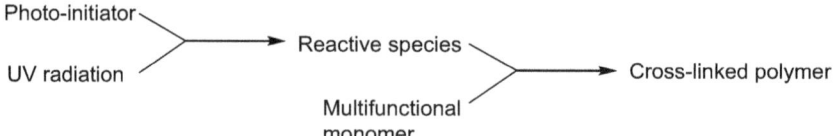

Figure 1.11 A typical process for the formation of cross-linked networks through UV curing.[7]

1.3.1 Photo-oxidation of Vegetable Oils for Direct Cross-linking

Drying oils are vegetable oils that are composed of mixtures of triglycerides. The high rate of unsaturation of these vegetable oils makes them sensitive to auto-oxidation under air. Drying oils are wildly applied as binders and film formers in paint and coating formulations because they can form polymer networks by auto-oxidation, peroxide formation and subsequent radical polymerization.[8,62–64]

Linseed oil is the most successful example of a drying oil, and the superior performance of linseed oil compared to other vegetable oils is mainly due to its faster drying.[65] Linseed oil is extensively used as a medium for paintings and elaborate linoleum, owing to its capacity to form a continuous thin layer easily, with good optical and mechanical properties within a reasonable time.[63]

The cross-linking mechanism (Figure 1.12)[9,62,69] was investigated in detail, and the formation of the lipidic network was attributed to the successive formation of radical species, isomerization, hydroperoxidation and cross-linking. The oxidation process, accelerated by UV irradiation using metal-based catalysts has also been studied.[63,66–69]

1.3.2 Photo-cured Polymer Networks based on Acrylated Vegetable Oils

Acrylated resins are the most widely used photo-curing systems, because of their high reactivity and the variety of available monomers and telechelic oligomers. A typical photo-curing formulation contains three basic components: (1) a photo-initiator which can generate free radicals by photolysis; (2) the acrylated functionalized oligomer which constitutes the backbone of the polymer network; and (3) the acrylated monomer which acts as a reactive diluent.[7] As shown in Figure 1.13, the photo-initiator plays a key role in the polymerization, and it governs both the rate of polymerization and the cure depth. The final degree of the polymerization and the physical and chemical properties of the photo-curing polymers are determined by the chemical structure and functionality of both the monomer and the oligomer.

Wuzella *et al.*[48] have studied the kinetic properties of acrylated epoxidized linseed oil monomers by UV-curing. They found that the photo-initiator

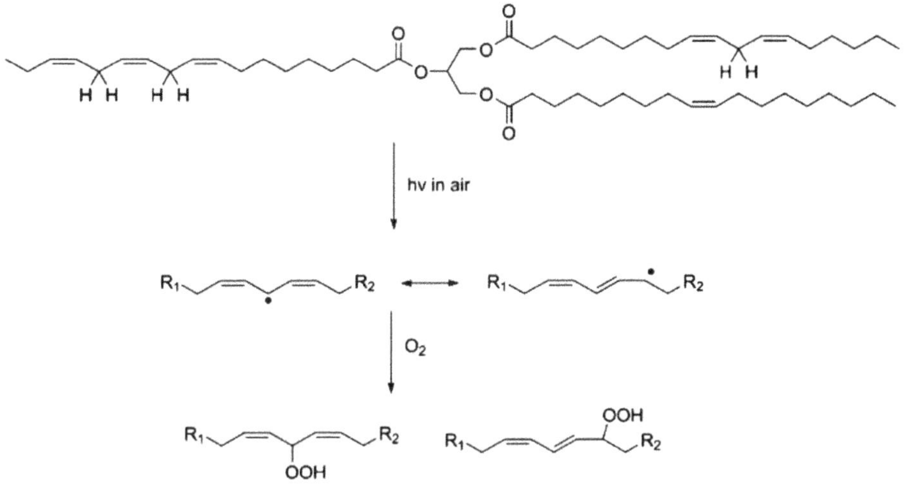

Figure 1.12   The auto-oxidation of drying oils in air.[9,62,69]

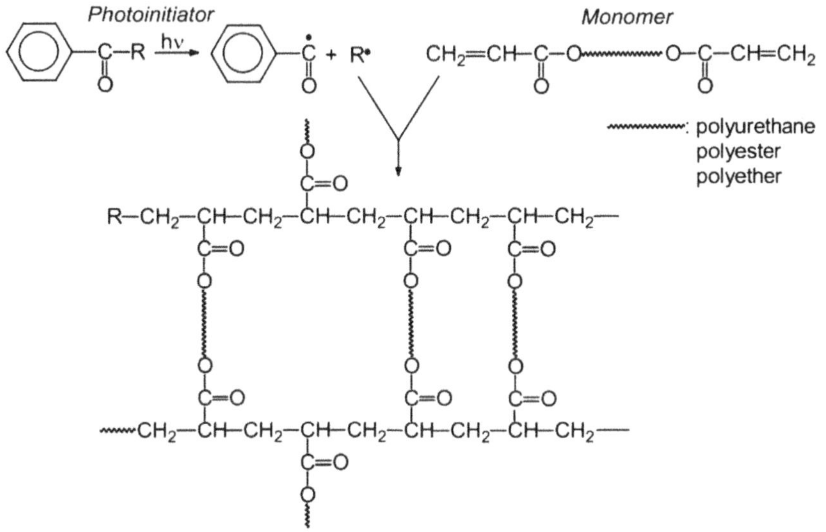

Figure 1.13   The reaction mechanism of photo-initiated radical polymerization.[7]

affects both the reaction rate and the final double-bond conversation. The structures of the monomer and photo-initiators are shown in Figure 1.14. The monomer with the photo-initiator HAC (2-hydroxy-2-methyl-1-phenyl-propan-1-one) showed the highest conversion rate and reached the highest level of double-bond conversion, followed by the mixtures with BP (benzo-phenone) TX (2,4-diethyl-9H-thioxanthen-9-one). The high conversion rate for HAC might be due to the quick generation of radical pairs through the

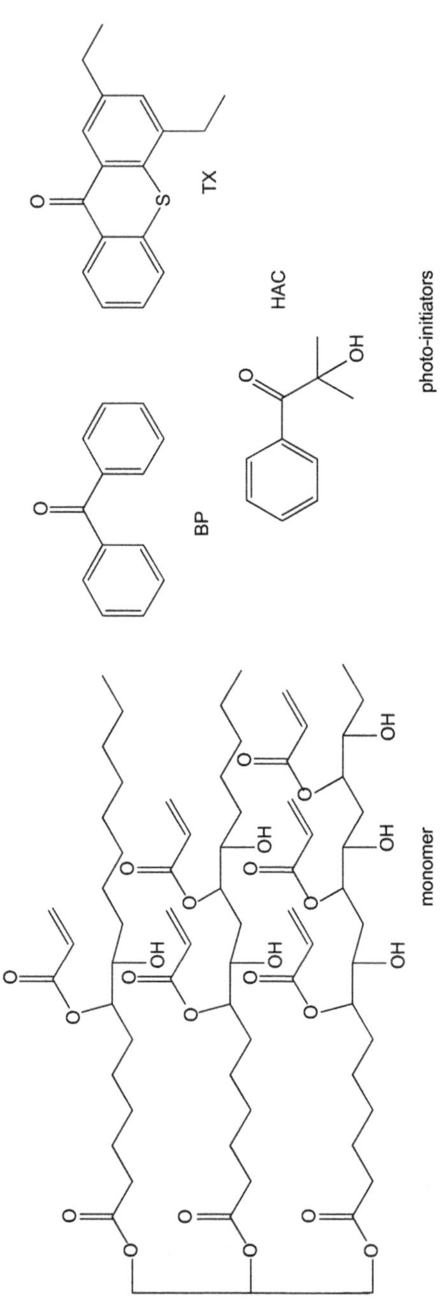

Figure 1.14 The structures of an acrylated epoxidized linseed oil monomer and three photo-initiators.[48]

efficient α-cleavage process of Type I photo-initiators. This UV-curable resin can be used for wood coating, as it contains sufficient cross-link density to withstand the solvent stress. Moreover, the polymer chains are flexible enough against scratches and exhibit good adhesion to wood substrates.

The influence of acrylate-reactive diluents on the photo-curing rate was investigated in detail, as well as the relationship between the number of acrylate functional groups on the oil backbone and the hardness of the resulting materials.[55,70,71] For example, the Patel group has prepared a novel binder system for UV-curing coatings based on tobacco seed (*Nicotiana rustica*) oil derivatives.[71] The UV-curing films of tobacco seed oil show good thermal stability at 100 °C, and the results of flexibility and adhesion tests revealed excellent performance. Higher functionalities of polyols, aromatic-type isocyanates, and lower oil ratios lead to poor adhesion and flexibility performance. Also, the aromatic nature of the isocyanate moiety further enhances the film hardness and toughness. Thus, the experimental sets based on higher polyols, higher functionality acrylate reactive diluents, and a lower proportion of oil gave better scratch hardness. Higher cross-linking densities showed better solvent and chemical resistance in the cured films.

Bio-degradable photo-cross-linked thin polymer networks based on acrylated hydroxy fatty acids have been reported. Di- and trimethacrylates,[71] or acrylated oligomers such as acrylated-PEG (polyethylene glycol) or acrylated-poly(ε-caprolactone) were used in co-polymerization. The bio-degradability of the resulting co-polymers was examined, and faster bio-degradation was observed for high-density cross-linking as a result of the low molecular weight between entanglements, that might otherwise block lipase attack sites.[72,73]

The Lecamp group[74] described a new synthesis process for vegetable-oil-based materials. It may provide potential opportunities to synthesize polymer materials from renewable resources by a clean and simple process that is totally transposable to lesser drying oils. First, linseed oil was thermally polymerized in bulk at 300 °C under an inert atmosphere. Then, the obtained stand oil was functionalized in a two-step one-spot process without solvent in order to graft onto it some photo-polymerizable groups. Finally, the materials were prepared by UV-curing of the modified linseed oil. The obtained materials were globally flexible, hydrophobic and non-bio-degradable. Compared to a naturally oxidized linseed-oil-based materials, the thermal stability and hydrophobicity remained unchanged.

1.3.3 Photo-cured Polymer Networks based on Epoxide Vegetable Oils

Multifunctional epoxide monomers can be converted into highly cross-linked polymer networks by UV irradiation in the presence of cationic photo-initiators. The cationic photo-polymerization of epoxidized oils is insensitive to oxygen, thus is highly attractive for many applications such as inks

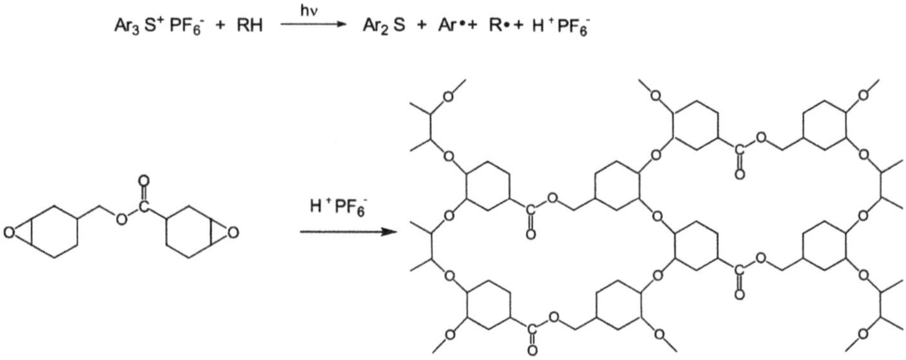

$$Ar_3 S^+ PF_6^- + RH \xrightarrow{h\nu} Ar_2 S + Ar\bullet + R\bullet + H^+ PF_6^-$$

Figure 1.15 The photo-initiated cationic polymerization of aliphatic dicycloepoxide.[7]

and adhesives.[7] The mechanism of photo-initiated cationic polymerization is illustrated in Figure 1.15.

The Soucek group[75] have reported the preparation and photo-polymerization of cyclohexene-derivatized linseed oil (CLO) and epoxycyclohexene-derivatized linseed oil (ECLO). The structures of CLO and ECLO and the photo-polymerization kinetics are shown in Figure 1.16. It was found that the addition of reactive or non-reactive diluents could reduce the viscosity of the formulations and increase the final epoxy conversion and the polymerization rate (Figure 1.17).[76] The properties of UV-curing hybrid films derived from epoxynorbornane-functionalized linseed oil and tetraethylorthosilane (TEOS) were also studied. The results indicated that the incorporation of TEOS can improve the performance of the films and enhance the tensile strength, thermal stability, fracture toughness, and general coating properties of epoxynorbornane-modified linseed oil.[77]

Ortiz's group studied the acceleration effect of substituted benzyl alcohols on the cationic photo-polymerization rate of epoxidized natural oils.[78] The Crivello group described the effect of the structure of both the cation and anion of diaryliodinium and triarylsulfonium photo-initiators on the polymerization rate of epoxidized triglycerides as renewable monomers. The obtained cured raw vernonia oil films based on this system exhibited a high degree of flexibility and impact strength.[79] Johansson's group has prepared UV-curable resins from a hydroxy functionalized hyperbranched polyether onto which an epoxy functional fatty acid, vernolic acid, had been grafted (Figure 1.18).[80] The Lalevée group[81] explored the cationic polymerization process of epoxidized soybean oil (ESO) using various photo-initiators under air and solar irradiation. A fundamental research study on photo-initiators was described for the case of the cationic photo-polymerization of triglycerides. Their newly developed photo-initiating systems are highly efficient under air upon solar irradiation, leading to 60% conversion, and form a completely tack-free and uncolored coating after 1 h.

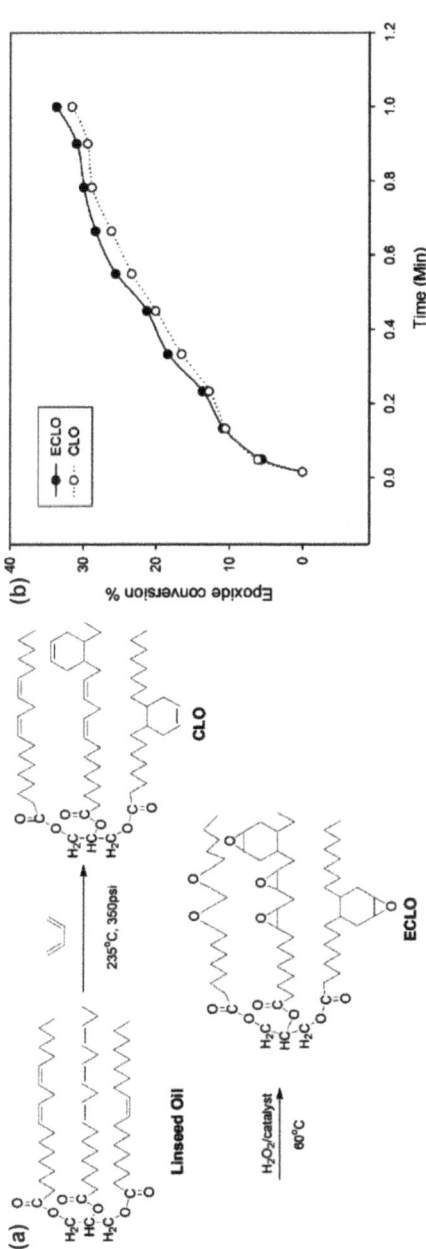

Figure 1.16 (a) The preparation of CLO and ECLO. (b) Conversion of the epoxy groups as a function of time. The concentration of the photo-initiator and light intensity are 4% and 60 mW cm^{-2}, respectively.[75]

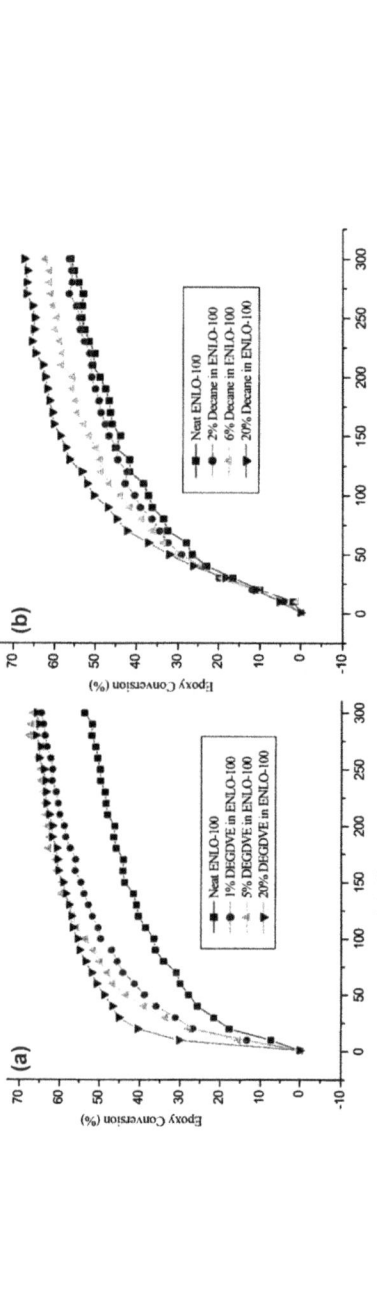

Figure 1.17 (a) Effect of the reactive diluent, DEGDVE (diethylene glycol divinyl ether), on the epoxy ring-opening polymerization of ENLO-100 (epoxy norbornane linseed oils) in the presence of 4 wt% OPPI as a function of the irradiation time. (b) Effect of the non-reactive diluent, DEGDEE (di(ethylene glycol) diethyl ether), on the epoxy ring-opening polymerization of ENLO-100 in the presence of 4 wt% OPPI ((4-octyloxyphenyl) phenyl iodonium hexafluoro antimonate) as a function of the irradiation time.[76]

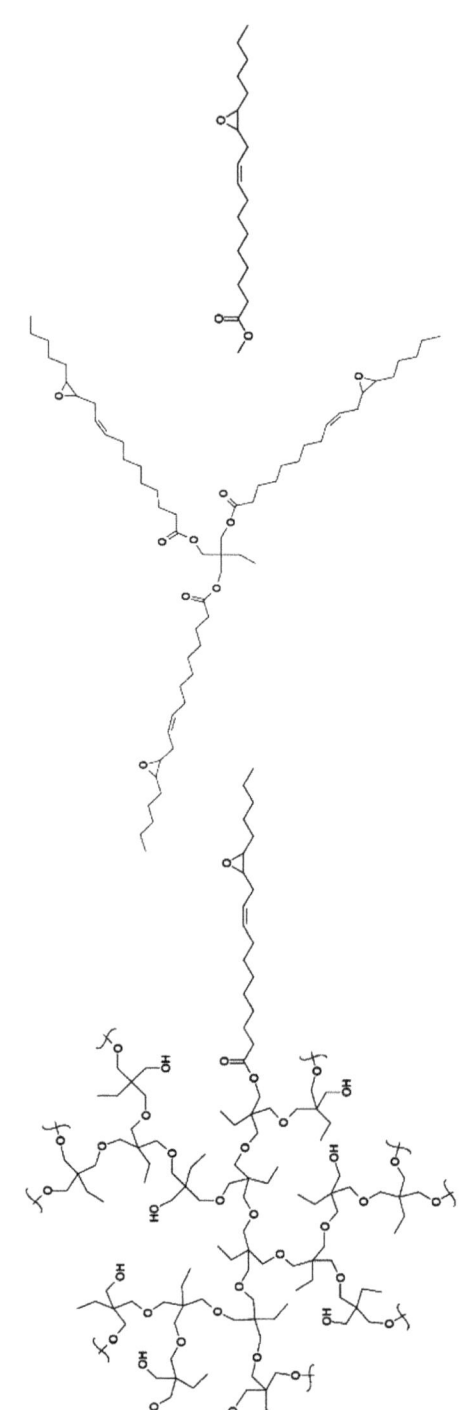

Figure 1.18 Structures of the hyperbranched polyether, poly-TMPO (hyperbranched poly-3-ethyl-3-(hydroxymethyl)oxetane)-vernoleate,

1.4 The Thiol-ene Reaction: A New Strategy for Photo-curable Materials from Vegetable Oils

The thiol-ene photo-reaction is a unique class of photo-polymerization that is advantageous with regard to the polymerization process and also the performance of the resulting polymers.[82–84] As shown in Figure 1.19, the thiol-ene reaction proceeds in a free-radical step-growth polymerization manner (Propagation A), and the excess acrylate continues to polymerize through homopolymerization (Propagation B).[51,84] In thiol-ene polymerizations a broad range of physical properties are achieved because a wide variety of enes such as acrylate, vinyl ether, allyl ether, vinyl acetate, and alkene are used. Thiol-ene photo-polymerization occurs through a step-growth reaction mechanism that offers many advantages such as delayed gel points, uniform structures, low polymerization shrinkage and reduced stress,[85–87] which make thiol-ene polymers high-performance materials,[88–90] especially for photo-curing coatings.[91,92]

The unsaturated nature of triglycerides obtained from vegetable oils make them good candidates for thiol-ene coupling. The fatty compounds can be transformed into polymeric networks by adding multifunctional thiols under UV or thermal radical conditions; the low thermal stability of lipidic compounds justifies the use of UV irradiation instead of thermal activation in the thiol-ene coupling reaction.[62,93–95] The thiol-ene photo-induced coupling reaction is a powerful method for the chemical modification of triglycerides or lipidic compounds to prepare vegetable oil derivates.[96,97] The Auvergne group[98] demonstrated an efficient thiol addition onto vegetable oils, leading to bio-based polyols (Figure 1.20). The most important feature of the reaction was the influence of the number of double bonds per chain in the vegetable oil on the thiol grafting yield. Indeed, disulfide formation and intermolecular recombination occurred. In the case of oleic acid functionalization, transesterification was also detected. Despite these side-reactions, byproducts were found to exhibit alcohol functional groups. Interestingly, the photo-reaction can therefore be performed under mild

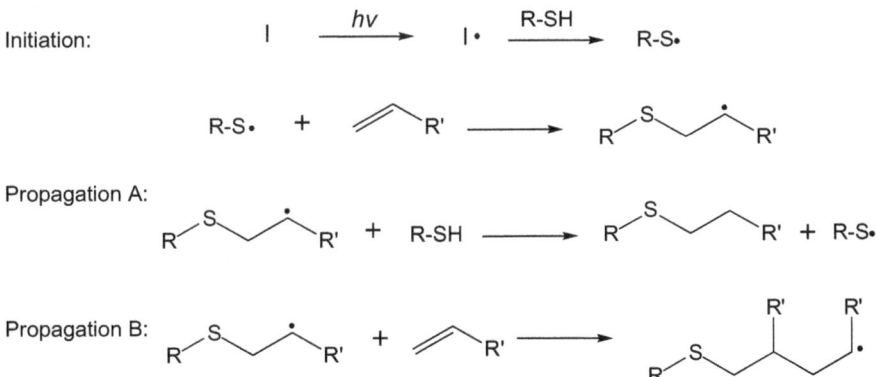

Figure 1.19 Proposed mechanism for the thiol-ene photo-reaction.[51]

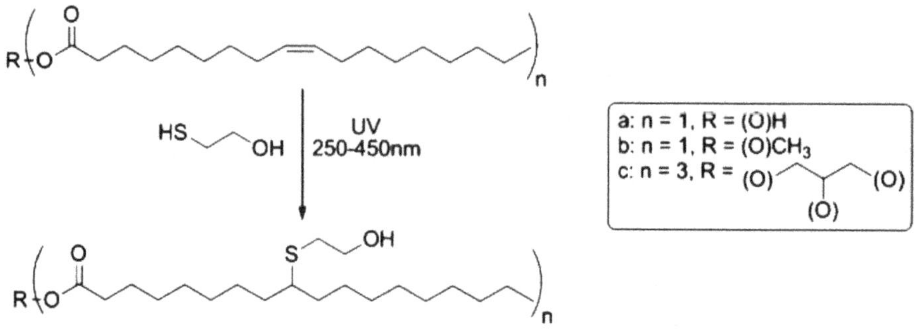

Figure 1.20 2-Mercaptoethanol photo-addition onto fatty compounds bearing one double bond per chain.[98]

conditions, requiring neither solvent nor photo-initiator. The crude product could also be purified by an easy procedure. This one-step route to produce fatty polyols represents a significant advance compared to the traditional epoxidation approach (that occurs in two steps) and further evidence of the synthetic usefulness of thiol addition in materials chemistry. Indeed, this synthetic method can readily incorporate different reactive functionalities onto vegetable oils and their derivatives, thus leading to functional precursors suitable for polymer synthesis. Through a similarly efficient coupling procedure, Meier[99] described the syntheses of materials derived from plant oils. By coupling both transesterification and thiol-ene addition, the Caillol group[100] synthesized a pseudo-telechelic diol from vegetable oil.

Many works focusing on UV-curable systems based on monomers derived from vegetable oils have been reported,[93–95,101] but UV-curable systems based on maleates, fumarates and acrylates have not been widely studied. Rawlins' group[101] reported the synthesis of thiol-ene UV-curable coatings based on vinyl ether, allyl, acrylate and derivatives of castor oil together with a multifunctional thiol. The films exhibited high solvent resistance and hardness as well as excellent adhesion and flexibility. Samuelsson *et al.* showed that coatings could be made by photo-induced thiol-ene polymerization using the internal double bonds of fatty esters (methyl oleate and methyl linoleate) and multifunctional thiols. The structure of both the unsaturated fatty derivatives and the thiol reagents affects the reactivity.[102]

Cádiz *et al.*[103] described polymer networks based on a kind of maleated soybean oil glyceride which was photo-polymerized with multifunctional thiols under very mild conditions (Figure 1.21). The mechanical properties of the polymer networks produced, resemble those of elastomeric materials. The materials exhibited a flexural modulus in the 240–340 MPa range and glass-transition temperatures below room temperature. Increasing the thiol functionality does not increase the rigidity of the materials due to structural factors. Bexell *et al.* reported the efficiency of a photo-induced reaction for coating aluminum surfaces with vegetable-oil-based films. In this case, linseed oil reacted with the thiol groups of mercaptosilane-treated aluminum.[104,105]

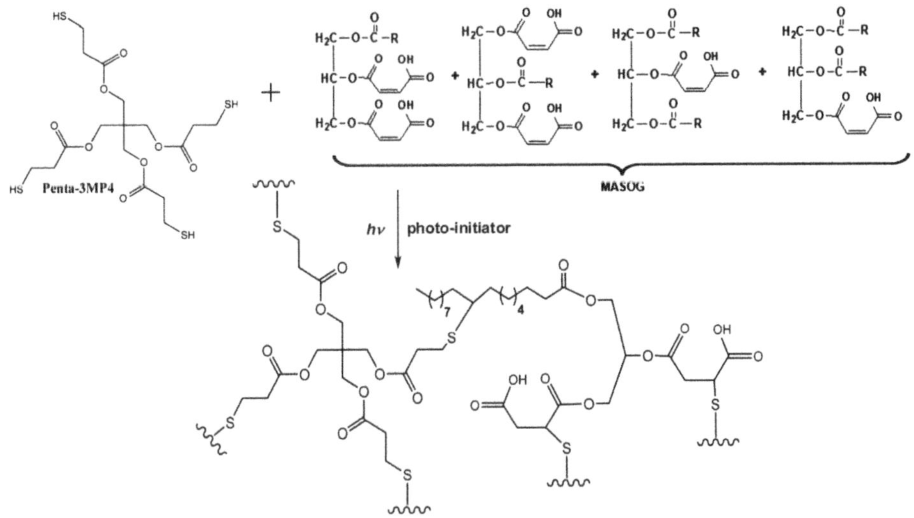

Figure 1.21 Simplified structure of the product obtained from the photo-polymerization of MASOG (maleated soybean-oil glycerides) with a tetrathiol.[103] Penta-3MP4 is pentaerythritol tetrakis(3-mercapto propionate).

Because of flexible thioether linkages and the inherent disadvantages of vegetable oil, it is difficult to obtain cross-linked materials with hardness, toughness and high glass-transition temperatures. Additionally, the typically unpleasant odors of low-molecular-weight thiols have also limited their commercial utilization. To overcome these disadvantages and obtain photocuring materials based on vegetable oils with enhanced performance, we first introduced polyhedral oligomeric silasesquioxanes (POSS) containing thiol groups into acrylated castor oil to develop a novel photo-cured hybrid material (Figure 1.22).[51] The obtained hybrid materials were transparent to visible light and no obvious phase separation was observed. The introduction of POSS can decrease the surface energy of the obtained hybrid materials. The thermal stability of the obtained hybrid materials increased with the increase of POSS content. The thermal decomposition temperature of the hybrid materials is between 250 and 300 °C, and the Young's modulus of the cured hybrid film is 230 MPa, which is higher than pure cured acrylated castor oil films. These characteristics give the obtained hybrid material potential applications in coating, and these studies provided a novel alternative approach to preparing hybrid materials from renewable sources.

1.5 Photo-initiators and other Functional Additives from Vegetable Oils

Due to their special structure and bio-compatibility, some vegetable oil derivates can be used as functional additives in resin formulations. Acrylated and epoxidized vegetable oils can be used as plasticizers to make photocuring epoxy or acrylate resin films with greater flexibility and toughness.[12]

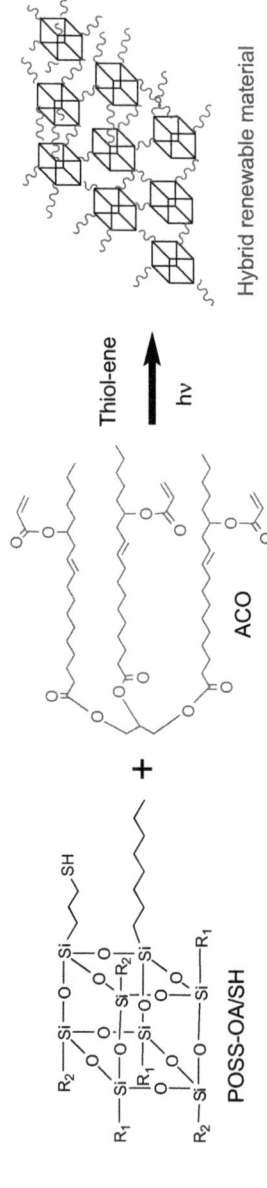

Figure 1.22 Photo-curing hybrid materials based on POSS-OA/SH and ACO.[51]

Due to their high adhesive properties, some photo-curing coating formulations may contain small proportions of vegetable oil derivates to enhance adhesion.[106,107] The Webster group[108] described the preparation of soybean-oil-based multithiol oligomers, which can be used as the cross-linking agent for photo-curing coating (Figure 1.23).

Kahraman's group[109] have prepared a kind of alkoxysilane-modified castor oil (Figure 1.24). They added the alkoxysilane-modified castor oil to a photo-curable resin formulation and obtained a hybrid UV-cured coating which is highly water repellent. Finally, highly hydrophobic and highly roughened coatings were prepared *via* a novel modification of an inexpensive and environmentally friendly bio-based renewable source. The most roughened coating showed a contact angle of 143°.

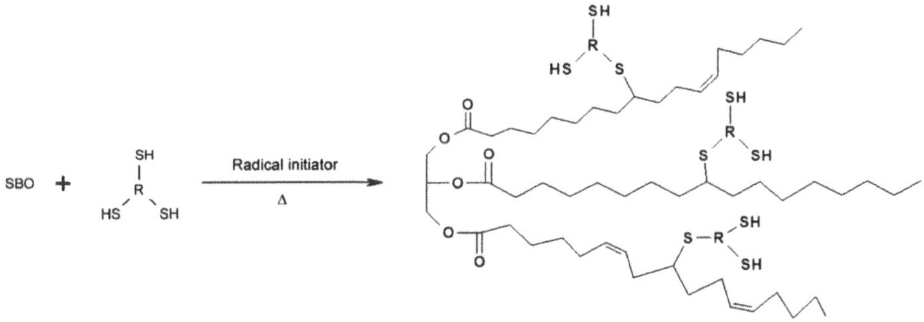

Figure 1.23 Thermal free-radical-initiated thiol-ene reaction between thiols and soybean oil (SBO).[108]

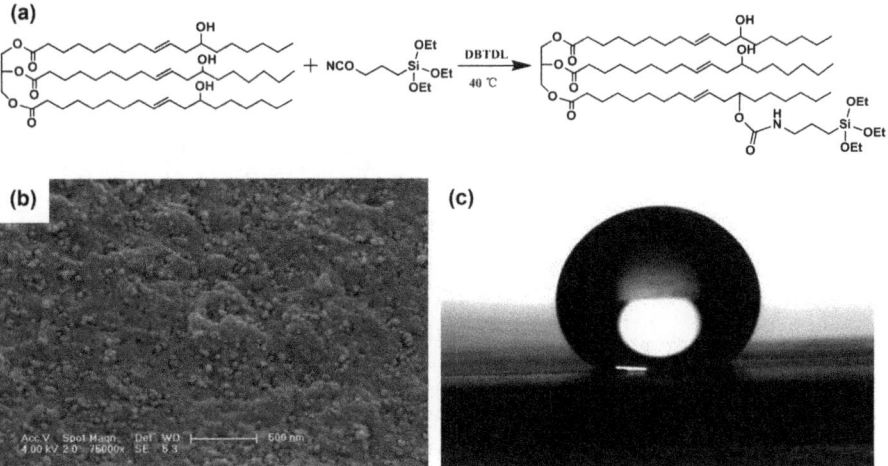

Figure 1.24 (a) Preparation of modified castor oil. (b) SEM image of the UV-cured coating surface. (c) A 5 μL liquid droplet on the UV-cured coating.[109]

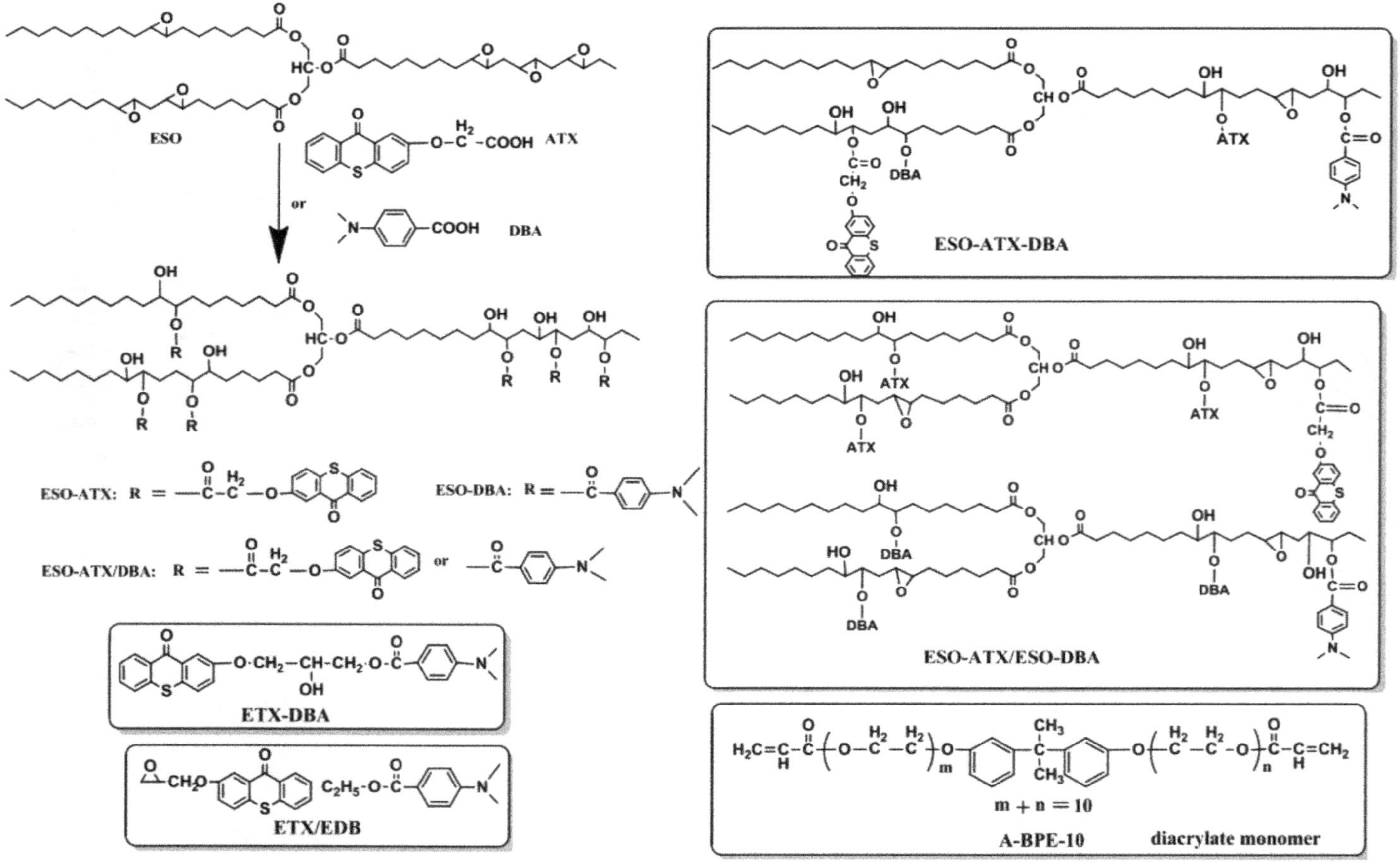

Figure 1.25   The synthesis of photo-initiators based on ESO, and the structures of four kinds of photo-initiator systems and the acrylate cross-linker A-BPE-10. ETX is 2-(2,3-epoxypropyloxy)thioxanthone; DBA is 4-(dimethylamino) benzoic acid; ATX is 2-acetic thioxanethone; EDB is ethyl-4-(dimethylamino) benzoate.

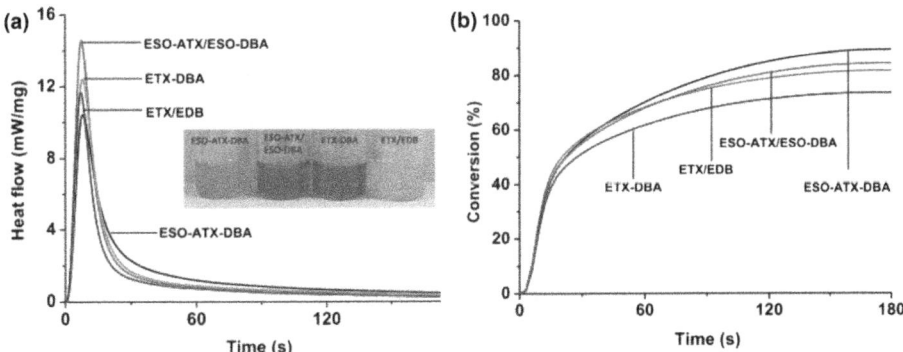

Figure 1.26 (a) Photo-DSC profiles for ESO-ATX-DBA, ESO-ATX/ESO-DBA, ETX-DBA and ETX/EDB, inset picture: dispersion of the four photo-initiators in A-BPE-10. (b) Conversion *versus* time curves for the polymerization of A-BPE-10 initiated by ESO-ATX-DBA, ESO-ATX/ESO-DBA, ETX-DBA and ETX/EDB, cured at 25 °C by UV light with an intensity of 45 mW cm^{-2} (all photo-initiator concentrations were 0.06 M wrt the TX moiety).

We synthesized a series of photo-initiators based on vegetable oil by introducing TX and co-initiator dimethylaminobenzene moieties into the ESO backbone through the reaction between carboxyl and epoxy groups (Figure 1.25). The obtained photo-initiators based on ESO exhibited high photo-efficiency in the photo-polymerization of acrylate monomers and low migration (Figure 1.26).[110]

1.6 Conclusions

Modifying vegetable oils can lead to new materials through a photo-polymerization process. The raw vegetable oils can be easily modified to obtain UV-curing monomers. Otherwise, modified vegetable-oil-based monomers can be introduced into polymeric materials to obtain specific properties, or form the basis of the polymers themselves. Consequently, photo-cured materials based on modified vegetable oils have a large variety of applications such as coatings, inks, bio-materials and so on.

Radiation-curing technology has developed very rapidly in recent years due to some obvious advantages. However, this technology retains some problems yet to be solved, such as the oxygen inhibition of radical polymerization, migration of photo-initiators and the low mechanical performance of the resulting materials. The development of non-hazardous photo-initiators which can be used safely in medicinal applications and in food contact coatings is a new opportunity. For industrial applications, new trends in the future will concern the development of ultrafast reactive monomers with low toxicity.

References

1. K. M. Schreck, D. Leung and C. N. Bowman, *Macromolecules*, 2011, **44**, 7520.
2. J. G. Kloosterboer, *Adv. Polym. Sci.*, 1988, **84**, 1.
3. C. Decker, *Polym. Int.*, 2002, **51**, 1141.
4. C. Decker, *Prog. Polym. Sci.*, 1996, **21**, 593.
5. C. Roffey, *Photogeneration of Reactive Species for UV-curing*, Wiley, New York, 1997.
6. Y. Yagci, S. Jockusch and N. J. Turro, *Macromolecules*, 2010, **43**, 6245.
7. C. Decker, *Pigm. Resin Technol.*, 2001, **30**, 278.
8. M. A. R. Meier, J. O. Metzger and U. S. Schubert, *Chem. Soc. Rev.*, 2007, **36**, 1788.
9. Y. Xia and R. C. Larock, *Green Chem.*, 2010, **12**, 1893.
10. F. Habib and M. Bajpai, *Chem. Chem. Tech.*, 2011, **5**, 317.
11. H. V. Patel, J. P. Raval and P. S. Patel, *Arch. Appl. Sci. Res.*, 2009, **1**, 294.
12. K. Prashnatha, V. Pai, B. Sherigara and K. S Prasanna, *Bull. Mater. Sci.*, 2001, **24**, 535.
13. S. Das and S. Lenka, *J. Appl. Polym. Sci.*, 2000, **75**, 1487.
14. V. Crivello and R. Ghoshai, *US Pat.*, 5 318 808, 1994.
15. D. Treybig, D. Wang, P. Sheih and L. Ho, *US Pat.*, 5 151 485, 1992.
16. R. W. Wan, R. Kumar, Z. S. Mek and M. M. Hilmi, *Eur. Polym. J.*, 2003, **39**, 593.
17. R. Raghavachar, G. Sarnecki, J. Baghdachi and J. Massingill, *J. Coat. Technol.*, 2000, **72**, 125.
18. S. Thames and H. Yu, *Surf. Coat. Technol.*, 1999, **115**, 208.
19. G. L. I. Puig, PhD thesis, University Rovira i Virgili, 2006.
20. R. Verhé, in *Renewable Bioresources: Scope and Modification for Non-food Applications*, ed. C. V. Stevens and R. Verhé, Wiley, West Sussex, 2004, pp. 208–250.
21. A. Remeikyte, J. Ostrauskaite and V. Grazuleviciene, *J. Appl. Polym. Sci.*, 2013, **129**, 1290.
22. K. Zou and M. D. Soucek, *Macromol. Chem. Phys.*, 2005, **206**, 967.
23. C. K. Mambo, P. M. Gitu, B. M. Bhaat, J. Chweya, S. Grinberg and D. Mills, *Bull. Chem. Soc. Ethiop.*, 1998, **12**, 121.
24. R. R. Kimwomi, MSc thesis, University of Nairobi, 1992.
25. P. Muturi, S. Dirlikov and P. M. Gitu, presented at the Inaugural Conference of the Kenya Chemical Society, Washington, DC, 1993.
26. U. Biermann, W. Friedt, S. Lang, W. Luhs, G. Machmuller, J. O. Metzger, M. Rüsch gen Klaas, H. J. Schafer and M. P. Schneider, *Angew. Chem., Int. Ed.*, 2000, **39**, 2206.
27. T. W. Findley, D. Swern and J. T. Scalan, *J. Am. Chem. Soc.*, 1945, **67**, 412.
28. S. Tayde, M. Patnaik, S. L. Bhagt and V. C. Renge, *Int. J. Adv. Eng. Technol.*, 2011, **2**, 491.
29. V. V. Goud, A. V. Patwardhan and N. C. Pradhan, *Bioresour. Technol.*, 2006, **97**, 1365.

30. S. Dinda, A. V. Patwardhan, V. V. Goud and N. C. Pradhan, *Bioresour. Technol.*, 2008, **99**, 3737.
31. C. Cai, H. Dai, R. Chen, C. Su, X. Xu, S. Zhang and L. Yang, *Eur. J. Lipid Sci. Technol.*, 2008, **110**, 341.
32. P. P. Meyer, N. Techaphattana, S. Manundawee, S. Sangkeaw, W. Junlakan and C. Tongurai, *Int. J. Sci. Technol.*, 2008, **13**, 1.
33. Z. S. Petrović, A. Zlatanic and C. C. Lava, *Eur. J. Lipid Sci. Technol.*, 2002, **104**, 293.
34. G. Du, A. Tekin, E. G. Hammond and L. Keith Wood, *J. Am. Oil Chem. Soc.*, 2004, **81**, 477.
35. A. E. Gerbase, J. R. Gregorio, M. Martinelli, M. C. Brasil and A. N. F. Mendes, *J. Am. Oil Chem. Soc.*, 2002, **79**, 179.
36. S. Bhattacharya, A. Drews, M. Kraume and M. Ansorge-Schumacher, *Chem. Ing. Tech.*, 2010, **82**, 1523.
37. I. Hilker, D. Bothe, J. Pruss and H. J. Warnecke, *Chem. Eng. Sci.*, 2001, **56**, 427.
38. M. Rüsch gen Klaas and S. Warwel, *Ind. Crops Prod.*, 1999, **9**, 125.
39. C. Orellana-Coca, S. Camocho, D. Adlercreutz, B. Mattiasson and R. Hatti-Kaul, *Eur. J. Lipid Sci. Technol.*, 2005, **107**, 864.
40. G. J. Piazza and T. A. Foglia, *J. Am. Oil Chem. Soc.*, 2005, **82**, 481.
41. T. Vlcek and Z. S. Petrović, *J. Am. Oil Chem. Soc.*, 2006, **83**, 247.
42. S. N. Khot, J. J. LaScala, E. Can, S. S. Morye, G. I. Williams, G. R. Palmese, S. H. Kusefoglu and R. P. Wool, *J. Appl. Polym. Sci.*, 2001, **82**, 703.
43. E. Sharmin, S. M. Ashraf and S. Ahmad, *Eur. J. Lipid. Sci. Technol.*, 2007, **109**, 134.
44. J. G. Homan, X. H. Yu, T. J. Connor and S. L. Cooper, *J. Appl. Polym. Sci.*, 1991, **43**, 2249.
45. M. A. Ali, T. L. Ooi, A. Salmiah, U. S. Ishiaku and Z. A. M. Ishak, *J. Appl. Polym. Sci.*, 2001, **79**, 2156.
46. D. Kolot and S. Grinberg, *J. Appl. Polym. Sci.*, 2004, **91**, 3835.
47. S. N. Khot, J. J. Lascala, E. Can, S. S. Morye, G. I. Williams, G. R. Palmese, S. H. Kusefoglu and R. P. Wool, *J. Polym. Sci., Part A: Polym. Chem.*, 2001, **82**, 703.
48. G. Wuzella, A. R. Mahendran, U. Muller, A. Kandelbauer and A. Teischinger, *J. Polym. Environ.*, 2012, **20**, 1063.
49. J. S. Nelson and T. H. Applewhite, *J. Am. Oil Chem. Soc.*, 1966, **43**, 542.
50. H. Pelletier and A. Gandini, *Eur. J. Lipid Sci. Technol.*, 2006, **108**, 411.
51. A. F. Luo, X. S. Jiang, H. Lin and J. Yin, *J. Mater. Chem.*, 2011, **21**, 12753.
52. M. Y. Cheong, T. L. Ooi, S. Ahmad, W. M. Z. W. Yunus and D. Kuang, *J. Appl. Polym. Sci.*, 2009, **111**, 2353.
53. T. Eren and S. H. Küsefoglu, *J. Appl. Polym. Sci.*, 2004, **91**, 2700.
54. H. V. Patel, J. P. Raval and P. S. Patel, *Arch. Appl. Sci. Res.*, 2009, **1**, 294.
55. J. G. Homan, X. H. Yu, T. J. Connor and S. L. Cooper, *J. Appl. Polym. Sci.*, 1991, **43**, 2249.
56. K. I. Patel, R. J. Parmar and J. S. Parmar, *J. Appl. Polym. Sci.*, 2008, **107**, 71.

57. E. Dzunuzovic, S. Tasic, B. Bozic, D. Babic and B. Dunjic, *Prog. Org. Coat.*, 2005, **52**, 136.
58. H. Pelletier, N. Belgacem and A. Gandini, *J. Appl. Polym. Sci.*, 2006, **99**, 3218.
59. H. Pelletier and A. Gandini, *Eur. J. Lipid. Sci. Technol.*, 2006, **108**, 411.
60. C. Decker, *Macromol. Rapid. Commun.*, 2002, **23**, 1067.
61. T. Scherzer and U. Decker, *Polymer*, 2000, **41**, 7681.
62. L. Fertier, H. Koleilat, M. Stemmelen, O. Giani, C. Joly-Duhamel, V. Lapinte and J. J. Robin, *Prog. Polym. Sci.*, 2013, **38**, 932.
63. M. Lazzari and O. Chiantore, *Polym. Degrad. Stab.*, 1999, **65**, 303.
64. V. Sharma and P. P. Kundu, *Prog. Polym. Sci.*, 2006, **31**, 983.
65. J. S. Mills and R. White, *Natl. Gallery Tech. Bull.*, 1980, **4**, 65.
66. A. Paramarta, X. Pan and D. C. Webster, *Polym. Prepr.*, 2011, **52**, 552.
67. S. Bovatzis, E. Ioakimoglou and P. Argitis, *J. Appl. Polym. Sci.*, 2002, **84**, 936.
68. A. C. Elm, *Ind. Eng. Chem.*, 1934, **26**, 386.
69. K. A. Tallman, B. Roschek and N. A. Porter, *J. Am. Chem. Soc.*, 2004, **126**, 9240.
70. N. Pietschmann and K. Stengel, *Farbe Lack*, 2004, **110**, 29.
71. K. I. Patel, R. J. Parmar and J. S. Parmar, *J. Appl. Polym. Sci.*, 2008, **107**, 71.
72. H. M. Kim, H. R. Kim and B. S. Kim, *J. Polym. Environ.*, 2010, **18**, 291.
73. H. M. Kim, H. R. Kim, C. T. Hou and B. S. Kim, *J. Am. Oil Chem. Soc.*, 2010, **87**, 1451.
74. O. Zovi, L. Lecamp, C. Loutelier-Bourhis, C. M. Lange and C. Bunel, *Green Chem.*, 2011, **13**, 1014.
75. K. Zou and M. D. Soucek, *Macromol. Chem. Phys.*, 2005, **206**, 967.
76. Z. Zong, M. D. Soucek, Y. Liu and J. J. Hu, *J. Polym. Sci., Part A: Polym. Chem.*, 2003, **41**, 3440.
77. Z. Zong, J. He and M. D. Soucek, *Prog. Org. Coat.*, 2005, **53**, 83.
78. R. A. Ortiz, D. P. Lopez, M. L. G. Cisneros, J. C. R. Valverde and J. V. Crivello, *Polymer*, 2005, **46**, 1535.
79. J. V. Crivello and R. Narayan, *Chem. Mater.*, 1992, **4**, 692.
80. J. Samuelsson, P. E. Sundell and M. Johansson, *Prog. Org. Coat.*, 2004, **50**, 193.
81. M. A. Tehfe, J. Lalevée, D. Gigmes and J. P. Fouassier, *Macromolecules*, 2010, **43**, 1364.
82. C. E. Hoyle, T. Y. Lee and T. J. Roper, *J. Polym. Sci., Part A: Polym. Chem.*, 2004, **42**, 5301.
83. D. P. Gush and A. D. Ketley, *Mod. Paint Coat.*, 1978, **68**, 58.
84. T. Y. Lee, Z. Smith, S. K. Reddy, N. B. Cramer and C. N. Bowman, *Macromolecules*, 2007, **40**, 1466.
85. J. A. Carioscia, J. W. Stansbury and C. N. Bowman, *Polymer*, 2007, **48**, 1526.
86. J. P. Phillips, N. M. Mackey, B. S. Confait, D. T. Heaps, X. Deng, M. L. Todd, S. Stevenson, H. Zhou and C. E. Hoyle, *Chem. Mater.*, 2008, **20**, 5240.

87. T. Y. Lee, J. Carioscia, Z. Smith and C. N. Bowman, *Macromolecules*, 2007, **40**, 1473.
88. K. Owusu-Adom, J. Schall and C. A. Guymon, *Macromolecules*, 2009, **42**, 3275.
89. L. M. Campos, T. T. Truong, D. E. Shim, M. D. Dimitriou, D. Shir, I. Meinel, J. A. Gerbec, H. T. Hahn, J. A. Rogers and C. J. Hawker, *Chem. Mater.*, 2009, **21**, 5319.
90. A. Bertin and H. Schlaad, *Chem. Mater.*, 2009, **21**, 5698.
91. M. Black and J. W. Rawlins, *Eur. Polym. J.*, 2009, **45**, 1433.
92. S. F. Thames and H. Yu, *Surf. Coat. Technol.*, 1999, **115**, 208.
93. A. Schwab, L. Gast and J. Cowan, *J. Am. Oil Chem. Soc.*, 1968, **45**, 461.
94. C. Lluch, J. C. Ronda, M. Galia, G. Lligadas and V. Cádiz, *Biomacromolecules*, 2010, **11**, 1646.
95. O. Turunc and M. A. R. Meier, *Green Chem.*, 2011, **13**, 314.
96. U. Bexell, M. Olsson, M. Johansson, J. Samuelsson and P. E. Sundell, *Surf. Coat. Technol.*, 2003, **166**, 141.
97. O. Turunc and M. A. R. Meier, *Macromol. Rapid Commun.*, 2010, **31**, 1822.
98. M. Desroches, S. Caillol, V. Lapinte, R. Auvergne and B. Boutevin, *Macromolecules*, 2011, **44**, 2489.
99. O. Trunc and M. A. R. Meier, *Eur. J. Lipid Sci. Technol.*, 2013, **115**, 41.
100. M. Desroches, S. Benyahya, V. Besse, R. Auvergne, B. Boutevin and S. Caillol, *Lipid Technol.*, 2014, **26**, 35.
101. M. Black and J. W. Rawlins, *Eur. Polym. J.*, 2009, **45**, 1433.
102. J. Samuelsson, M. Jonsson, T. Brinck and M. Johansson, *J. Polym. Sci., Part A: Polym. Chem.*, 2004, **42**, 6346.
103. D. A. Echeverri, V. Cádiz, J. C. Ronda and L. A. Rios, *Eur. Polym. J.*, 2012, **48**, 2040.
104. U. Bexell, R. Berger, M. Olsson, T. M. Grehk, P. E. Sundell and M. Johansson, *Thin Solid Films*, 2006, **515**, 838.
105. U. Bexell, M. Olsson, M. Johansson, J. Samuelsson and P. E. Sundell, *Surf. Coat. Technol.*, 2003, **166**, 141.
106. M. S. Kroll, M. Acevedo, J. C. Camemlh, T. F. Kauffman, J. S. Lindquist, E. R. Simmons, D. B. Malcolm and K. A. Coleman, *US Pat.*, 6 579 915 B2, 2003.
107. R. S. Kosiorek and L. R. Gatechair, *US Pat.*, 4 649 062, 1987.
108. J. F. Wu, S. Fernando, D. Weerasinghe, Z. Chen and D. C. Webster, *ChemSusChem*, 2011, **4**, 1135.
109. Y. Mulazim, E. Cakmakc and M. V. Kahraman, *J. Vinyl Addit. Technol.*, 2013, **19**, 31.
110. A. F. Luo, X. S. Jiang and J. Yin, *Polymer*, 2012, **53**, 2183.

CHAPTER 2

Chemical Synthesis of Carbonates, Esters, and Acetals from Soybean Oil

KENNETH M. DOLL

USDA, ARS, National Center for Agricultural Utilization Research, Bio-Oils Research Unit, 1815 N. University Street, Peoria, IL 61604, USA[†]
Email: kenneth.doll@ars.usda.gov

2.1 Use of Natural Oils

Lack of petroleum[1] and life cycle[2] are two of the phenomena that you faithful readers have encountered and will undoubtedly encounter repeatedly in this book. Mercifully, they will not be mentioned again, at least in this chapter. On the surface, the search for "Green Materials from Plant Oils" seems to be an easy task. Just pull a leaf off of the tree in your yard or the plant in your office, and squeeze. There you have it, green material. Now, if one wishes to find a useful green material, well, that is a bit more of a story.

A good starting point originated about a decade ago when the "Principles of Green Chemistry"[3,4] were laid out in a manner suitable for all to read. Some of these such as: the avoidance of unnecessary derivatization, the production of substances of low toxicity, and use of catalytic reagents

---

[†]Mention of trade names or commercial products in this publication is solely for the purpose of providing specific information and does not imply recommendation or endorsement by the U.S. Department of Agriculture. USDA is an equal opportunity provider and employer.

RSC Green Chemistry No. 29
Green Materials from Plant Oils
Edited by Zengshe Liu and George Kraus
© The Royal Society of Chemistry 2015
Published by the Royal Society of Chemistry, www.rsc.org

are found throughout this chapter. However, the main driver in much of this research is the principle that the raw material feedstock in a process should be renewable. That thread is common throughout the use of plant oils in industry, with even the USA federal government in agreement.[5] To this end, there are many good reviews available, some of them specifically on the uses of natural oils.[6,7] That material will not be reproduced here, where instead you, friendly reader, can move on to a tale of the twisting of soybean oil into carbonates, acetals, and branched-chain esters.

2.2 Epoxidation of Soybean Oil or Alkyl Esters of Soybean Oil

In its natural state, refined, bleached, and deodorized soybean oil consists of a glycerol structure with three fatty chains. A common reaction, used in bio-diesel synthesis (Scheme 2.1), involves the esterification of the three chains of the glycerol structure producing a family of alkyl esters. Many of these chains, both in triacylglycerols and the alkyl esters, will contain one or more double bonds that are important for two reasons. First, it is usually these unsaturated groups which chemists can utilize in the modification of these fatty chains. Second, due to a variety of effects, the locations adjacent to the double bonds are highly susceptible to different reactions such as hydrogen abstraction and oxidation.

Whether on the full triacylglycerol, or on the ester, these double bonds will be utilized in this chapter to form epoxides, then further to form cyclic carbonates, esters, and acetals (Figure 2.1). This is not a comprehensive list of possible modifications; elsewhere ethers,[8,9] amines,[10,11] azides,[12] thiols,[13,14] hydroxides,[15] and many others have been described.

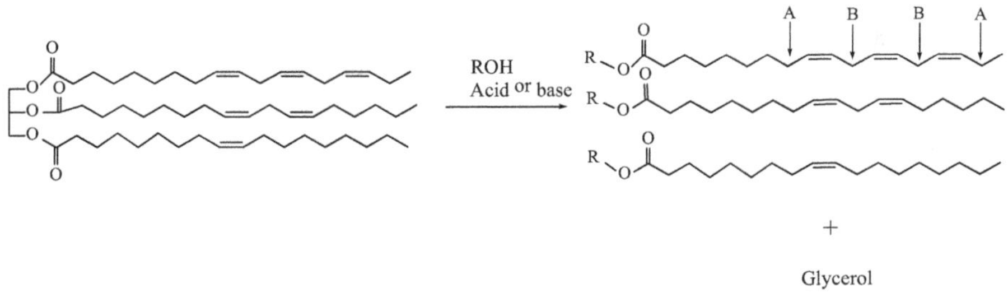

Scheme 2.1   A soybean oil structure and its esterification to alkyl esters. A series of non-systematic names describe the fatty chains which naturally occur in the *cis*-configuration. For example, the methyl esters would be called "methyl stearate", "methyl oleate", "methyl linoleate", and "methyl linolenate" for 18-carbon fatty chains with 0, 1, 2, and 3 double bonds, respectively. The allylic (A) and bis-allylic (B) positions, which are especially susceptible to oxidative reactions, are shown on the linoleic chain.

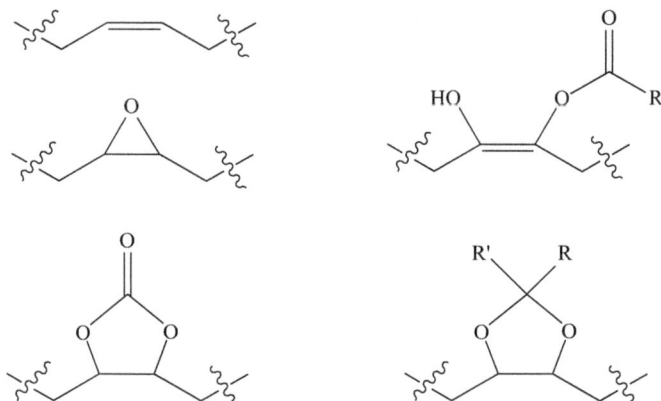

Figure 2.1 The alkene, epoxide, cyclic carbonate, (left side), mid-chain ester, and acetal (right side) structures discussed in this chapter. The acetal structure is also sometimes referred to as a "ketal".

The epoxidation of soybean oil involves the synthesis of a three-membered ring which contains an oxygen group, called an "oxirane". There are several methods for the synthesis of this material, which has been used as a plasticizer in large volumes.[16,17] It also has properties which make it good for use in lubricants[18–21] and as a starting material in fuel additives,[8,22] polymers,[23–25] hydrogels,[26,27] coatings,[28] resins,[29] and surfactants.[30] The epoxide structure itself has long been known, although the best chemical representation of the moiety was a past controversy.[31] Its primary reactivity involves ring opening, which can be accomplished readily under acidic conditions.

2.2.1 Overview of Methods

The epoxidation of soybean oil directly with oxygen over a metal catalyst such as silver, as is done when converting ethene to ethylene oxide, would be ideal. Unfortunately, the active allylic hydrogens in soybean oil are a problem for this process, and, along with the added steric hindrance at double bonds, render this type of process ineffective. What has been done for the last half century is to use hydrogen peroxide as the oxygen source, along with some type of catalyst.

The use of carboxylic acids, such as formic or acetic acids,[32,33] has been shown to be effective (Scheme 2.2). In this reaction, a peracid is formed which then reacts with the substrate and regenerates the carboxylic acid. If hydrogen peroxide is used, the reaction is bi-phasic, which makes is difficult to study. However, it has proven effective, and a slight modification of this procedure was the method of choice used for the generation of epoxides studied here, as well as to get the intermediate materials for follow-up syntheses.[34]

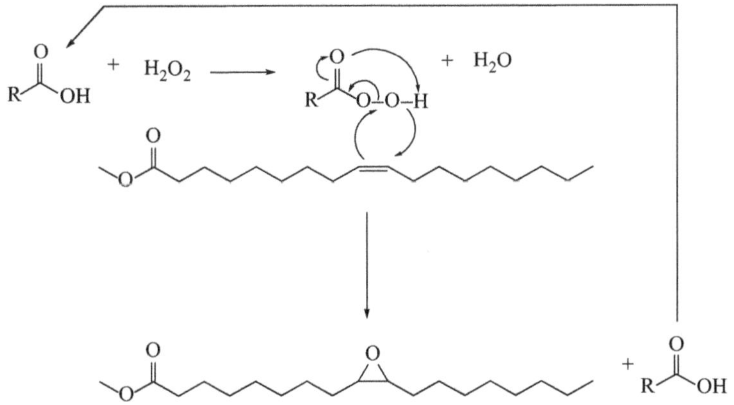

Scheme 2.2 A mechanistic scheme showing the epoxidation of methyl oleate with hydrogen peroxide using a carboxylic acid catalyst. The stereochemistry of the epoxide groups, not specified in the scheme, is the expected, with the hydrogens on the same side of the chain. Many epoxides show multiple stereoisomers with respect to the relative positions of the epoxide groups.

Instead of the use of the carboxylic acids, heterogeneous catalysts have also been employed. Unmodified alumina showed the ability to catalyze the reaction in yields up to 54% using hydrogen peroxide.[35] Treating the alumina with zinc, tin, zirconium, or titanium did not improve the reaction. In different studies,[36–39] titanium silicates demonstrate high reactivities with up to 95% conversion when using *t*-butyl hydroperoxide on the *trans*-isomer of methyl oleate.

Others have used enzymes to effect the transformation.[40–42] The use of oat seed lipase and *t*-butyl hydroperoxide were shown to yield over 60% epoxidation under optimized conditions.

2.2.2 Properties of Epoxidized Oils

One of the primary reasons to perform the epoxidation reaction is to improve the oxidation stability of the material. Systematic study of alkenes and epoxidized materials[43,44] has shown that the oxidation onset temperature, as measured by pressurized differential scanning calorimetry, shows an increase of up to 42 °C (Table 2.1). However, there are some tradeoffs. The viscosity of the material increases by a large amount, especially for the un-saturated materials with multiple double bonds. This can be a good thing or a bad thing, depending on the desired formulation. The pour point, the lowest temperature at which a liquid is pourable, is also increased. This is almost always considered a drawback in lubrication applications.

Overall, the epoxidized soybean oil and epoxidized methyl esters are successful products. Their combination of lubricity and better oxidation

Table 2.1 Physical properties of methyl oleate and methyl linoleate and their corresponding epoxides. Oxidation onset temperature, pour point, and kinematic viscosity are reported.

Sample	Oxidation onset/°C	Kinematic viscosity at 40 °C/mm s^{-1}	Pour Point/°C
Soybean oil	155	~32	−9
Methyl oleate	177	3.7	−27
Methyl linoleate	138	3.2	−48
Epoxidized soybean oil	199	175	3
Epoxidized methyl oleate	190	8.0	0
Epoxidized methyl linoleate	180	145	−2

stability make them effective as plasticizers and in grease applications. However, their high pour point and viscosity open the door for further chemical modifications.

2.3 Carbonates

Carbonates are materials that contain a carbon double bonded to an oxygen, and with single bonding to two other oxygen atoms (Figure 2.1). If these atoms are linked by a carbon chain, forming a ring, then it is a special case called a "cyclic carbonate". These cyclic carbonates have found industrial uses as solvents, dielectric fluids, and diluents for many decades.[45] They can be reacted with amines, alcohols, or thiols to form polyurethanes and polycarbonates of varying physical properties. Ethylene carbonate, propylene carbonate, and glycerol carbonate are three of the common organic carbonates that are well known for their bio-degradability, low toxicities, and high flash points.

Cyclic carbonates are generally synthesized by either the reaction of a chlorohydrin, the reaction of a diol with phosgene or, most promisingly, the insertion of carbon dioxide into an epoxide.[46] Early examples of the phosgene method[47,48] were used on methyl hydroxy stearates, and the chlorohydrin method has also been shown to produce satisfactory products.[49] However, following the principles of green chemistry, avoidance of chlorinated intermediates can be accomplished by the use of epoxides, easily produced as described in the previous section.

2.3.1 Previous Syntheses at Lower Pressure

There have been several reports on the transformation of oils into cyclic carbonates including epoxidized soybean oil,[50,51] vernonia oil,[52] and epoxidized cottonseed oil.[53] One interesting report also produced a carbonated bio-diesel by first subjecting epoxidized soybean oil to an esterification reaction, then performing the carbonation.[54] These reports all have a couple of things in common. Firstly, they utilize a ring-opening catalyst such as tetrabutylammonium bromide with carbon dioxide at relatively low pressure. Secondly, they all take a long time to accomplish the conversion.

2.3.2 Carbonation using Supercritical Carbon Dioxide

Supercritical fluids have been shown to be valuable tools in chemical syntheses of almost all types due to their possession of both liquid-like, and gas like-properties.[55] In the case of the formation of a carbonate, the use of supercritical carbon dioxide has a double effect. That is, because in this reaction, it is not only the solvent, serving both the reagents and catalyst together, it is also a reactant.

Commercially available epoxidized soybean oil was reacted with supercritical carbon dioxide under a pressure of 10.3 MPa (Scheme 2.3). In order to facilitate the ring-opening reaction, tetrabutylammonium bromide was used as a catalyst at a concentration of 5 mol% wrt to oxirane oxygen atoms. Under these reaction conditions, 82% conversion was noted in only 10 hours of reaction time, and complete conversion could be achieved in only

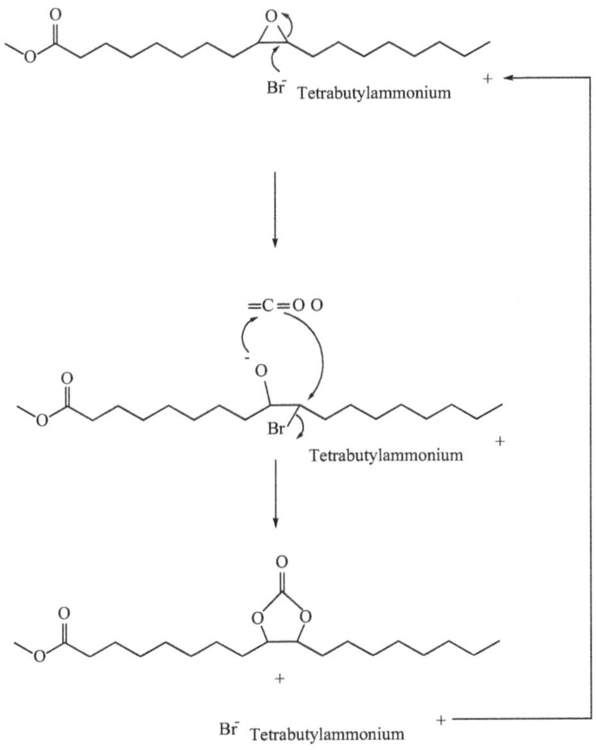

Scheme 2.3 The insertion of carbon dioxide into an oxirane, catalyzed by tetrabutylammonium bromide.

40 hours.[56] Other catalysts were also employed including potassium bromide, lithium bromide, and tetrabutylammonium hydroxide, but none were observed to be particularly efficacious.

The catalyst could be removed from the product in a couple of different ways. The most interesting one takes advantage of the Hoffmann reaction where tetrabutylammonium bromide can be decomposed to tributylamine, butene, and hydrobromic acid. Because of the stability of carbonated soybean oil, a temperature was found at which the catalyst would decompose and volatilize but the product remained unharmed. The volatiles from the sample were collected and found to contain the expected amine, and the product material was found to be identical to that made using the normal method of catalyst removal, washing with sonication.

Turning to smaller vegetable-oil-based materials, the same reaction utilizing supercritical carbon dioxide could also be performed on epoxidized methyl oleate, epoxidized methyl linoleate, and epoxidized methyl linolenate, *i.e.* materials with 1, 2, and 3 oxirane groups, respectively.[57] Individual epoxides of these esters were not available commercially, but the formic acid catalyzed epoxidation was viable. In order to afford each of the materials in high yield, it was necessary to directly follow the epoxidation reaction using gas chromatography in order to avoid ring-opened side-products. A mixture of the epoxides of the 2-ethylhexyl esters and epoxidized 2-ethylhexyl soyate was available commercially, so it was also carbonated for comparison.

2.3.3 Selected Properties of Vegetable-oil-based Carbonates

A comparison of the physical stabilities of some vegetable-oil-based carbonates (Table 2.2) shows a large increase in the temperature it takes for half of the sample to volatilize. Another interesting property is the very high viscosity, 170 Pa·s at 25 °C. This is over 400 times that of the epoxidized material, and over 3000 times that of ordinary soybean oil.

Table 2.2 Physical stability of carbonates based on vegetable oil, and some comparative samples.

Sample	50% Weight loss by thermogravimetric analysis/°C
Methyl oleate	187
Epoxidized methyl oleate	211
Carbonated methyl oleate	260
Methyl linoleate	189
Epoxidized methyl linoleate	231
Carbonated methyl linoleate	271
Carbonated 2-ethylhexyl soyate	301
Carbonated soybean oil	>350

2.4 Branched-chain Esters

Ring opening the epoxide with materials that are not carbon dioxide can also be straightforward or complicated. Ring opening multiple-epoxide compounds does not form polyhydroxy compounds in high yields, as much of the literature would suggest. Instead, complicated furan structures dominate the observed product mixtures.[58,59] However, in materials with only a single epoxide in a fatty chain, the reaction is well behaved, forming the branched esters (Scheme 2.4). It was studied thoroughly with smaller carboxylic acids, despite the complication of a bi-phasic system.[60] However, the first catalyst-free system developing an entire family of these compounds was only demonstrated recently.[61]

2.4.1 Simple Catalyst-free Synthesis

The ring-opening reaction of an epoxide with a carboxylic acid can be performed in a solvent which will readily solvate both reactive species, such as butanol. What works more efficiently is the use of a larger carboxylic acid which can function both as the solvent and the reactant. Using this methodology, the reaction can be run at temperatures from 80–120 °C,

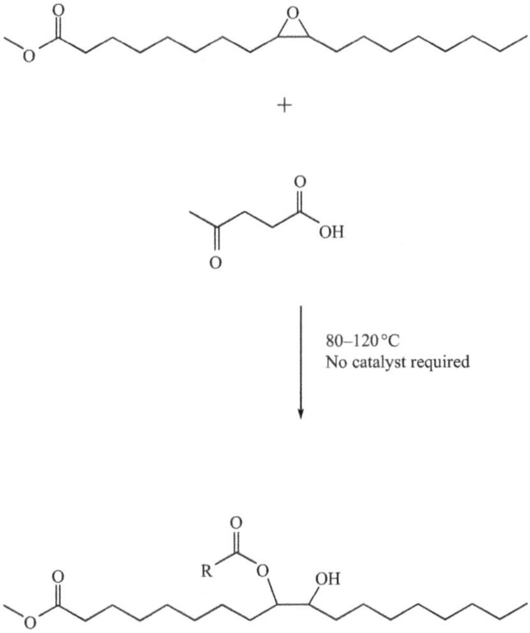

80–120 °C
No catalyst required

Scheme 2.4 The ring opening of an epoxide with a carboxylic acid to form a mid-chain ester. The ester and hydroxy groups of the mid-chain ester are shown in the 10 and 9 positions of the chain, respectively, but the opposite isomer is also produced in equal quantities.

Table 2.3 The physical properties of branched-chain esters including pour point and oxidation onset temperature.

Sample	Oxidation onset/°C	Pour Point/°C
Methyl oleate	177	−27
Epoxidized methyl oleate	190	0
Propanoic acid branched	175	−15
2-Ethylhexanoic acid branched	166	−33
Octanoic acid branched	160	−24

giving complete conversion in only a few hours at the higher temperatures.[62] Propanoic, octanoic, and 2-ethylhexanoic acids were employed.

2.4.2 Good Lubricant Properties

As mentioned earlier, the epoxides, while showing good oxidation stability, suffer from high pour points. This new family of compounds shows some success in the battle between these two properties (Table 2.3). While not quite up to epoxy material stability, the new compounds are close. Further, the longer chains even surpass the very favorable pour point of the alkene. Other lubricant and fuel properties,[63] such as surface tension and friction reduction, are also favorable, and the versatility and ease of the methods lend them to building an even larger family of compounds.[22]

2.5 Acetals

The final structure discussed in this chapter will be the acetal, also sometimes referred to as a "ketal" (Figure 2.1). A vegetable-oil-based acetal was synthesized from a polyhydroxy compound and acetone, utilizing acid catalysis.[64] This chemistry is a familiar protecting-group reaction, where both stable and unstable regions could be envisioned through control of the acid and base chemistry.

2.5.1 Catalyst Control and the Amazing Levulinic Acid Case

Structures of this same type are even more conveniently available through the use of an epoxide. Using 2-pentanone and phosphoric acid, acetals can be conveniently formed at only 50 °C.[65] This temperature regime is important in a special case, that of using levulinic acid as a ketone.

Levulinic acid is available from biological sources, and is often considered a platform commodity in the quest to transform cellulose into industrial chemicals.[66] It is also bi-functional, containing the two reactive groups that were just discussed earlier in this section and in the previous section. Because the carboxylic acid reaction does not require a catalyst, but does require high temperatures, and the acetal reaction requires the opposite, a

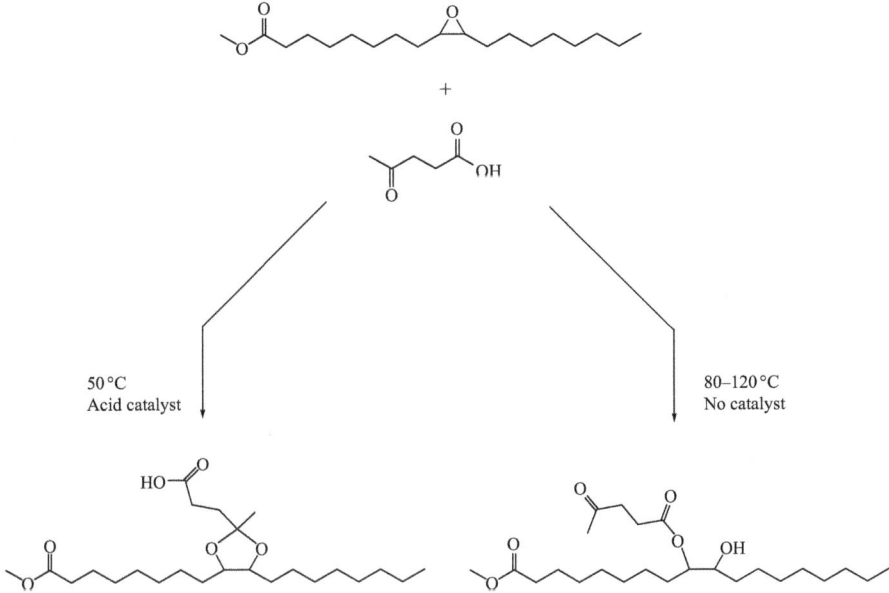

Scheme 2.5 The synthesis of an acetal (left) or branched-chain ester (right) from epoxidized methyl oleate and levulinic acid.

situation where the formation of products is conditional was reached (Scheme 2.5). In other words, different products from the same reactants can be formed in the system. Up to 87% selectivity for branched esters, and 64% selectivity for acetals is possible.

2.6 Conclusions

The friendly reader has just learned about a few of the things that can be done with vegetable oils. The reactive points are at the glycerol backbone, which can be changed with esterification, and at the double bonds. This work focused on the epoxidation of the double bonds, which produces useful material in its own right or gives further possibilities. Just a few of those possibilities are mentioned here which, with luck, can be part of the new world of *Green Materials from Plant Oils*.

Acknowledgements

This work was primarily funded through the Agricultural Research Service, the in-house research arm of the United States Department of Agriculture. For work on these projects, acknowledgement goes to Dr Brajendra K. Sharma, Donna I. Thomas, Jennifer R. Koch, Dr Karl E. Vermillion, Erin L. Walter, and Daniel A. Knetezer.

References

1. P. van Arnum, *Chem. Mark. Rep.*, 2004, **266**, 1.
2. L. Gustafsson and P. Borjesson, *Int. J. Life Cycle Assess.*, 2007, **12**, 151.
3. J. C. Warner, A. S. Cannon and K. M. Dye, *Environ. Impact Assess. Rev.*, 2004, **24**, 775.
4. P. T. Anastas and M. M. Kirchhoff, *Acc. Chem. Res.*, 2002, **35**, 686.
5. K. Collins, *Fed. Regist.*, 2005, **70**, 1792.
6. U. Biermann, W. Friedt, S. Lang, W. Lühs, G. Machmüller, J. O. Metzger, M. Rüsch gen Klaas, H. J. Schäfer and M. P. Schneider, *Angew. Chem., Int. Ed.*, 2000, **39**, 2206.
7. U. Biermann, U. Bornscheuer, M. A. R. Meier, J. O. Metzger and H. J. Schäfer, *Angew. Chem., Int. Ed.*, 2011, **50**, 3854.
8. B. R. Moser and S. Z. Erhan, *J. Am. Oil Chem. Soc.*, 2006, **83**, 959.
9. B. R. Moser and S. Z. Erhan, *Eur. J. Lipid Sci. Technol.*, 2007, **109**, 206.
10. A. Biswas, A. Adhvaryu, S. H. Gordon, S. Z. Erhan and J. L. Willett, *J. Agric. Food Chem.*, 2005, **53**, 9485.
11. A. Biswas, B. K. Sharma, K. M. Doll, S. Z. Erhan, J. L. Willett and H. N. Cheng, *J. Agric. Food Chem.*, 2009, **57**, 8136.
12. A. Biswas, B. K. Sharma, J. L. Willett, A. Advaryu, S. Z. Erhan and H. N. Cheng, *J. Agric. Food Chem.*, 2008, **56**, 5611.
13. M. Desroches, S. Caillol, V. Lapinte, R. M. Auvergne and B. Boutevin, *Macromolecules*, 2011, **44**, 2489.
14. G. B. Bantchev, J. A. Kenar, G. Biresaw and M. G. Han, *J. Agric. Food Chem.*, 2009, **57**, 1282.
15. K.-W. Dietrich, T. Heinemann and M. Dietrich, *US Pat.*, 5 886 062, 1999.
16. M. Rüsch gen Klaas and S. Warwel, *Ind. Crops Prod.*, 1999, **9**, 125.
17. V. Goud, N. Pradhan and A. Patwardhan, *J. Am. Oil Chem. Soc.*, 2006, **83**, 635.
18. H.-S. Hwang and S. Z. Erhan, *J. Am. Oil Chem. Soc.*, 2001, **78**, 1179.
19. H.-S. Hwang, A. Adhvaryu and S. Z. Erhan, *J. Am. Oil Chem. Soc.*, 2003, **80**, 811.
20. B. K. Sharma, A. Adhvaryu, Z. Liu and S. Z. Erhan, *J. Am. Oil Chem. Soc.*, 2006, **83**, 129.
21. A. Adhvaryu and S. Z. Erhan, *Ind. Crops Prod.*, 2002, **15**, 247.
22. B. R. Moser, B. K. Sharma, K. M. Doll and S. Z. Erhan, *J. Am. Oil Chem. Soc.*, 2007, **84**, 675.
23. Z. Liu and D. A. Knetzer, *Green Mater.*, 2013, **1**, 87.
24. M. Ionescu, Z. Petrović and X. Wan, *J. Polym. Environ.*, 2007, **15**, 237.
25. J. Xu, Z. Liu, S. Z. Erhan and C. J. Carriere, *J. Am. Oil Chem. Soc.*, 2002, **79**, 593.
26. Z. Liu and S. Erhan, *J. Am. Oil Chem. Soc.*, 2010, **87**, 437.
27. Z. Liu, K. M. Doll and R. A. Holser, *Green Chem.*, 2009, **11**, 1774.
28. S. F. Thames and H. Yu, *Surf. Coat. Technol.*, 1999, **115**, 208.
29. Z. S. Liu and S. Z. Erhan, *J. Appl. Polym. Sci.*, 2002, **84**, 2386.
30. Z. Liu and G. Biresaw, *J. Agric. Food Chem.*, 2011, **59**, 1909.

31. R. E. Parker and N. S. Isaacs, *Chem. Rev.*, 1959, **59**, 737.
32. T. W. Findley, D. Swern and J. T. Scanlan, *J. Am. Chem. Soc.*, 1945, **67**, 412.
33. W. R. Schmitz and J. G. Wallace, *J. Am. Oil Chem. Soc.*, 1954, **31**, 363.
34. S. P. Bunker and R. P. Wool, *J. Polym. Sci., Part A: Polym. Chem.*, 2002, **40**, 451.
35. P. A. Z. Suarez, M. S. C. Pereira, K. M. Doll, B. K. Sharma and S. Z. Erhan, *Ind. Eng. Chem. Res.*, 2009, **48**, 3268.
36. M. Guidotti, N. Ravasio, R. Psaro, G. Ferraris and G. Moretti, *J. Catal.*, 2003, **214**, 242.
37. M. Guidotti, N. Ravasio, R. Psaro, E. Gianotti, L. Marchese and S. Coluccia, *Green Chem.*, 2003, **5**, 421.
38. M. Guidotti, N. Ravasio, R. Psaro, E. Gianotti, S. Coluccia and L. Marchese, *J. Mol. Catal. A: Chem.*, 2006, **250**, 218.
39. M. Guidotti, R. Psaro, N. Ravasio, M. Sgobba, E. Gianotti and S. Grinberg, *Catal. Lett.*, 2008, **122**, 53.
40. G. J. Piazza and T. A. Foglia, *J. Am. Oil Chem. Soc.*, 2005, **82**, 481.
41. G. J. Piazza and T. A. Foglia, *J. Am. Oil Chem. Soc.*, 2006, **83**, 1021.
42. A. Kockritz and A. Martin, *Eur. J. Lipid Sci. Technol.*, 2008, **110**, 812.
43. B. K. Sharma, K. M. Doll and S. Z. Erhan, *Green Chem.*, 2007, **9**, 469.
44. K. M. Doll, B. K. Sharma and S. Z. Erhan, *Clean: Soil, Air, Water*, 2008, **36**, 700.
45. J. H. Clements, *Ind. Eng. Chem. Res.*, 2003, **42**, 663.
46. K. M. Doll, J. A. Kenar and S. Z. Erhan, *Chem. Today*, 2007, **25**, 7.
47. W. I. Riedeman, *US Pat.*, 2 858 286, 1958.
48. W. I. Riedeman, *US Pat.*, 2 826 591, 1958.
49. J. A. Kenar and I. D. Tevis, *Eur. J. Lipid Sci. Technol.*, 2005, **107**, 135.
50. B. Tamami, S. Sohn, G. L. Wilkes and B. Tamami, *J. Appl. Polym. Sci.*, 2004, **92**, 883.
51. G. L. Wilkes, S. Sohn and B. Tamami, *US Pat.*, 7 045 577, 2006.
52. N. Mann, S. K. Mendon, J. W. Rawlins and S. F. Thames, *J. Am. Oil Chem. Soc.*, 2008, **85**, 791.
53. L. Zhang, Y. Luo, Z. Hou, Z. He and W. Eli, *J. Am. Oil Chem. Soc.*, 2014, **91**, 143.
54. R. A. Holser, *J. Oleo Sci.*, 2007, **56**, 629.
55. C. A. Eckert, B. L. Knutson and P. G. Debenedetti, *Nature*, 1996, **383**, 313.
56. K. M. Doll and S. Z. Erhan, *Green Chem.*, 2005, **7**, 849.
57. K. M. Doll and S. Z. Erhan, *J. Agric. Food Chem.*, 2005, **53**, 9608.
58. G. J. Piazza, A. Nunez and T. A. Foglia, *J. Am. Oil Chem. Soc.*, 2003, **80**, 901.
59. H. Benecke, B. R. Vijayendran and J. Cafmeyer, *US Pat.*, 2010/0029523 A1, 2005.
60. A. Campanella and M. A. Baltanas, *Chem. Eng. J.*, 2006, **118**, 141.
61. S. Z. Erhan, K. M. Doll and B. K. Sharma, *US Pat.*, 8 173 825, 2012.
62. K. M. Doll, B. K. Sharma and S. Z. Erhan, *Ind. Eng. Chem. Res.*, 2007, **46**, 3513.

63. B. K. Sharma, K. M. Doll and S. Z. Erhan, *Bioresour. Technol.*, 2008, **99**, 7333.
64. J. Filley, *Bioresour. Technol.*, 2005, **96**, 551.
65. K. M. Doll and S. Z. Erhan, *Green Chem.*, 2008, **10**, 712.
66. J. J. Bozell, L. Moens, D. C. Elliott, Y. Wang, G. G. Neuenscwander, S. W. Fitzpatrick, R. J. Bilski and J. L. Jarnefeld, *Resour. Conserv. Recycl.*, 2000, **28**, 227.

CHAPTER 3

Preparation of Bio-polymers from Plant Oils in Green Media

ZENGSHE LIU

USDA, ARS, National Center for Agricultural Utilization Research, Bio-Oils Research Unit, 1815 N. University Street, Peoria, IL 61604, USA[†]
Email: kevin.liu@ars.usda.gov

3.1 Introduction

About 100 years ago, chemicals were commonly derived from plant-based materials. However, the availability of low-cost fossil resources completely changed the landscape of the chemical industry to the point that almost all chemicals are now based on petroleum feedstock. As a result of the tremendous pressure put on fossil resources by a growing global population's needs, it is understood that, sooner or later, the era of low-cost, easily available fossil resources will come to an end.[1] It is notable that over the past 20 years, the price of crude oil has increased approximately 10-fold.[2] With recent social emphasis on the environment and resource renewability, utilizing natural materials as renewable resources for industrial products is attracting great attention. Therefore, the widest possible usage of renewable raw materials will significantly contribute to sustainable development for the 21st Century. Furthermore, the utilization of renewable raw materials can, in some cases, meet other principles of green chemistry.[2]

[†]Mention of trade names or commercial products in this publication is solely for the purpose of providing specific information and does not imply recommendation or endorsement by the U.S. Department of Agriculture. USDA is an equal opportunity provider and employer.

RSC Green Chemistry No. 29
Green Materials from Plant Oils
Edited by Zengshe Liu and George Kraus
© The Royal Society of Chemistry 2015
Published by the Royal Society of Chemistry, www.rsc.org

Among bio-based products from agricultural resources, such as plant oils, polysaccharides (mainly cellulose and starch), sugars, wood, and others, plant oils make up the greatest portion of the current consumption of renewable raw materials in the chemical industry. Plant oils are non-toxic, biodegradable, non-polluting, and relatively harmless to the environment. In addition to the above-mentioned advantages, plant oils are liquids at room temperature and can be chemically modified because of their excellent solubility in most reaction media. Approximately 80% of the global fat production is vegetable oil, and 20% is of animal origin (share decreasing).[3] In 2009, soybean accounted for about 23.4% of planted acreage in the USA, just behind corn, which accounts for about 26.1%, and ahead of wheat, which accounts for about 18.0%. About three billion bushels of soybean are grown annually in the USA, of which the current market demand is about 2.9 billion. Developing new uses for surplus soybean oil (SBO) is important to prevent price depression due to oversupply.

SBO has a rich history of use as mechanical lubricants and soaps, ranging from ancient times in East Asia to paints, fuel and biodiesel currently worldwide. The most recent large-scale industrialized use of SBO and other vegetable oils is in the area of alternative fuels for diesel engines. Polymeric materials prepared from vegetable oils have become increasingly important because of their low cost, readily availability and bio-degradability. Polymerized SBOs have been employed in printing inks and paints.[4,5] Nowadays, there is a growing interest in producing various green materials from plant oils. For example, polyurethanes from vegetable-oil-based polyols have been widely studied by the Petrović group and others.[6,7] Since the start of the decline in petroleum supplies, preparation of polymers from renewable sources has become more important due to their attractive properties, such as bio-degradability and, in some cases, they are cheaper than petroleum polymers. Particularly, in the USA, SBO is one of the most abundant renewable resources.

SBO is a triglycerol with saturated and unsaturated fatty acids, 80–85% being unsaturated fatty acids. SBO has around 4.6 double bonds from oleic (C18:1), linoleic (C18:2) and linolenic (C18:3) acids, as shown in Figure 3.1. Unfortunately, the internal 1,2-disubstituted non-conjugated double bonds are of low reactivity and polymerize with difficulty.[6] However, these double bonds may be converted into more reactive oxirane (or epoxide) moieties by reaction with peracids or peroxides. Epoxidized soybean oil (ESO) used as a raw material for the synthesis of new polymers has been reported by our group.[8,9] Figure 3.2 shows the structure of ESO. Euphorbia oil (EuO) is

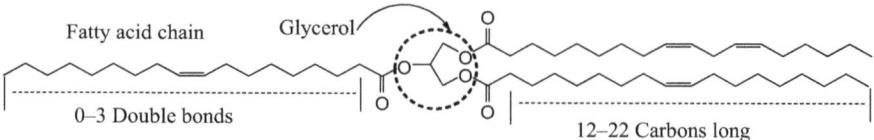

Figure 3.1 The structure of soybean oil (SBO).

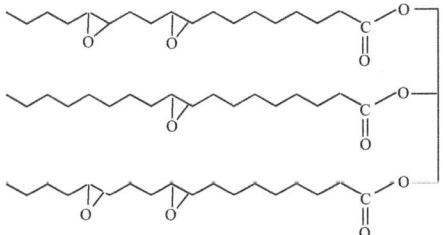

Figure 3.2   The structure of epoxidized soybean oil (ESO).

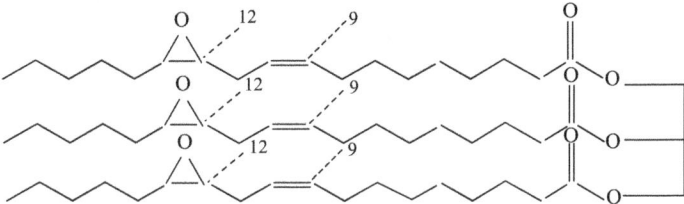

Figure 3.3   The structure of euphorbia oil (EuO).

natural epoxy oil from *Euphorbia resinifera* seeds, and its structure is shown in Figure 3.3. The advantage of polymerization of EuO is that the epoxidation reaction of plant oils is avoided. The polymerization of EuO has been reported by Liu and co-workers.[10]

Liquid carbon dioxide was first used as a solvent for polymerization in 1960,[11] but it is only recently that there has been an explosion of interest in the use of supercritical carbon dioxide (scCO$_2$). This is due to its low toxicity of CO$_2$ and the lack of solvent residues in the final product, as well as it being an environmentally acceptable alternative to conventional solvents.[12–14] Also, CO$_2$ is inexpensive, readily available and non-flammable. Supercritical CO$_2$ is a green medium for organic reactions and polymerizations. The critical parameters of CO$_2$ are a P_c (critical pressure) of 73.8 bar and a T_c (critical temperature) of 31.1 °C. Although the T_c is low, obtaining the high P_c may not be easy. However, CO$_2$ in the liquid state provides many of the advantages of the supercritical system, but at a lower pressure and temperature. We wish to review here the work we carried out on the polymerization of plant oils such as SBO, ESO and EuO using supercritical and subcritical CO$_2$ as reaction media. The catalyst used for the polymerization was a Lewis acid, boron trifluoride diethyl etherate, BF$_3 \cdot$ OEt$_2$. The formed polymers are referred to "PSBO", "RPESO" and "RPEuO" for SBO, ESO and EuO, respectively. The thermal properties of the resulting polymers have been determined. Also, the conversion of RPESO into polysoap (HPESO) and its surface and interface properties have been studied in order to explore their applications in various areas. Here, P means polymerized, RP means ring-opening polymerized, H means hydrolyzed.

3.2 Materials

Soybean oil (SO-5) was purchased from Purdue Farms Inc., Refined Oil Division, (Salisbury, MD). ESO was obtained from Elf Atochem Inc. (Philadelphia, PA) and used as received. The euphorbia seeds (51% oil content) were provided by Dr Richard Roseberg, Oregon State University. The oil from the seeds was extracted using a hydraulic press (Model 3851, Carver, Inc., Wabash, IN) fitted with cage equipment (Carver Catalog No. 2094). Purified and redistilled $BF_3 \cdot OEt_2$, was obtained from Sigma–Aldrich (St. Louis, MO). Sodium hydroxide (97.5%) was obtained from Fisher Scientific (Fair Lawn, NJ), and potassium hydroxide (ACS reagent, 88.3%) was obtained from J. T. Baker (Phillipsburg, NJ). Triethanolamine (98%) was obtained from Sigma–Aldrich. Deionized water was purified to a conductivity of 18.3 MΩ for the preparation of aqueous polysoap solutions for surface and interfacial tension measurements. Carbon dioxide (>99.8%) was obtained from Linde Gas LLC (Independence, OH). Sodium bicarbonate (certified ACS grade) was obtained from Fisher Scientific (Fair Lawn, NJ).

3.3 Experimental

3.3.1 Polymerization Procedure in CO_2

Polymerization of plant oils was carried out in a 300 mL or 100 mL high-pressure reactor. The reactor was a Parr (Moline, IL) 4560 mini benchtop unit equipped with a Parr 4843 controller and thermocouple. A schematic diagram of the experimental set-up is depicted in Scheme 3.1.

The reactor was attached to an Isco Model 260D high-pressure syringe pump used to charge the reactor with CO_2. In a typical experiment of the

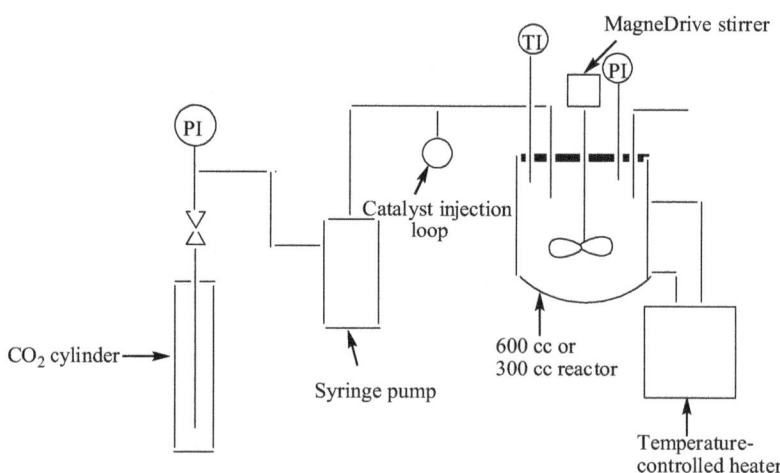

Scheme 3.1 A schematic diagram of the experimental set-up.

ring-opening polymerization of ESO, 30 g of ESO was added to the reactor, which was then sealed. N_2 was purged into the reactor for 5 min. CO_2 was pumped in until the reactor pressure, 62.1 bar, was reached. A controller (Parr 4843) was used to control the temperature. Once the reactor was brought to the appropriate temperature (generally 25 °C), $BF_3 \cdot OEt_2$ was charged into the reactor by using a Rheodyne injector. Then CO_2 was pumped in to clear the injection loop at a pump pressure of 65.5 bar. After reaction for 2 h, 2 mL of ethanol/H_2O (1 : 1) was added to the reactor to deactivate the catalyst. The white-colored polymers (RPESO) were washed sequentially with 5% aqueous sodium bicarbonate, and H_2O. The RPESO polymer was dried in a vacuum at 60 °C. About 28–30 g of the RPESO sample was obtained. Soxhlet extraction with hexane as the refluxing solvent was used to extract the soluble substance from the RPESO samples for analysis. The details of the preparation of PSBO, RPESO and RPEuO were reported by Liu and co-workers.[15–17]

3.3.2　Analysis

3.3.2.1　*Fourier-transform Infrared Spectroscopy*

FT-IR spectra were recorded on a Thermo Nicolet Nexus 470 FT-IR system (Madison, WI) coupled with a Smart ARK accessory for liquid samples in a scanning range of 650–4000 cm^{-1} for 32 scans at a spectral resolution of 4 cm^{-1}. Solid samples were recorded on this FT-IR system coupled with the Smart Orbit accessory.

3.3.2.2　*Nuclear Magnetic Resonance Spectroscopy*

^{1}H-NMR and ^{13}C-NMR spectra for the extracted soluble substances from the RPESO samples were recorded using a Bruker (Rheinstetten, Germany) ARX-500 NMR spectrometer operating at a frequency of 500.13 and 125.77 MHz, respectively, using a 5 mm inverse Z-gradient probe in $CDCl_3$ (Cambridge Isotope Laboratories, Andover, MA). Solid-state ^{13}C-NMR spectra for extracted insoluble substances were recorded using a Bruker ARX-300. NMR spectrometer operation at a frequency of 300.12 and 75.46 MHz.

3.3.2.3　*Gel Permeation Chromatography*

GPC profiles were obtained on a Waters HPLC system including a 1515 isocratic HPLC pump, 717 plus automated injector, column heater, and controlled with the Breeze software obtained from Waters Corporation (Milford, MA). The columns used for separation were a pair of PLgel 3 μm MIXED-E, 300×7.5 mm columns and a PLgel 5 μm guard, 50×7.5 mm column (part numbers PL1110-6300, PL1110-1520, respectively) from Polymer Laboratories (Varian, Inc., Amherst, MA). Signals generated from a mini-DAWN TREOS triple-angle light-scattering detector and Optilab rEX refractive index (RI) detector obtained from Wyatt Technology Corporation

(Santa Barbara, CA) were processed using the Astra V macromolecular characterization software also from Wyatt Technology Corporation. THF was used as the mobile phase at a flow rate of 1 mL min^{-1} and columns were maintained at 40 °C. The liquid-phase samples were brought into solution with THF stabilized with butylated hydroxytoluene from Fisher Scientific (Suwanee, GA) at a known concentration near 4.00×10^{-3} g mL^{-1}. The Waters' autosampler was used to make 100 μL injections from a 1 mL sample vial. Linear polystyrene standards (Polymer Laboratories), $M_n = 580$–100 K, $M_w/M_n = 1$ were used for the calibration of the molecular weights of all polymers of RPEO. The Astra V software was used to calculate the molecular weights.

3.3.2.4   *Differential Scanning Calorimetry*

DSC thermograms of the test samples were recorded using a TA Instruments (New Castle, DE) Q2000 model DSC with an autosampler. Typically, about 10 mg of the RPESO sample was accurately weighed in an aluminum pan and sealed with pin-perforated lids. The DSC oven was ramped at 10 °C per min to 110 °C per min to eliminate thermal history and possible moisture. A refrigerated cooling system was used to equilibrate the samples at −60 °C, from 110 °C, at a rate of 5 °C per min. Data was recorded while the oven temperature was increased from −60 °C to 150 °C at a rate of 5 °C per min. The DSC method applied an inert atmosphere by purging the oven with nitrogen at 50 mL min^{-1}. Thermal Advantage and Universal Analysis software provided by TA Instruments were used for data analysis.

3.3.2.5   *Thermogravimetric Analysis*

A TA Instruments Q500 thermogravimeter with an autosampler was used to measure the weight loss of the SA-RPESO (super acid catalyst used) samples under a flowing nitrogen atmosphere. Generally, 20 mg of an RPESO sample was used in the TGA. The samples were heated from 30 to 600 °C at a heating rate of 10 °C per min and the weight loss was recorded as a function of temperature.

3.3.3   **Hydrolysis of RPESO Polymers**

A solution of 2.5 g of RPESO in 50 mL of 0.4 N NaOH was refluxed for 24 h. The solution was then filtered with a filter paper and cooled to room temperature. The resulting gel was precipitated with 80 mL of 1.0 N HCl, followed by several washings with water, and finally with two more washings with 10% (v/v) aqueous acetic acid. The resulting polymer was dried overnight at 80 °C in an oven. The sample was further dried under vacuum at 70 °C to a constant weight. The procedure yielded around 2.1 g (about 84% yield) of product, and is referred to as "HPESO polyacid".

3.3.4 Preparation of HPESO and HPSO Polysoaps with Different Counterions

Procedures for the preparation of polysoaps (HPESO) were as follows: Typically, 1.0 g of a HPESO polyacid sample was weighed into a 50 mL beaker and the required amount of NaOH or KOH (to neutralize all the carboxylic acid groups) was dissolved in 10 mL water (18.3 MΩ). The beaker containing the HPESO sample and base solution was then placed in a 75 °C water bath, and stirred with a glass rod until the HPESO sample had dissolved. The resulting solution was then transferred to a 100 mL volumetric flask. The beaker was rinsed three times with 10 mL of water each and added to the volumetric flask. The solution was then cooled to room temperature and filled with water to the 100 mL mark. The aqueous stock solution of the HPESO-TEA salt was prepared by using a 2.5 : 1 ratio of triethanolamine (TEA) to the carboxylic acid groups in HPESO.

3.3.5 Dynamic Surface and Interfacial Tension

Dynamic surface and interfacial tension measurements were conducted using the axisymmetric drop shape analysis (ADSA) method[19] on an FTA-200 automated goniometer (First Ten Angstroms, Portsmouth, VA) equipped with the FTA-32 v2.0 software. All dynamic surface and interfacial tension measurements were conducted at room temperature (23 ± 2 °C). Details of the data processing were reported in ref. 18.

3.4 Results and Discussion

3.4.1 Effect of Molecular Structure on the Polymerization

The results of study of the effects of variable conditions such as catalyst concentration, reaction time, initiator concentration, monomer concentration, *etc.* on the molecular weights and glass-transition temperatures (T_g) of the formed polymers are reported in the following tables: Tables 3.1 and 3.2 report the results for PSBO; Tables 3.3–3.5 report the results for RPESO; and Tables 3.6–3.8 report the results for RPEuO. It can be seen that the polymerization of plant oils with epoxy groups is much easier than the polymerization of SBO. The conditions for the polymerization of ESO and

Table 3.1 Effect of catalyst amount on the molecular weight of PSBO.

Sample	SO/g	P/bar	T/°C	Reaction time/h	Catalyst/mol	M_w/g mol^{-1}
1 (control)	100	110	140	2	0	1300
2	100	110	140	2.5	0.014	24 582
3	100	110	140	2.5	0.018	82 608
4	100	110	140	4.0	0.014	85 624
5	100	110	140	4.0	0.018	110 360

Table 3.2 Effect of reaction time on the molecular weight of PSBO.

Sample	SO/g	P/bar	T/°C	Reaction time/h	Catalyst/mol	M_w/g mol^{-1}
1 (control)	100	110	140	2	0	1300
2	100	110	140	2	0.018	21 842
3	100	110	140	2.5	0.018	82 608
4	100	110	140	3.0	0.018	118 300
5	100	110	140	4.0	0.018	110 360

Table 3.3 The glass-transition temperatures of RPESO polymers prepared at various initiator loadings.

Sample	Polym. temperature/°C	Initiator loading/mmol	T_g/°C
RPESO-V	25	1.87	− 13.8
RPESO-II	25	2.79	− 15.9
RPESO-VI	25	4.75	− 15.7
RPESO-VII	25	7.72	− 24.1

Table 3.4 Data related to RPESO polymers.

Sample	Polym. temperature/°C	Initiator loading/ mmol	Monomer loading/ mol	M_w (soluble)/ g mol^{-1}	Soxhlet extraction/ wt%	
					Soluble	Insoluble
RPESO-I	20	2.79	0.03	1040	21	79
RPESO-II	25	2.79	0.03	1937	5	95
RPESO-III	30	2.79	0.03	1429	16	84
RPESO-IV	35	2.79	0.03	1237	17	83
RPESO-V	25	1.87	0.03	2364	11	89
RPESO-VI	25	4.75	0.03	3606	3	97
RPESO-VII	25	7.72	0.03	3318	7	93
RPESO-VIII	25	2.79	0.04	4578	24	91
RPESO-IX	25	2.79	0.05	3066	20	94
RPESO-X	25	2.79	0.06	3349	17	83

Table 3.5 Thermal stability data for the RPESO polymers.

Sample/Temperature	200–220 °C	240–320 °C	340–450 °C
RPESO-I	5 wt% loss	13 wt% loss	82 wt% loss
RPESO-II	stable	7 wt% loss	91 wt% loss
RPESO-III	5 wt% loss	14 wt% loss	81 wt% loss
RPESO-IV	3 wt% loss	12 wt% loss	82 wt% loss
RPESO-V	5 wt% loss	10 wt% loss	88 wt% loss
RPESO-VI	stable	6 wt% loss	91 wt% loss
RPESO-VII	stable	5 wt% loss	90 wt% loss
RPESO-VIII	5 wt% loss	15 wt% loss	82 wt% loss
RPESO-IX	2 wt% loss	12 wt% loss	83 wt% loss
RPESO-X	stable	10 wt% loss	85 wt% loss

Table 3.6 The glass-transition temperatures of RPEuO prepared at various initiator concentrations.

Sample	Polym. temperature/°C	Initiator loading /mmol	T_g/°C
RPEuO-I	25	1.87	− 22.7
RPEuO-II	25	2.79	− 15.9
RPEuO-III	25	4.66	− 15.0

Table 3.7 Data related to the extracted soluble substances from RPEuO polymers.

Sample	Polym. temperature/°C	Initiator loading/mmol	Monomer loading/g	Soxhlet extraction/ wt% Soluble	Insoluble
RPEuO-I	25	1.87	30	30	70
RPEuO-II	25	2.79	30	29	71
RPEuO-III	25	4.66	30	14	86
RPEuO-IV	25	4.66	40	22	78
RPEuO-V	25	4.66	50	35	65

Table 3.8 Thermal stability data for the RPEuO polymers.

Sample/Temperature[a]	200–220 °C	240–320 °C	340–450 °C
RPEuO-I	stable	2 wt% loss	87 wt% loss
RPEuO-II	stable	3 wt% loss	85 wt% loss
RPEuO-III	stable	4 wt% loss	85 wt% loss
RPEuO-IV	stable	3 wt% loss	84 wt% loss
RPEuO-V	stable	3 wt% loss	88 wt% loss

[a]Roman numbers here are only to identify the samples prepared at different conditions, such as temperature, catalyst amounts and initiator amounts.

EuO are mild, such as at room temperature, low pressure and shorter reaction times. The polymerization of SBO needs harsh reaction conditions, such as temperature more than 100 °C, pressures of 110 bar and longer reaction times. This is because the internal 1,2-disubstituted non-conjugated double bonds create difficulties in the polymerization reaction. When those double bonds in SBO are converted into oxirane (or epoxide) moieties, they show more reactivity.

3.4.2 Characterization of the Polymers

Confirmation of the resulting polymer structures was accomplished by a number of different techniques, ^{1}H-NMR, ^{13}C-NMR, FT-IR, solid-state NMR and GPC. Figure 3.4 shows the ^{1}H-NMR spectra of SBO and PSBO.

The signals at 5.40 ppm are characteristic of olefinic hydrogens and the signals at 5.1–5.3 ppm represent the methine proton of $-CH_2-CH-CH_2-$ (the glycerin backbone). The signals at 4.0–4.4 ppm are from the methylene protons of $-CH_2-CH-CH_2-$. The peak at 2.80 ppm corresponds to the protons in

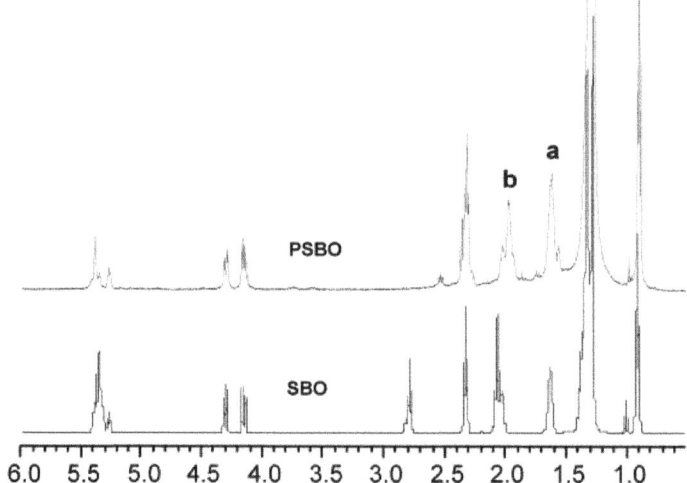

Figure 3.4 ^{1}H-NMR spectra of SBO and PSBO.

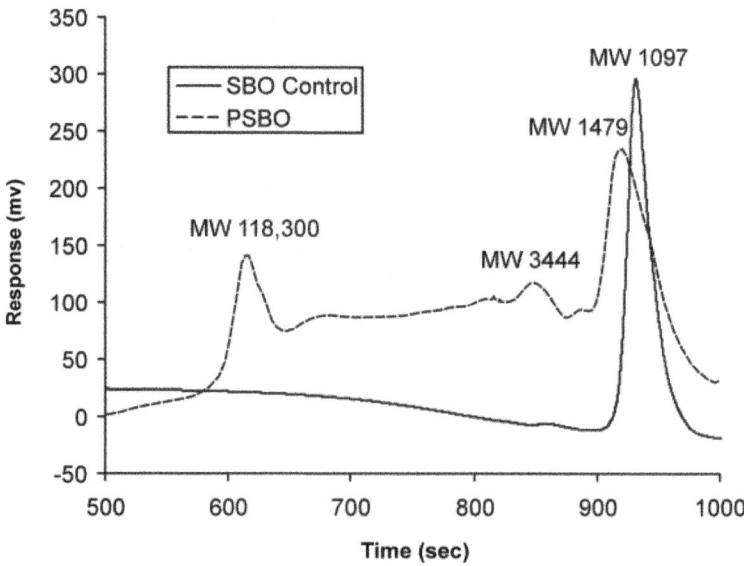

Figure 3.5   Overlays of SBO (control) and PSBO obtained from the RI detector of the GPC.

the CH_2 groups between two C=C bonds. The signals at 2.10 ppm are the α-methylene protons CH_2 adjacent to the C=C bonds. It can be seen from Figure 3.4 that the peaks at 5.1–5.4 ppm and 2.10 ppm for PSBO are greatly decreased compared to those for SBO. The peak at 2.80 ppm for PSBO has disappeared. These observations indicate that the polymerization of SBO has occurred, and the number of C=C bonds has greatly reduced. Figure 3.5

shows the GPC profile of SBO and PSBO. From the GPC profiles, it is of note that between the SBO (control) peak (1097) and the high-molecular-weight peak of PSBO (118 300), there are only two peaks at (1479) and (3444), corresponding to a dimer and a trimer of the SBO molecule, respectively.

ESO, a liquid oil at room temperature, shown in Figure 3.6 (bottle), was polymerized under mild conditions to form a white-colored polymer (RPESO) shown in Figure 3.6 (dish). Figure 3.7 shows FT-IR spectra of ESO, the hexane-extracted soluble substances, and the insoluble portion of an RPESO sample. It can be clearly seen that absorption at 838 cm^{-1} for the oxirane group has disappeared in the FT-IR spectra of insoluble RPESO, indicating that ring-opening polymerization of ESO has taken place. The FT-IR data also show the oxirane absorption band from ESO is not present in the hexane-extracted soluble substances. The results suggest the hexane-extracted soluble substances are the ring-opening products with low molecular weights.

Figure 3.8 shows the solid-state ^{13}C-NMR spectrum of the insoluble substances obtained after the extraction of the RPESO sample. The spectrum clearly shows the presence of ester carbonyls (δ 167 ppm) from the oil triglyceride structure. There is no signal at δ 52 ppm, which shows that the C–C epoxy bond has disappeared. This indicates that ESO was polymerized through ring-opening polymerization and that highly cross-linked polymers were formed.

Figure 3.9 shows the ^{1}H-NMR spectra of EuO and the hexane-extracted RPEuO soluble fraction. The peaks in the δ 2.7–2.9 ppm region related to

Figure 3.6 Photos of an RPESO polymer (dish) and the starting material, ESO (bottle).

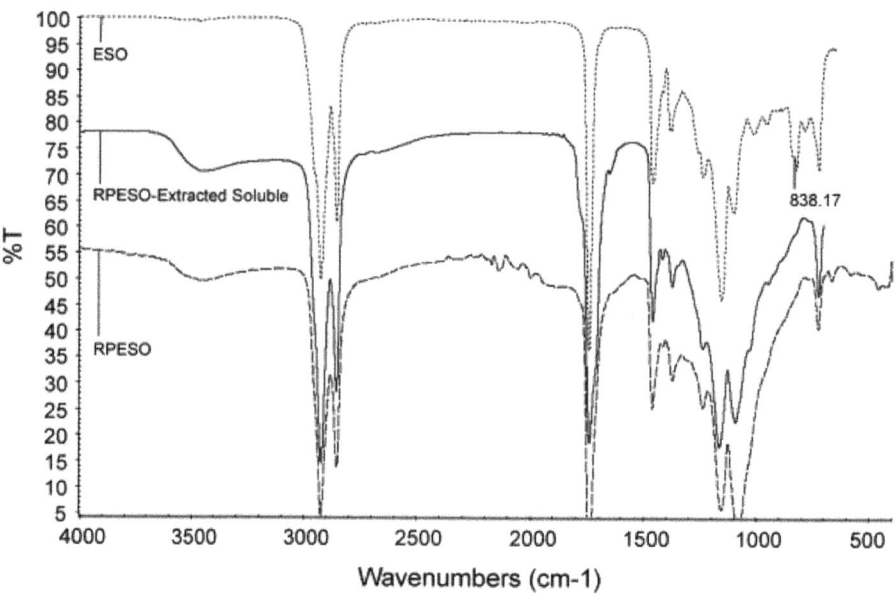

Figure 3.7   FT-IR spectra of an RPESO polymer (soluble and insoluble fractions) and
the starting material, ESO.

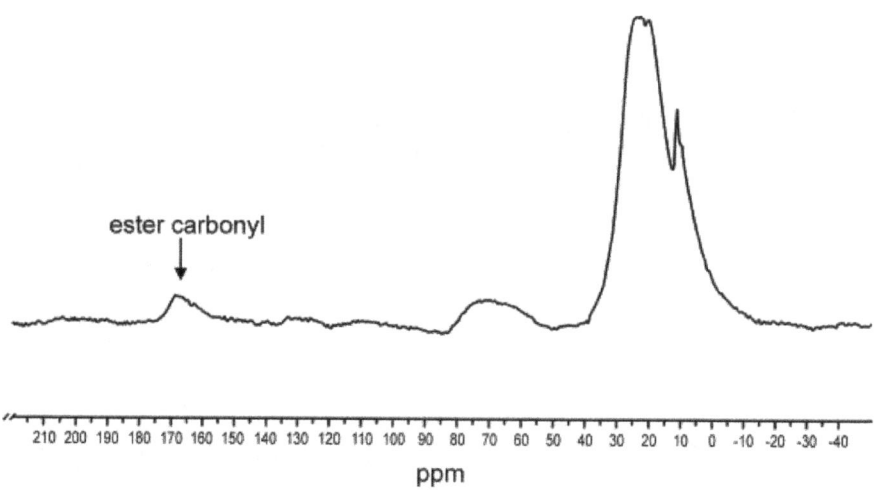

Figure 3.8   Solid-state ^{13}C-NMR spectrum of RPESO.

epoxy protons are apparent in both spectra. The methine proton –CH$_2$–CH–
CH$_2$– of the glycerol backbone at δ 5.1–5.3 ppm, and the methylene
protons –CH$_2$–CH–CH$_2$– of the glycerol backbone at δ 4.1–4.3 ppm are
observed, which means the triglyceride structure of EuO is not disturbed.
The solid-state ^{13}C-NMR spectrum of RPEuO clearly shows (Figure 3.10)

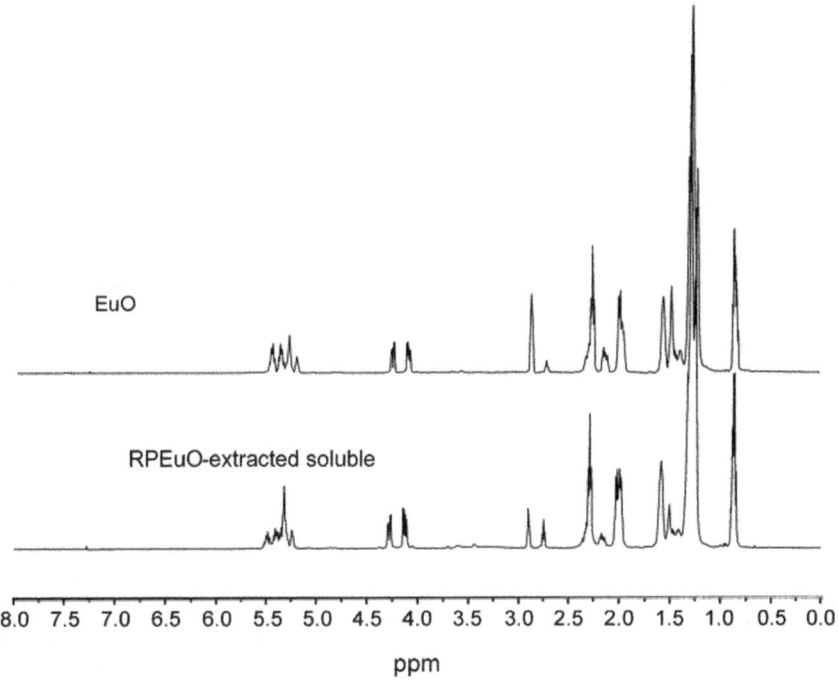

Figure 3.9 ¹H-NMR spectra of the EuO and the RPEuO extracted soluble fraction.

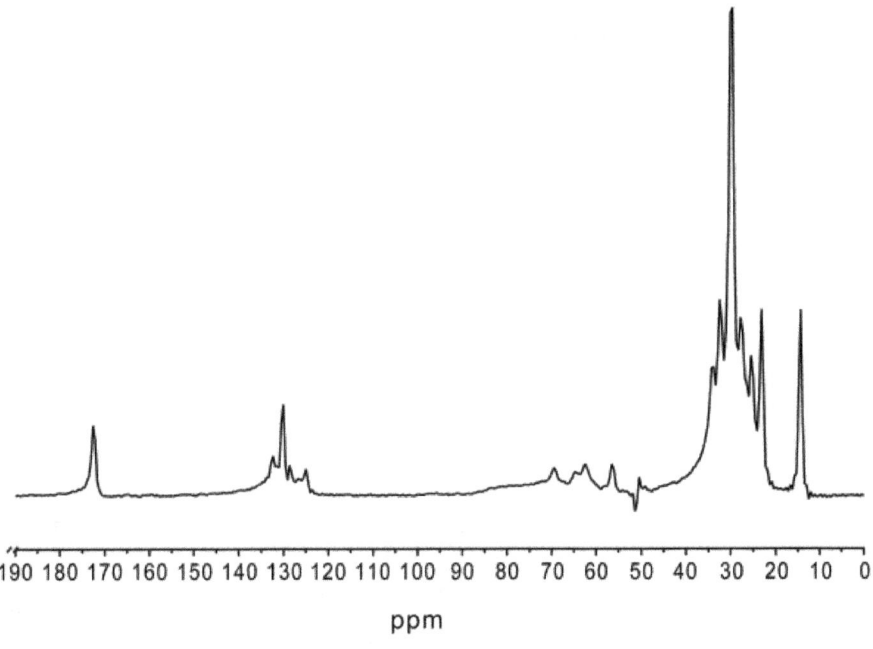

Figure 3.10 Solid-state ¹³C-NMR spectrum of RPEuO.

the presence of ester carbonyls (δ 173 ppm) from the oil triglyceride structure. There is no signal at δ 52 ppm which indicates that the C–C epoxy bond has been converted to an ether, and confirms that the EuO has been polymerized through ring-opening polymerization and that cross-linked polymers were formed.

3.4.3   Thermal Stability of the Polymers

To better understand the thermal properties of the resulting polymers, TGA was used to investigate their thermal decomposition behavior under a nitrogen atmosphere. Figure 3.11 shows the TGA curve of RPEuO-I (the non-soluble fraction after extraction). It can be seen that RPEuO-I appears to be thermally stable at temperatures below 200 °C. Two distinct temperature regions are observed where samples experienced weight loss (220–320 °C) and (340–437 °C). The material slowly loses about 2% of its weight at temperatures between 220 and 320 °C, followed by an abrupt weight loss of 80% between 340 and 437 °C. TGA measurements revealed a total 98% weight loss observed at temperatures between 240 and 437 °C. For the other RPESO samples obtained at various temperatures and initiator concentrations, their TGA curves showed similar behavior to RPEuO-I. Some samples showed less than 5 wt% loss around 220 °C. Table 3.5 summarizes the thermal stability

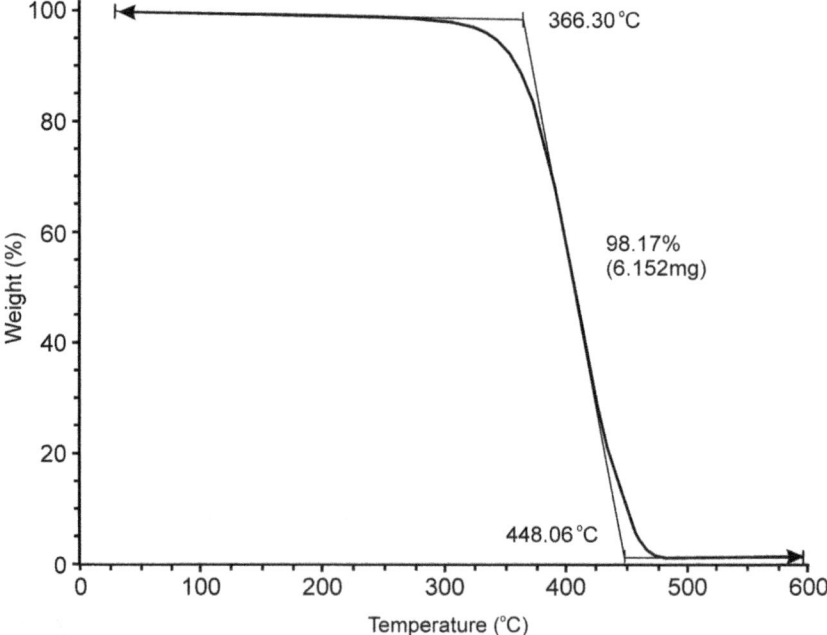

Figure 3.11   TGA thermogram of RPEuO-I, weight loss *vs.* temperature (under a N$_2$ atmosphere).

results of the RPEuO polymers studied by TGA. It can be seen that in the first temperature region, about 15 wt% loss occurs and 79–91 wt% of the weight is lost in the second temperature region for all the RPEuO polymers. Table 3.8 summarizes the thermal stability results of RPEuO polymers studied by TGA.

3.4.4 Surface Tension of Aqueous HPESO Polysoaps

A series of aqueous solutions of the HPESO polysoaps with Na^+, K^+ and TEA^+ counterions were prepared, as mentioned above, and their dynamic surface tensions were investigated. Table 3.9 lists the concentrations investigated for the HPESO polysoaps. The molecular weights were either 2.6 or 3.2 kg mol^{-1}.

Typical dynamic surface tension data for an aqueous HPESO salt are illustrated in Figure 3.12. The data in Figure 3.12 shows triplicate measurements on an aqueous HPESO salt of $M_w = 3.2$ kg mol^{-1}, with a TEA^+

Table 3.9 Polysoaps investigated in this work.

Polysoap	Counterion[a]	M_w/kg mol^{-1}
HPESO-004K	K^+	2.615
HPESO-004A	TEA^+	2.615
HPESO-003N	Na^+	3.219
HPESO-003K	K^+	3.219
HPESO-003A	TEA^+	3.219

[a]TEA^+ = Triethonalamonium.

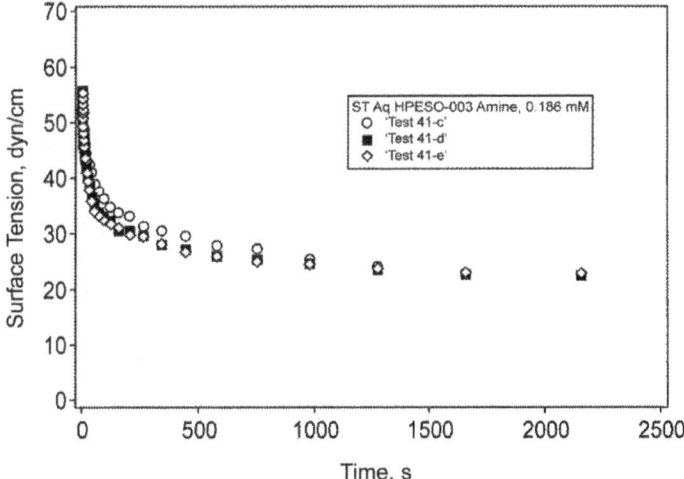

Figure 3.12 Typical data from repeat measurement of the dynamic surface tension of an aqueous polysoap on an automated pendant drop goniometer. The polysoap had a M_w of 3.2 kg mol^{-1} with the TEA^+ counterion.

Table 3.10   Effect of HPESO polysoaps on the minimum equilibrium surface tension of water.

Polysoap	M_w/kg mol^{-1}	Surface tension/dyn cm^{-1}
HPESO-004K	2.6	19.9 ± 0.6
HPESO-004A	2.6	22.9 ± 0.4
HPESO-003N	3.2	21.6 ± 0.5
HPESO-003K	3.2	19.9 ± 1.1
HPESO-003A	3.2	23.9 ± 1.4

counterion, and 0.186 mM concentration. As can be seen from Figure 3.12, the surface tension decreases sharply with time initially, then displays a gradual decrease, and finally levels off to a more-or-less constant value over a long period of time. This profile is consistent with the generally accepted mechanism of amphiphile diffusion from the droplet–bulk to the droplet–air interface. Initially, the concentration of polysoap at the interface is low, and, as a result, the surface tension is high. This causes rapid diffusion of poly-soap molecules to the interface, causing a rapid increase in the surface concentration and a corresponding rapid decrease of surface tension. As the concentration of the polysoap molecules at the air–water interface ap-proaches the equilibrium value, the diffusion slows and so the rate of surface tension declines. After a very long time, the concentration of polysoap at the interface reaches the equilibrium value and the surface tension become constant and independent of time. A detailed discussion related to the surface tension of aqueous HPESO polysoaps was reported in ref. 18.

Table 3.10 shows a summary of HPESO polysoaps on the minimum equilibrium surface tension of water. It can be seen that the minimum equilibrium surface tension values of the various aqueous HPESO polysoaps are in a narrow range of 20–24 dyn cm^{-1}, indicating that the polysoaps have similar surface energies.

3.4.5   Interfacial Tension of HPESO Polysoaps between Water and Hexadecane

Figure 3.13 shows the dynamic interfacial tension between water and hexa-decane in the presence of a HPESO polysoap. The profile of the time *versus* interfacial surface tension data is similar to the time *versus* surface tension data discussed earlier. In the presence of a HPESO polysoap in the water, the water–hexadecane interfacial tension shows an initial fast drop, followed by a gradual drop and, finally a constant equilibrium value for a very long time. It should be noted that the equilibrium interfacial tension value is a function of the concentration of HPESO in the water phase. As demonstrated in Figure 3.13 the equilibrium interfacial tension decreases with increasing HPESO concentration in the water phase. Table 3.11 is a summary of the minimum equilibrium interfacial tension between aqueous HPESO poly-soaps and hexadecane. From the study of the surface tension and interfacial

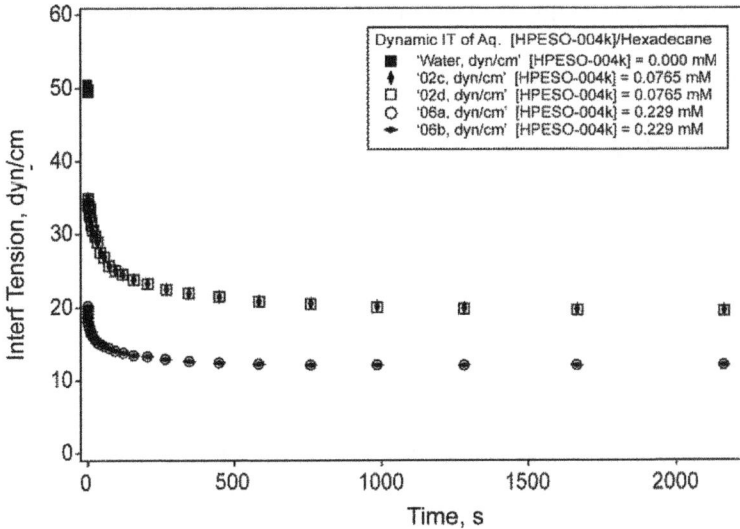

Figure 3.13 Typical dynamic interfacial tension data between water and hexadecane in the presence of a HPESO polysoap. The polysoap had a M_w of 2.6 kg mol^{-1} with the K$^+$ counterion.

Table 3.11 Minimum equilibrium interfacial tension between aqueous HPESO polysoaps and hexadecane.

Polysoap	M_w/kg mol^{-1}	Interfacial tension/dyn cm^{-1}
HPESO-004K	2.6	11.9 ± 0.1
HPESO-004A	2.6	14.2 ± 0.1
HPESO-003N	3.2	13.0 ± 0.04
HPESO-003K	3.2	12.7 ± 0.2
HPESO-003A	3.2	16.9 ± 0.2

tension of HPESO polysoaps, it can be concluded that they are very active bio-surfactants and higher surface activity than some conventional synthetic surfactants. They have potential applications in various areas.

3.5 Conclusions

SBO, ESO and EuO have been polymerized using a BF$_3 \cdot$OEt$_2$ catalyst in subcritical (liquid) or supercritical CO$_2$. Polymerization of ESO and EuO were possible under mild conditions such as room temperature, with short reaction times and low CO$_2$ pressures. However, for the polymerization of SBO, harsher conditions were needed. The formed polymers were found to be typical cross-linked polymers. They have T_g values ranging from -14 to -24 °C. TGA analysis showed that all of the polymers appear to be thermally stable at temperatures lower than 200 °C, and decomposition temperatures were found to be mainly above 340 °C. HPESO polysoaps were effective at lowering the surface tension of water and the interfacial tension of water–hexadecane, and displayed minimal tension values in the range of 20–24 and 12–17 dyn cm^{-1}, respectively, at concentrations of 200–250 μM.

Acknowledgements

The author gratefully acknowledges Mr Daniel Knetzer for his help with the GPC, DSC, and TGA experiments, Mrs Erin Walter for help with the oil physical properties study, and Dr Karl Vermillion for collecting the NMR spectra.

References

1. C. Jokers and H. van Bekkum, *Green Chem.*, 1999, **1**, 107.
2. M. Eissen, J. O. Metzger, E. Schmidt and U. Schneidewind, *Angew. Chem., Int. Ed.*, 2002, **41**, 414.
3. J. O. Metzger and U. Bornscheuer, *Appl. Microbiol. Biotechnol.*, 2006, **71**, 13.
4. S. Z. Erhan and M. O. Bagby, *J. Am Oil Chem. Soc.*, 1991, **68**, 635.
5. A. A. Foster and J. Viscoine, *US Pat.*, 5 320 670, 1994.
6. Z. Petrović, *Polym. Rev.*, 2008, **48**, 109.
7. R. Gu and M. M. Sain, *J. Polym. Environ.*, 2013, **20**, 30.
8. Z. S. Liu, B. K. Sharma and S. Z. Erhan, *Biomacromolecules*, 2007, **8**, 233.
9. Z. S. Liu and S. Z. Erhan, *J. Am. Oil Chem. Soc.*, 2010, **87**, 437.
10. Z. S. Liu, S. Shah, R. Evangelista and T. Isbell, *Ind. Crop. Prod.*, 2013, **41**, 10.
11. R. H. Biddulph and P. H. Plesch, *J. Chem. Soc.*, 1960, **82**, 3913.
12. R. Butler, C. M. Davies and A. I. Cooper, *Adv. Mater.*, 2001, **13**, 1459.
13. A. I. Cooper, *Adv. Mater.*, 2003, **15**, 1049.
14. C. D. Wood, B. Tan, H. Zhang and A. I. Cooper, *Thermodynamics, Solubility and Environmental Issues*, ed. T. Letcher, Elsevier, Amsterdam, Netherlands, 2007, ch. 21, p. 383.
15. Z. S. Liu and S. Z. Erhan, *J. Polym. Environ.*, 2010, **18**, 243.
16. Z. S. Liu, K. M. Doll and R. A. Holser, *Green Chem.*, 2009, **11**, 1774.
17. Z. Liu, S. Shah, R. Evangelista and T. Isbell, *Ind. Crop. Prod.*, 2013, **41**, 10.
18. G. Biresaw, Z. S. Liu and S. Z. Erhan, *J. Appl. Poly. Sci.*, 2008, **108**, 1976.
19. Y. Rotenberg, L. Boruvka and A. W. Neumann, *J. Colloid Interface Sci.*, 1983, **93**, 169.

CHAPTER 4

Thiol-ene and H-Phosphonate-ene Reactions for Lipid Modification

GRIGOR B. BANTCHEV,* STEVEN C. CERMAK, GIRMA BIRESAW, MICHAEL APPELL, JAMES A. KENAR AND REX E. MURRAY

USDA, ARS, National Center for Agricultural Utilization Research, Bio-Oils Research Unit, 1815 N. University Street, Peoria, IL 61604, USA[†]
*Email: grigor.bantchev@ars.usda.gov

4.1 Introduction

This chapter provides an overview of bio-based materials obtained from unsaturated lipids reacted with H–E, where E is either a thiyl or a phosphorus-centered group. The lipid modifications are carried out through the free-radical mechanism (see Scheme 4.1). Non-radical addition of H–E to a C=C double bond has been reported, when the groups R_1 or R_2 are electron-withdrawing groups, which is not the case for fatty acids (FAs).

The compounds synthesized from fatty materials and thiols have been shown to be good for polymers and lubricant additives (mostly anti-oxidative and metal removal applications). The compounds synthesized using

---

[†]Mention of trade names or commercial products in this publication is solely for the purpose of providing specific information and does not imply recommendation or endorsement by the U.S. Department of Agriculture. USDA is an equal opportunity provider and employer.

RSC Green Chemistry No. 29
Green Materials from Plant Oils
Edited by Zengshe Liu and George Kraus
© The Royal Society of Chemistry 2015
Published by the Royal Society of Chemistry, www.rsc.org

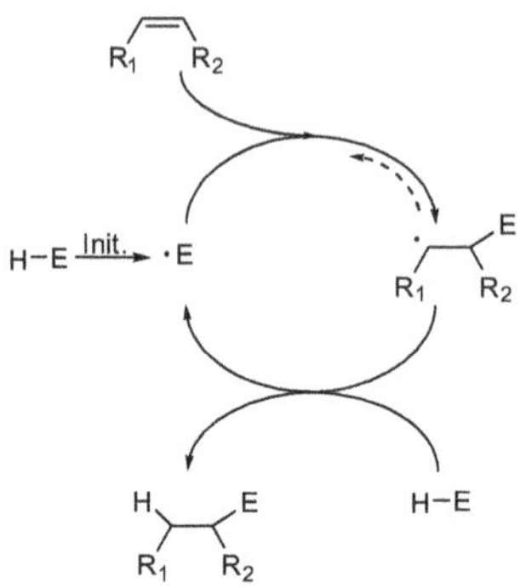

Scheme 4.1 Overall radical addition of a thiol or H-phosphonate to a double bond.

H-phosphonates have shown to be good as additives in plastics (mostly as plasticizers), lubricants (anti-wear additives) and anti-microbial agents.

4.2 Thermodynamics of the Reaction

The overall reaction involves breaking one C=C bond, one H–S (or H–P) bond and the formation of a C–S (or C–P) bond, and a C–H bond. Some literature data for the bond strengths are presented in Table 4.1. The data should be evaluated with care, since the experimental values are for molecules in the gas phase and the quantum computation data is for small molecules.

Nevertheless, it can be seen that the E–H bond is stronger than the E–C bond. The driving force for the reaction is the strong C–H bond that is formed. The energetics of a thiyl radical attaching to a double bond $(RS^{\bullet} + C=C \rightarrow RS\text{-}C\text{-}C^{\bullet})$ can be approximated by $D(S\text{-}C) + D(C\text{-}C) - D(C=C)$. Using the values from Table 4.1, a close-to-zero value can be obtained: $347 + 272 - 614 = 619 - 614 = +5$ kJ mol^{-1}. Based on the bond dissociation energies (BDEs), the addition of a thiyl radical to a double bond is predicted to be a reversible process. The energetics of the second step (chain transfer): $RS\text{-}C\text{-}C^{\bullet} + HSR \rightarrow RS\text{-}C\text{-}C\text{-}H + {}^{\bullet}SR$ can be evaluated by $D(C\text{-}H) - D(S\text{-}H) = 405 - 356 = +49$ kJ mol^{-1}. This second step is expected to be an irreversible process.

4.3 The Thiol-ene Reaction

4.3.1 Historical Overview

One feature of the thiol-ene reaction is that the addition of the thiyl radical to the double bond is reversible. After the thiyl radical attacks the double bond, a carbon-centered radical is formed. If there is a low concentration of hydrogen-atom donors (thiol molecules) in the system, this carbon-centered radical can decompose back to a double bond and a thiyl radical. The carbon-centered radical intermediate has a freely rotating single bond, and

Table 4.1 Bond dissociation energies (BDEs). BDEs in the current work were calculated using structures that had been geometry optimized with the Parameterized Model number 3 (PM3) and Unrestricted Hartree-Fock/PM3 methods. The BDEs are calculated in heat of formation energies.

Bond	$D/\text{kJ mol}^{-1}$	Reference
Me(Et)(H)C–H	405	1
Allylic C–H	344	2
	363	1
Bis-allylic C–H	305	3
Bis-allylic C–H	313	2
MeS–H	360	1
BuS–H	356	Current work
H_2P–H	345	1
$(MeO)_2(O)$P–H	366	Current work
C=C $\rightarrow$ C: :C	614	4
C–C	347	4
BuS–CH$_3$	272	Current work
$(MeO)_2(O)$P–CH$_3$	266	Current work

the overall decomposition process leads to isomerization of the double bonds to their thermodynamic equilibrium *cis–trans* mixture of $\sim 30/70\%$.

The importance of thiol molecules in biological systems (the most noteworthy being cysteine) has generated significant academic interest in the impact of this reaction, especially since it can lead to *in vivo* formation of *trans*-FAs from *cis*-FAs. Sivertz *et al.*[5] seem to be the first to report that the apparent activation energy of the reaction of a thiol with an alkene is negative (*i.e.* the rate is faster at lower temperatures) for some solvents. From this observation, they deduced that Step 1 of the cycle (the addition of the thiyl radical to the double bond) is reversible. Later, Walling and Helmreich[6] and Sivertz[7] reported experimental data for the *cis*-2-butene isomerization to *trans*-2-butene in the presence of thiyl radicals. Neureiter and Bordwell[8] published a series of experiments with the addition of thiol-acetic acid to 2-chloro-2-butene. From the ratio of the erythro- to threo-addition products, it was concluded the thiol radical adds to the double bond without forming a bridged radical intermediate.

Kircher[9] was the first to investigate the isomerization of lipids in the presence of thiols. The reaction was faster under direct sunlight, and influenced by the presence of oxygen, which strongly indicated a radical mechanism. The isomerization rate was independent of the oleate concentration in most of the measurements. Isomerization of double bonds was faster for methyl oleate and olive oil than for methyl linoleate and methyl linolenate.

Schwab *et al.*[10–13] were the first to use the thiol-ene reaction to obtain bio-based products. They showed that under UV light irradiation, H_2S adds more slowly to linseed oil than to methyl oleate.[10] Later,[11] they showed the thiol-ene reaction to be zero-order with respect to the double bond concentration and the reaction rate was oleic $> \sim 2 \times$ linoleic $> \sim 3 \times$ linolenic.

Oils with conjugated FAs showed lower reactivity than oils without conjugated FAs. The same group[13] demonstrated the reaction between H_2S and olefinic bonds in fatty compounds can be carried out not only under radical initiation conditions, but also under nucleophilic catalytic conditions. Nevertheless, the radical version of the thiol-ene reaction has been used exclusively for bio-based material synthesis.

4.3.2 Kinetics of the Thiol-ene Reaction

4.3.2.1 Kinetics of the Thiol-ene Reaction (Single Ene)

An in-depth study of the mechanism and kinetics of the isomerization of lipids in the presence of thiyl radicals was carried by Chatgilialoglu et al.[14-19] Since their focus was on understanding a live biological system, they carried out their experiments with low thiol concentrations where the addition reaction is not favored. A summary of their rate constants is shown in Table 4.2. A schematic of the reactions, whose rate constants are included in Table 4.2, is shown in Scheme 4.2. The table also includes related data from other authors.[20-26]

Biermann et al.[27] reported that the thiol and alkene can form an electron-donor–electron-acceptor complex, which reacts with another thiol molecule to generate radicals that initiate the thiol-ene chain.

Other studies investigated the kinetics of the reaction, without reporting rate constants. Claudino et al.[28] investigated the reaction between a trifunctional thiol (2-ethyl-(hydroxymethyl)-1,3-propanediol trimercapto propionate) and methyl oleate and methyl elaidate. They monitored the reaction with time-resolved FT-IR, Raman and NMR spectroscopy. They determined that the reaction of the addition of the thiyl radical to the double bond is fast and reversible, leading to cis–trans isomerization of the double bond, and that the

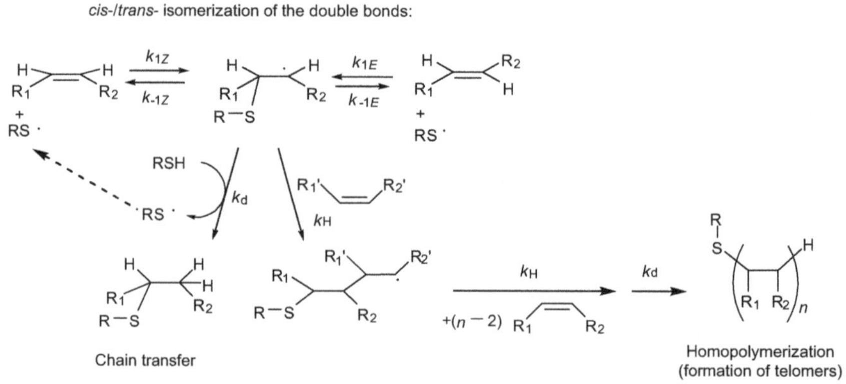

Scheme 4.2 The propagation reactions of the thiol-ene reaction. R_1 and R_2 are the remainder of the FA (most often oleic, $R_1 = n\text{-}C_8H_{17}$, $R_2 = -CH_{27}COOR'$ or vice versa).

Table 4.2 Rate constants from the literature for the reactions from Scheme 4.2.

Constant	Value	Units	Notes	Ref.
k_{1Z}	1.5×10^5	$M^{-1} s^{-1}$	At 25 °C, in *tert*-butanol	15, 16
k_{-1Z}	2.2×10^7	s^{-1}	At 25 °C, in *tert*-butanol	
k_{1E}	2.0×10^5	$M^{-1} s^{-1}$	At 25 °C, in *tert*-butanol	
k_{-1E}	1.5×10^8	s^{-1}	At 25 °C, in *tert*-butanol	
$k_{1Z} = k_{1E}$	4.5×10^6	$M^{-1} s^{-1}$	Liposomes in water pH $= 6$	20
k_{1Z}	3.5×10^6	$M^{-1} s^{-1}$	—	21
k_{1Z}	7.0×10^7	$M^{-1} s^{-1}$	—	21[a]
$k_{-1Z} = k_{-1E}$	3.0×10^5	s^{-1}	—	20, 21
k_d	$(0.2–6) \times 10^7$	$M^{-1} s^{-1}$	At 30 °C. Depends on solvent polarity.	22
k_d	1.1×10^7	$M^{-1} s^{-1}$	At 22 °C in 1.2 M pyridine in benzene.	23
k_d	0.8×10^7	$M^{-1} s^{-1}$	For primary radical + *t*-BuSH, claims the rate for secondary and tertiary radicals is close	24
$k_1{}^c$	3.0×10^9	$M^{-1} s^{-1}$	For PhS• radicals	16[b]
k_1	1.0×10^5	$M^{-1} s^{-1}$	Dinorbornene	25
k_d	1.0×10^5	$M^{-1} s^{-1}$	—	
k_t	1.0×10^6	$M^{-1} s^{-1}$	—	
k_1	$(0.5–2.2) \times 10^5$	$M^{-1} s^{-1}$	Divinyl ether, acrylate, vinyl silazane	25
k_d	$(0.17–2.5) \times 10^5$	$M^{-1} s^{-1}$		
k_H	2.5×10^4	$M^{-1} s^{-1}$	Acrylic homopolymerization	
Bis abstraction	$(0.3–3.1) \times 10^7$	$M^{-1} s^{-1}$	—	26
Bis abstraction	$(1.2–1.6) \times 10^7$	$M^{-1} s^{-1}$		20
k_d	$\sim 1 \times 10^5$	$M^{-1} s^{-1}$	(for P–H + •CH$_2$R) at 80 °C	18

[a]Ref. 21 attributed the authorship of the value to ref. 26.
[b]k_t is the termination rate constant, of the reaction of forming of a molecule from two radicals. Despite the value for k_t being for a thiophenyl radical, refs. 15 and 16 use it successfully to fit the data for other thiyl radicals.
[c]k_1 is an effective rate constant, depending on k_{1Z}, k_{-1Z}, k_{1E} and k_{-1E}.

subsequent reaction of hydrogen abstraction is a slow, rate-determining step. They obtained high thiol-ene conversions without side reactions.

Lee *et al.*[29] investigated the reaction of vinyl acrylate with thiols under UV light (365 nm) and demonstrated that, in this compound, the acrylate double bond homopolymerizes, while the thiyl radical exclusively attacks the vinyl double bonds. It was proposed that the selectivity was due to the electrophilic nature of the thiyl radical and the low electron density of the acrylic double bond in the vinyl acrylate. When catalyzed by primary amines, instead of UV light, the acrylate double bond reaction occurred so rapidly it could not be monitored under normal mixing conditions. The thiol-ene reaction is not stopped by the presence of oxygen, because peroxy radical can abstract a hydrogen atom from a thiol group.

Roper *et al.*[30] reported that the reactivity of alkenes follows the order monosubstituted > disubstituted > trisubstituted. Norbornene (bicyclo[2.2.1]-hept-2-ene) reacted faster than 1-hexene; the authors explained this with the release of ring strain when the double bond is attacked. The higher reactivity of the norbornene double bond compared to 1-octene was also reported by other authors.[31] The rate-limiting step is the hydrogen transfer (chain transfer),[25,30] indicating there is no apparent dependence on the ene concentration, but a dependence on the thiol concentration. *cis*-2-Hexene is much more reactive than *trans*-2-hexene. The higher reactivity of a primary alkene compared to internal double bonds led to the use of undecylenic acid and its esters for the bio-based thiol-ene reaction. Undecylenic acid is a product of the pyrolysis of ricinoleic acid, the main FA in castor oil.

Ferreri *et al.*[17] showed that the isomerization of the double bonds in vesicles depends on the proximity of the double bonds to the vesicle–solvent interface.[17] The hydrophilic $HOCH_2CH_2S^{\bullet}$ radical, used in their study, isomerized all the double bonds (oleic, linoleic, or arachidonoic) at the same rate, independent of their position, when the FAs were freely dissolved. When the lipid chains were within large unilamellar vesicles, the double bonds, closer to the vesicle–water interface (closer to the head $-COO^{-}$ group) were isomerized much faster than those further from it.

Bantchev *et al.*[32] reacted butanethiol with corn and canola oil under UV light. One of the observations was that the reaction is faster at lower temperatures – an observation that is in agreement with the negative activation energy for the thiol-ene reaction reported by Sivertz *et al.*[5] Again, the authors explained the observation with the reversibility of the first step of the reaction and with shifting the equilibrium of the first step towards the unreacted materials with the increase in temperature.

4.3.2.2   The Effect of bis-Allylic Hydrogens

The reaction of thiyl radicals with compounds having two double bounds (like linoleic or linolenic acid) can be influenced by the side-reaction of a thiyl radical abstracting a bis-allylic radical. The kinetics of this reaction were first studied by Schoneich *et al.*[26] They reported the kinetic rates to be on the order of $(0.3–3.1)\times10^7$ M^{-1} s^{-1}. The more lipophilic thiols showed a higher rate and the rate constants were higher in polar solvents.

The abstraction of bis-allylic hydrogens by thiyl radicals can change the thiyl radical concentration and, correspondingly, the reaction rate. Samuelsson *et al.*[33] and Desroches *et al.*[34] observed that linoleate reacts much more slowly than oleate when they react separately with thiols. Bantchev *et al.*[32] observed that the oleate double bond and the linoleate double bond in corn oil react at the same rate. These two observations can be reconciled using the results for phospholipids produced by Ferreri *et al.*:[14] the oleic double bonds reacted with the thiyl radical more quickly than the linoleic double bonds did, when there were no linoleic groups present. When both oleic and linoleic moieties were present simultaneously in the system, the rate of reaction of the linoleic moiety did not change, but the rate of reaction of the oleic moiety decreased significantly, to a rate corresponding to the same rate of reaction per double bond as the linoleic moiety.

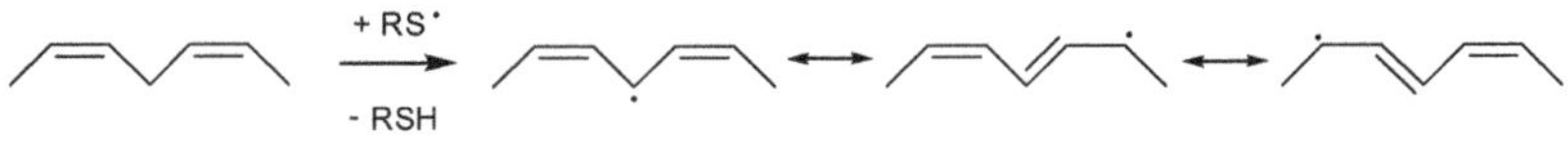

Scheme 4.3 Abstraction of bis-allylic hydrogen by thiyl, and the resonance stabilization of the resulting pentadienyl radical.

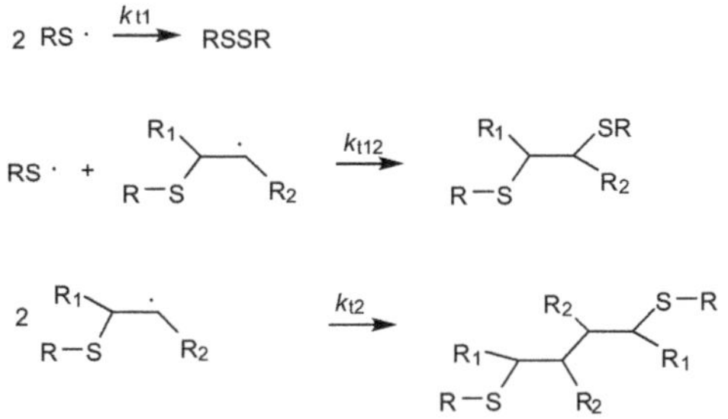

Scheme 4.4 Some of the possible termination reactions. In Chatgilialoglu *et al.*[14–19] and Bantchev *et al.*[32] the first termination reaction (between two thiyl radicals) is the dominant termination reaction.

Ferreri *et al.*[14] explained the observation with the presence of bis-allylic hydrogens in the linoleic moiety, which act as inhibitors; these hydrogens can be abstracted (see Scheme 4.3), and this reduces the concentration of thiyl radicals. The pentadienyl radical formed is resonance-stabilized and, correspondingly, has a low reactivity. Ferreri *et al.*[14] reported that they detected dimeric FAs, products of reactions with these pentadienyl radicals. Bantchev *et al.*[32] and Samuelsson *et al.*[33] did not report dimeric FAs, nor the appearance of conjugated double bonds, possibly due to an inherent lack of sensitivity in their analytical methods.

4.3.2.3 Termination Reactions

The termination reaction occurs when any two radicals from the initiation or propagation steps combine to form a non-radical species (see Scheme 4.4). Nevertheless, kinetic studies have led to the conclusion that the dominant termination reaction for the reaction of FAs with thiols is recombination of the thiyl radicals to form disulfides. The termination reaction is also assumed to be diffusion controlled. Our work on the reaction of vegetable oil with butanethiol supports this observation, since butanedisulfide was the only termination product identified. Nevertheless, it should not be forgotten that the thiols in these studies were small molecules. In polymer synthesis, larger thiol molecules are frequently utilized and, in the course of the reaction, attachment to polymer chains and thiyl radicals can occur. The result is slower diffusion and the possibility that other modes of radical recombination become dominant.

Cramer *et al.*[25] investigated the kinetics of the thiol-ene reaction for the synthesis of non-bio-based polymers. Their work focused mostly on highly active double bonds: acrylates, vinyl esters and related species. The data was modeled with k_1/k_d ratios of 13:1 to 0.2:1, while Chatgilialoglu *et al.*[16,19] reported ratios of 0.02:1 (see Table 4.2). Chatgilialoglu *et al.*'s results indicate the thiyl attack on the double bond is much slower than the subsequent chain-transfer abstraction of hydrogen by the carbon-centered radical. Cramer *et al.*'s result showed a faster or slightly slower rate of thiyl attack. This discrepancy can be attributed to the fact that Cramer *et al.*[25] used a simplified model which neglected the reversibility of Step 1. Other observations can be attributed to the fact that Chatgilialoglu *et al.*[16,19] worked with small molecules with large diffusion coefficients, while Cramer *et al.*[25] investigated polymeric systems with more restricted diffusion. The termination steps of these reactions can be diffusion controlled and influence the overall reaction rates.

4.4 Bio-based Materials made with the Thiol-ene Reaction

4.4.1 Low-molecular-weight Materials

Obtaining polymeric products with the thiol-ene reaction requires the coupling of compounds with at least two thiol groups to compounds with at least two ene groups. Otherwise, low-molecular-weight compounds will be obtained. This reaction scheme is utilized in the synthesis of low-molecular-weight compounds for lubricants.

The products of reacting of oleic acid, undecanoic acid and docosenoic acid with mercaptoacetic and 3-mercaptopropionic acids were patented as lubricating base oils by Gadd *et al.*[35] They used a thermo-initiator (lauroyl peroxide) for the reaction. They reported the kinematic viscosity at 100 °C as being between 3.14 and 7.28 $mm^2 s^{-1}$, the viscosity index as being between 123 and 183, and the Pour Point as being below −9 °C for some compounds and below −60 °C for others. The oxidative stability, measured as an onset temperature to exothermic oxidative peak in a pressurized differential scanning calorimetry (DSC) experiment, was substantially higher compared to the corresponding polyol ester and synthetic hydrocarbon compounds.

Bantchev *et al.*[32] added butanethiol to corn and canola oils and evaluated the products as lubricants and lubricating additives (see Scheme 4.5). The resulting oils had better cold flow properties (Cloud Point and Pour Point) than the starting vegetable oils; however, weak effects were observed when the materials were used as additives. The sulfide-modified corn oil showed improved oxidative stability and good anti-oxidative properties as an additive.[36,37]

In a very recent report by the same group, both sulfide-modified corn oil and canola oil were also shown to effectively remove heavy metals from water (see Scheme 4.6). The extraction of Ag^+ from a water solution was used as a

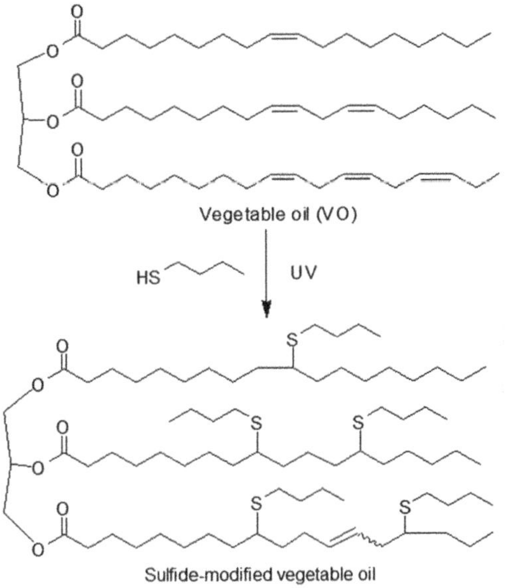

Scheme 4.5 The synthesis of sulfide-modified vegetable oil.[32]

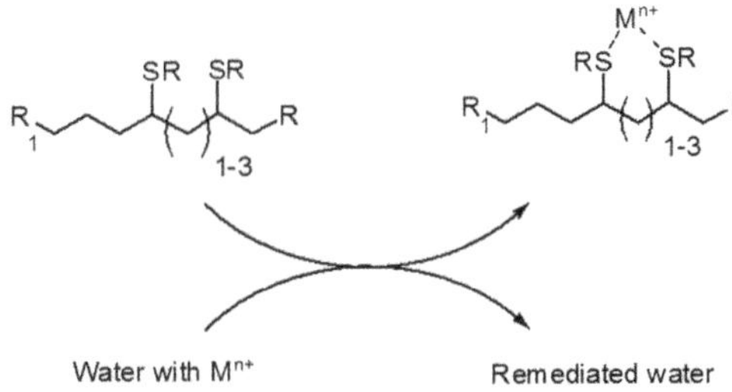

Scheme 4.6 The use of sulfide-modified vegetable oils for the remediation of water contaminated with heavy metals.[38]

model example for the remediation of water contaminated with heavy metals.[38] Thus, a new technological use for thiol-ene products (in this case ligand-modified vegetable oils) in a broad realm of applications, encompassing metal coordination, was demonstrated.

Epoxidized soybean oil (ESBO) was a feedstock for a similar reaction with thiols using perchloric acid as the catalyst[39] (see Scheme 4.7). BF$_3$ in hexane (a Lewis acid) was reported to be ineffective as a catalyst for the reaction. *n*-Butyl-, *n*-decyl-, *n*-octadecyl-, and cyclohexyl-thiols were successfully reacted with ESBO to yield the corresponding alkyl thioethers of the hydroxy vegetable oil (BTHV, DTHV, OTHV, and CTHV). The products were later evaluated as anti-wear additives in soybean oil (SBO) and toluene.[40] In a four-ball test (15 min, 40 kg, 1200 rpm, 25 °C) 5% BTHV in soybean oil (SBO) decreased the coefficient of friction to lower values than comparative

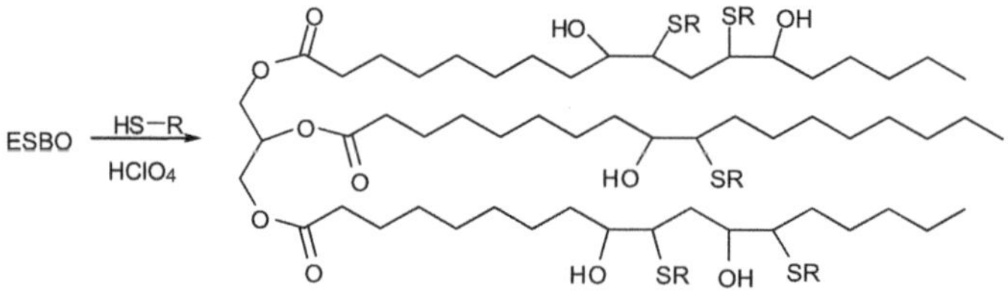

Scheme 4.7 The synthesis of an alkyl thioether of hydroxy vegetable oil. R is an alkyl group.[39]

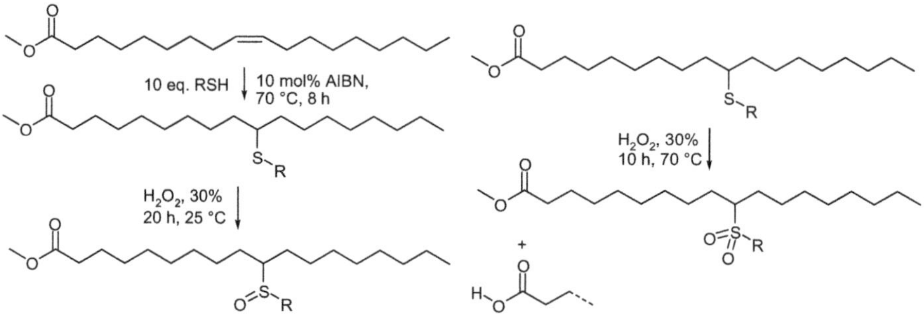

Scheme 4.8 Scheme of the synthesis of a sulfoxide and a sulfone. The harsher conditions, applied for oxidation to the sulfone, led to partial hydrolysis of the carboxylic ester group. R = ethyl, butyl or hexyl.[40] The 9-regioisomer is not shown.

commercial additives. While the BTHV reduced the wear scar diameter, it was less effective than most of the commercial additives. The anti-wear properties of BTHV, DTHV, OTHV, CTHV, and commercially comparable products were evaluated as additives in toluene on a ball-on-disk instrument. The wear-track widths for BTHV, DTHV, OTHV, and CTHV were smaller than the results for the commercial additives. The effect was explained by the affinity of the hydroxyl and thioether groups towards metal surfaces.

de Espinosa *et al.*[41] synthesized thioethers, sulfoxides and sulfones from oleic acid thiols and hydrogen peroxide. The reactions are shown in Schemes 4.8 and 4.9.

The oleic acid was either technical grade oleic acid (94.9% oleic acid, 2.8% linoleic acid and 2.3% stearic acid) or a FA mixture, obtained from the hydrolysis of high-oleic canola oil (82.3% oleic, 12.4% linoleic, 3.9% palmitic, and 1.3% stearic acids). They report that the reaction time for the thiol-ene coupling was essentially the same for both grades of oleic acid, which indicates that the increase of linoleic acid from ~3 to ~12% has little effect on the reactivity.

The products of the reaction were considered for use as polyvinylchloride (PVC) plasticizers. Thioethers and sulfoxides were deemed inappropriate

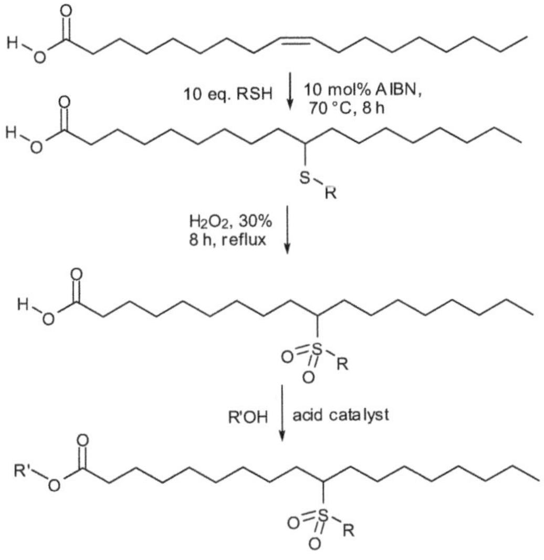

Scheme 4.9 The synthesis of sulfone derivatives of oleic acid. R = ethyl, butyl or hexyl, R′ = methyl or butyl.[40]

because of the unpleasant odor of the products and their decomposition products. The sulfoxides decomposed above 135 °C, whereas the sulfones were thermally stable to 250 °C.

Six of the bio-based sulfones synthesized and two commercial plasticizers were compared as plasticizers in a PVC composition containing 100 parts by weight PVC, 50 parts plasticizer, 3 parts ESBO as a co-stabilizer, and 2 parts Ca/Zn-base stabilizer. The glass transition temperature (T_g) of the sulfones was between −9.6 and −30.0 °C, which was higher than −33.6 and −43.6 °C for the commercial plasticizers. The volatility of the sulfones after 4 min at 200 °C was comparable to the volatility of one commercial plasticizer and higher than other commercial plasticizers. The worst volatility was observed for the product from canola oleic acid. The authors attributed the high volatility to the presence of sulfones from linoleic acid, which they thought were less stable thermally. The complex viscosity at 25 °C, which is preferred to be low, was about an order of magnitude higher than the complex viscosity of the composition with commercial plasticizers. The PVC formulations with sulfones had reduced gelation temperatures, compared to the commercial formulations, indicating that the bio-based formulations could be easier to process. The thermal stability of the bio-based formulations at 200 °C was lower than that of the commercial additives; the PVC darkened due to the release of hydrochloride from the backbone within 4–5 min for the formulation with sulfones, while the formulations with the commercial plasticizers darkened after 7 min. The Shore hardness was ~10% lower for the bio-based sulfones, thus showing better effectiveness in plasticizing the polymer. The bio-based plasticizers induced higher water uptake and higher plasticizer losses after drying.

4.4.2 Pre-polymers

The thiol-ene reaction was successfully used to introduce other functional groups in place of the double bonds. Most often, it is used to introduce primary –OH groups, but also –COOH, –COOCH$_3$, –NH$_2$, or –SH groups. The authors, reporting on these syntheses, do not report the functional group, which has been introduced, to have a strong (negative) effect on the thiol-ene reactivity. The only exception was the amine group, which decreased the miscibility and reaction rate, but this problem was overcome by protecting it as a hydrochloride salt during the reaction. Afterwards, the products were utilized for the synthesis of various polymeric materials.

Most often, the double bonds were reacted with thiols containing –OH groups[34,42–47] and the product was used either in polyurethane (PU) or in polyester synthesis. The thio-alcohol was usually 2-mercaptoethanol (2ME), but 1-thioglycerol was also utilized to introduce two hydroxyl groups per double bond.[42]

Reacting the double bonds with 3-mercaptopropionic acid[48,49] or 2-mercaptoacetic acid methyl ester[43,50] permits the transformation to carboxylic acids and esters.

Amino groups were introduced using cysteamine hydrochloride, and consecutive washes with a solution of Na$_2$CO$_3$ (ref. 51) or K$_2$CO$_3$ (ref. 52). When direct reaction with cysteamine was attempted, lower yields were observed.[52]

Thiol groups can be introduced several ways. One is direct addition of excess H$_2$S[10–13,53] or polyfunctional thiols[54] to double bonds. Caution should be used when performing this method, as the formed thiol groups can also react with double bonds and form sulfides and H$_2$S is a highly toxic gas. An indirect method can be used, utilizing the thiol-ene reaction with thioacetic acid and consecutive transesterification (see Scheme 4.10).[55,56]

Gonzalez-Paz et al.[44] obtained diols by reacting the double bonds of undecylenic and oleic acids with 2ME. Two routes were successfully tested to add a second hydroxyl group to the compound. One was esterifying the acid group with allyl alcohol, whose olefinic double bond reacted afterwards with the 2ME. The other route was esterifying the acid group with methanol and, afterwards, reducing it with LiAlH$_4$ (see Scheme 4.11).

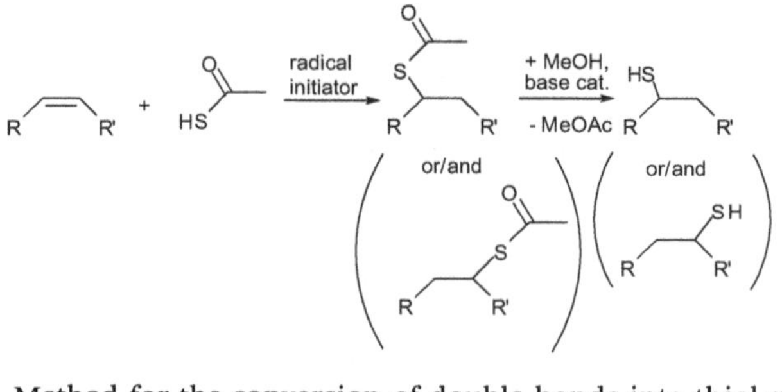

Scheme 4.10   Method for the conversion of double bonds into thiol groups.[56]

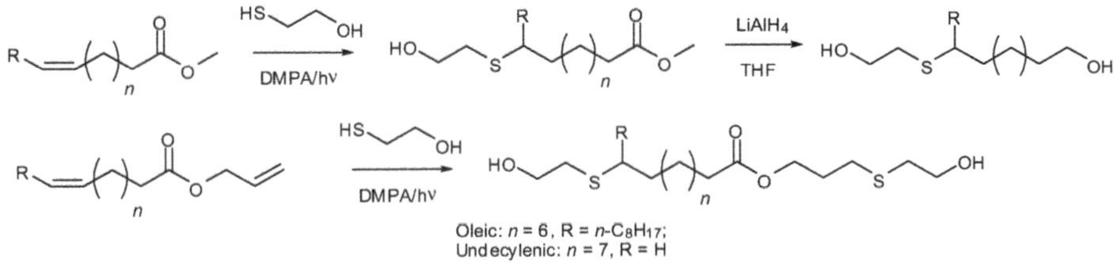

Scheme 4.11 Preparation of diols from oleic and undecylenic acids.[44] In the case of oleic acid, only one of the two isomers is shown. THF = tetrahydrofuran, DMPA = 2,2-dimethoxy-2-phenylacetophenone, a photoinitiator.

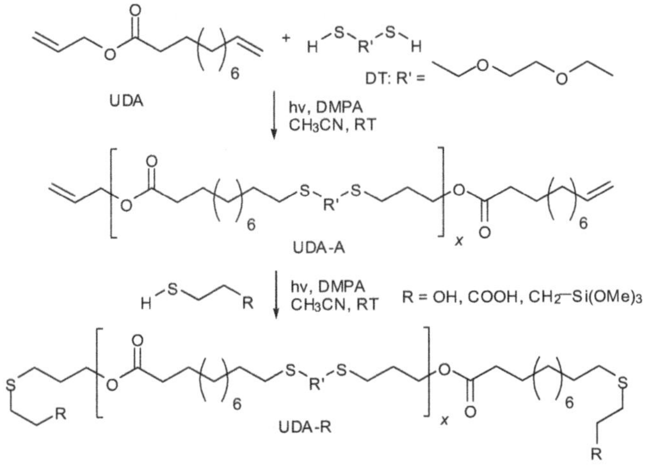

Scheme 4.12 The synthesis of telechelic polymers.[42] DT is 3,6-dioxa-1,8-octanedithiol.

The resulting diols were polymerized with 4,4′-methylenebis(phenylisocyanate), in a dimethylformamide (DMF) solution using tin(II) 2-ethylhexanoate as the catalyst. The synthesized PUs utilizing the LiAlH$_4$ reduction, had higher T_g values than the ones utilizing the allylic alcohol route. The PUs synthesized from undecylenic acid had higher T_g values than the ones from oleic acid. The PUs synthesized from undecylenic acid showed stress–strain curves, indicating the presence of rigidity increasing crystalline domains. The PUs synthesized from oleic acid showed behavior similar to lightly cross-linked rubbers.

Lluch *et al.*[42] reported on the synthesis of telechelic polymers (polymers where both ends possess the same reactive functionality). As a starting material, they utilized allyl 10-undecenoate (UDA). UDA has a high reactivity towards thiols, since both olefinic double bonds are primary. The two-step reactions (Scheme 4.12) were successfully carried out in one pot. Telechelics

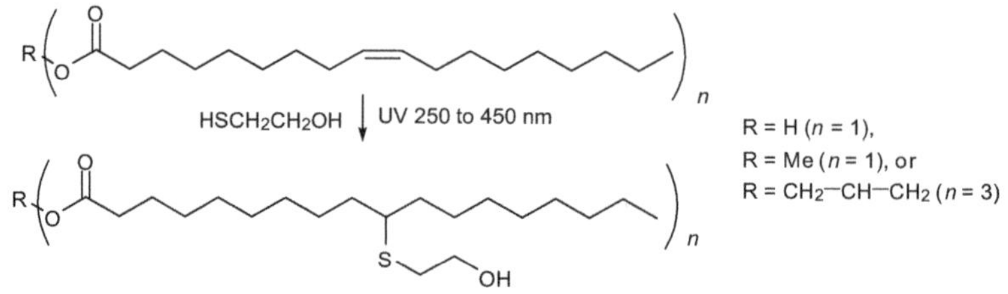

Scheme 4.13 Reaction of unsaturated materials with 2ME.[34] In the case of canola oil, the formed polyol was successfully used to make PU.

with three end-groups (hydroxyl, carboxyl and trimethoxysilane) were reported. One of the end-groups synthesized by this method, macrodiol, was used for the formation of poly(ester urethane) with 4,4'-methylenebis-(phenylisocyanate) and 1,4-butanediol (as a chain extender). Two T_g values (-45 and $+55$ °C) and two melting temperatures (-9 and 190 °C) were observed, suggesting the formation of block polymers with a phase-separated morphology. A further study of analogous poly(ester urethane)s as drug-release agents was reported by the same group.[49]

Desroches *et al.*[34] reacted 2ME with oleic acid, methyl oleate, methyl linoleate, and canola oil with irradiation with a UV light (250–450 nm) without an initiator, with or without solvent (ethanol). The obtained products contained the primary alcohol group from 2ME (Scheme 4.13) and were used successfully to react with isocyanates to form elastomeric PUs. The PUs, obtained from 2ME-modified canola oil and isocyanates, had thermal properties (T_g, mass loss at 200 °C and residual mass at 500 °C) that are close to PUs from the same isocyanates and commercial polyols.

Next, the group investigated the same thiol-ene reaction of canola oil methyl esters with 2ME, followed (or preceded) by an esterification of the carboxylic ester group with ethylene glycol.[46] The product mixture obtained by this strategy had a hydroxyl group functionality close to two; correspondingly, the PU formed from it had high fractions soluble in DMF (76%) and THF (38%).

In a subsequent report, the same group[45] used soybean methyl esters to form diols. The double bonds were converted to primary –OH groups using the thermal thiol-ene reaction with 2ME (Schemes 4.14 and 4.15). The carboxylic group was either transesterified with a diol or amidified with diamine or amine-alcohol. Details of the SBO used for the synthesis were not described. A typical SBO contains mostly linoleic acid. Linoleic acid can react with two moles 2ME, and in the following transesterification would yield a product with 3 –OH groups per molecule. The reported relatively low numbers (2.01 to 2.63 –OH groups per molecule) imply that either the high-oleic variety of SBO was used or close to 50% of the olefinic bonds reacted with 2ME.

When the diols were reacted with Voramer 2093, a series of soft PU pre-polymers of methylene diphenyl-4,4'-diisocyanate, the product started

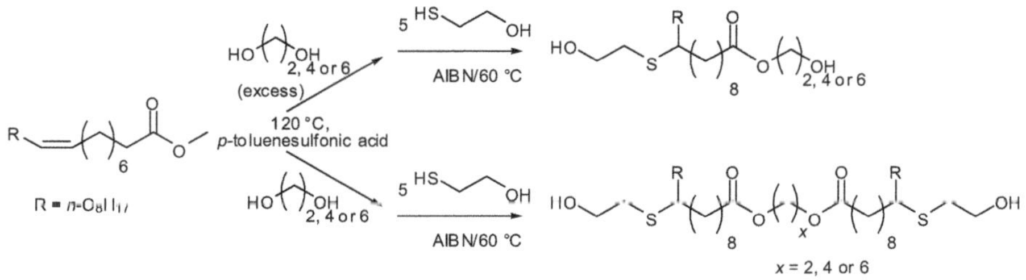

Scheme 4.14 The synthesis of diols using 2ME. The actual synthesis utilized soybean methyl esters, which contain linoleic acid methyl ester as their main component.[45]

decomposing thermally in nitrogen at 261–272 °C. The T_g values of the PUs were T_g (monoamido diols) > T_g (diamido diols) > T_g (monoester diols) > T_g (diester diols). Mostly, increasing the number of CH_2 groups in the diols led to lower T_g values. The amide groups probably catalyzed the reactions, since the PUs from amides had shorter gel times than the PUs from esters.

Lluch *et al.*[48] coupled a thiol with oleic, 10-undecylenic acid and 10-undecylenic acid triglyceride with 3-mercaptopropionic acid to obtain compounds with two or three carboxylic acid groups (Scheme 4.16). Next, the carboxylic groups were converted to mixed anhydrides with acetic anhydride, followed by melt condensation (see Scheme 4.17), to obtain polymeric anhydrides. The formed anhydrides were cast and their degradation in pH 7.4 buffer was observed. A fast degradation of the polymer (more than 50% in 24 h) was observed. In addition, polymers were prepared with a hydrophobic dye (rhodamine B) cast in the polymer as a model drug. Most of the drug was released within 60 h in the pH 7.4 buffer.

More *et al.*[50] reported the successful addition of 2-mercaptoacetic acid methyl ester to undecylenic acid methyl esters under mild conditions (35 °C, or UV light at room temperature). The ester groups were successfully converted to diisocyanate (1-isocyanato-10-(isocyanatomethyl)thio]decane (DITD), Scheme 4.18) and reacted with diols to form PU. Compared to 1,8-diisocyanatooctane, DITD was less stable when exposed to air. This observation was explained by the sulfur atom increasing the reactivity of the nearby isocyanate group. PU formed with DITD, showed several transition peaks in a DSC thermogram, which were explained by the enhanced flexibility of the hydrocarbon chain due to the presence of the sulfur atoms.

Türünç and Meier[43] synthesized several derivatives of 10-undecylenic acid and 10-undecylenic alcohol using the thiol-ene reaction (Scheme 4.19). Next, a polycondensation reaction was carried out using the triazabicyclodecene (TBD) base catalyst to form several polyester compounds. The so-formed polymers had number-average molecular weights (M_n) of 3.9 to 9.4 kDa, and polydispersities (M_w/M_n) of 1.87 to 3.37. The DSC thermogram showed sharp endothermic peaks, indicating that the polymers were semi-crystalline.

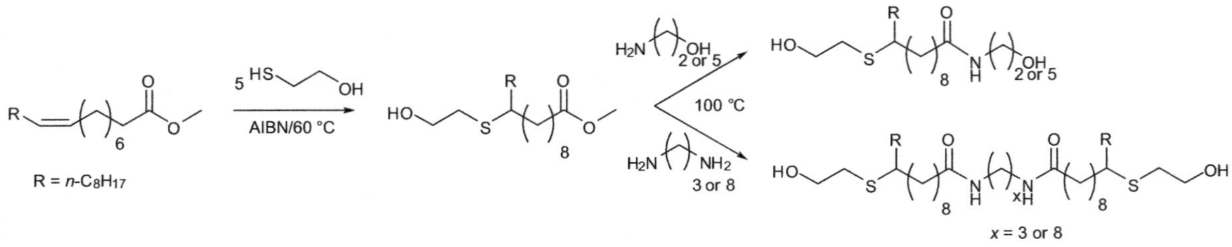

Scheme 4.15 The synthesis of diols using 2ME and diamines or amine-alcohols.[45] AIBN = azobisisobutyronitrile.

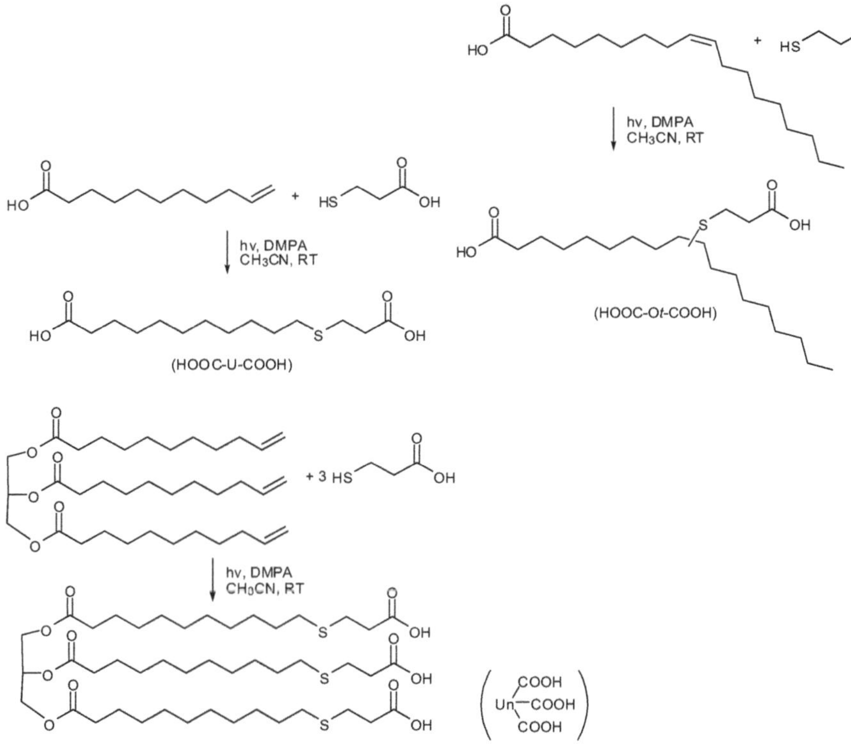

Scheme 4.16 Using the thiol-ene reaction to synthesize di- and trifunctional carboxylic acids.[48]

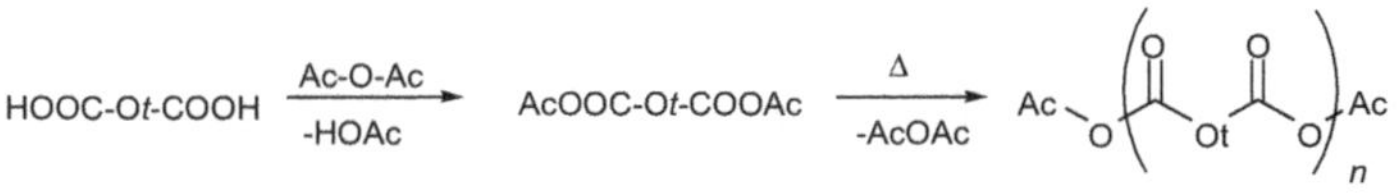

Scheme 4.17 Formation of an anhydride polymer by acetylation and melt condensation. Several polymeric anhydrides were prepared, using the di- and tricarboxylic acid compounds from Scheme 4.16.[48]

The melting points were around 50–70 °C, but they depended on the pretreatment (annealing). The decomposition (5 wt% loss) started at 300 °C or higher.

The same group also synthesized amine-esters using the thiol-ene click reaction.[52] The solubility of the cysteamine salt was better than the solubility of cysteamine, and the yields were better. The undecylenic acid methyl ester gave an 85% isolated yield of the product, without use of an initiator. The less reactive methyl oleate and methyl erucate gave 75% and 66% isolated yields, and the reaction needed to be initiated with UV light and DMPA (see Scheme 4.20).

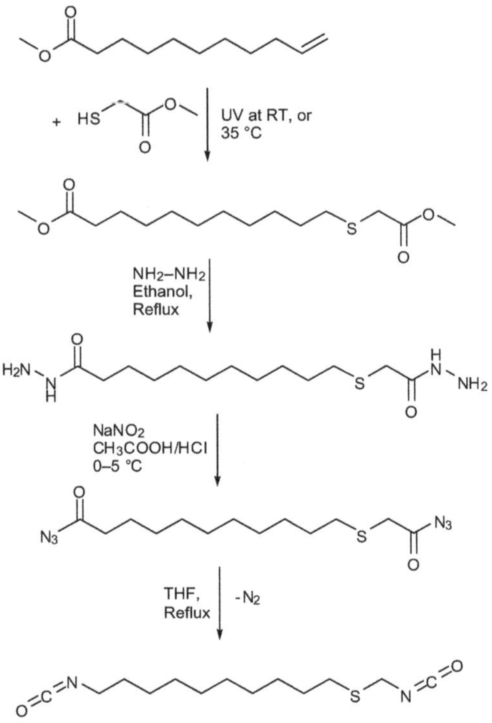

Scheme 4.18   Introduction of ester groups, and conversion to isocyanate groups. The sulfur atom changes the reactivity of the isocyanate groups.[50]

The synthesized amine-esters were reacted using a base catalyst (TBD) to form polyamides. The molecular weight of the polyamides was highest for the undecylenic-derived polyamide. The undecylenic-derived polyamide had the highest melting point (138 °C), with those for the oleic- and erucic-derived polyamides being below −50 °C, and +43 °C, respectively. The undecylenic-derived polyamide was soluble in hexafluoroisopropanol, while the oleic- and erucic-derived polyamides were soluble in THF. The molecular weight was increased by co-condensation with dimethyl adipate and 1,6-hexamethylene diamine (monomers, used for nylon-6,6 synthesis). Co-condensation also allowed changing of the melting points.

Stemmelen *et al.*[51] introduced amino groups into grapeseed oil (which contains mostly linoleic FA), using similar chemistry to that employed by Türünç *et al.*[52] (Scheme 4.21). Using 10 mol% DMPA and UV irradiation for 8 h, they succeeded in reacting 100% of the double bonds. There was some homopolymerization, since the addition of amine was estimated to be only 87%. The formed aminated grapeseed oil was reacted successfully with epoxidized linseed oil, to obtain a 3D polymeric network. The enthalpy of cross-linking was relatively low, 50 J g^{-1}; the time for cross-linking,

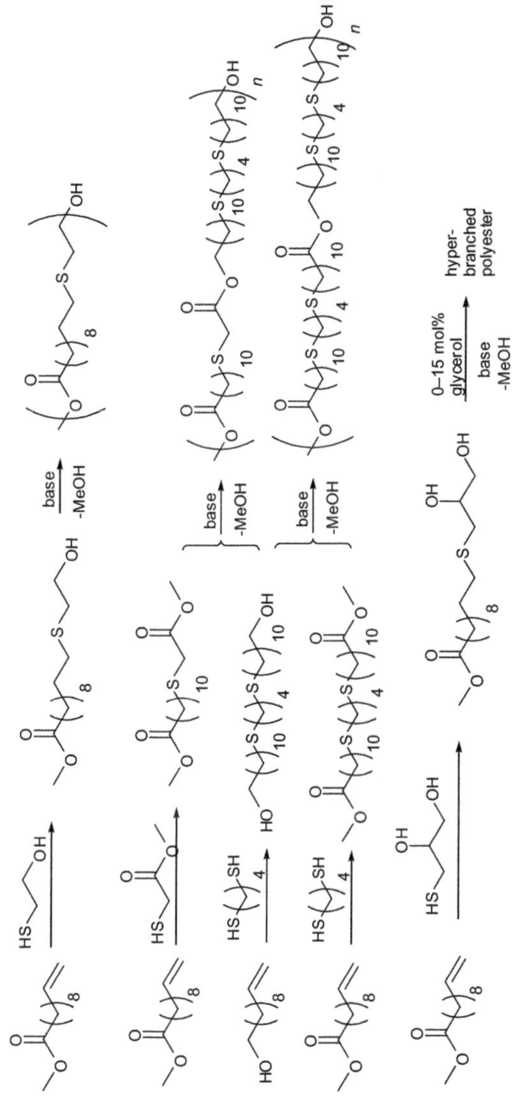

Scheme 4.19 The synthesis of alcohol-esters, used for formation of polyesters.[43]

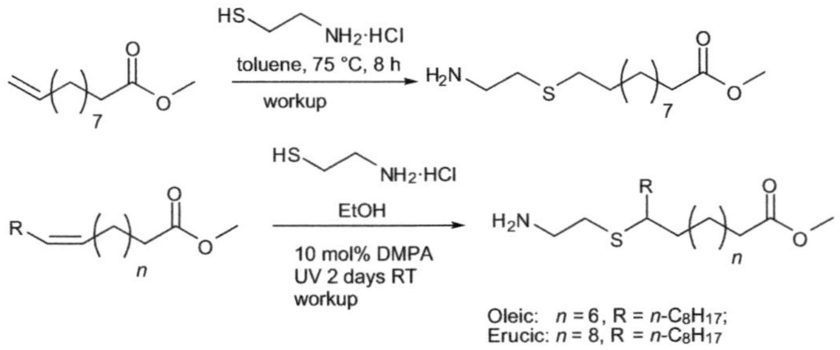

Scheme 4.20 Introduction of amine functionalities, using the thiol-ene reaction. The other regioisomers for methyl oleate and methyl erucate derivatives are not shown.[52]

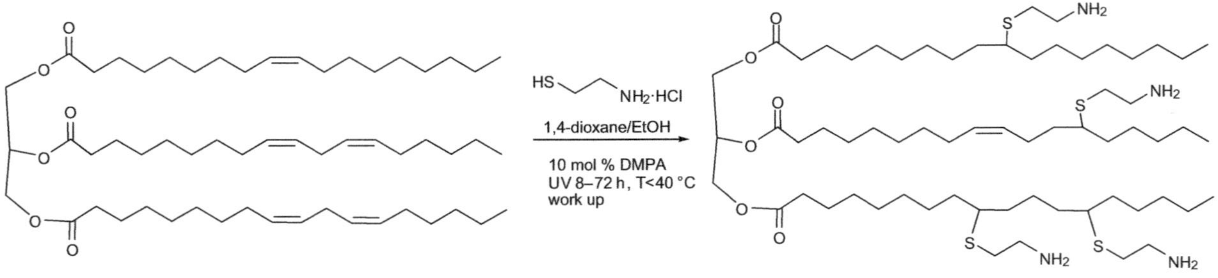

Scheme 4.21 The synthesis of aminated grapeseed oil.[51]

measured as stabilization of the elastic modulus at 100 °C, was 14 h. The thermomechanical profile resembled those of thermoset amorphous polymers. The T_g was −38 °C.

The same group reported interesting chemistry for the synthesis of a star polymer. Methyl oleate or grapeseed oil was used as the core with the double bonds utilized to introduce primary hydroxyl groups by reaction with 2ME.[47] The best conversion of double bonds (63%) was achieved after 8 h of irradiation (high-pressure Hg lamp, maximum intensity at 365 nm) with 0.02 mol DMPA per double bond.

Byers *et al.*[53] made sulfur-containing SBO derivatives by the addition of pressurized H_2S to SBO with a medium-pressure Hg lamp (see Scheme 4.22). Good thiol-to-cyclic sulfur compound ratios were reported for the system. The resulting bio-based thiols were reacted with isocyanates or sulfur to synthesize polythiourethanes or polysulfide-cross-linked materials, which were used as coatings for fertilizers. The coated fertilizers had desirable slow-release properties.

4.4.3 Polymers

4.4.3.1 Grafting to Polymers using the Thiol-ene Reaction

Ates *et al.*[57] used a bio-based large-cycle lactone (globalide) to form an unsaturated polyester polyglobalide using ring-opening polymerization (Scheme 4.23). The double bond in the polyglobalide was reacted with three thiols using AIBN as an initiator, and a minimal amount of THF as a solvent to improve miscibility. The reaction temperature was 80 °C, (above the melting temperature of polyglobalide). The conversion of the double bonds was above 95% with mercaptohexanol and *N*-acetylcysteamine, and >75% with butyl-3-mercapto propionate. DSC showed that the crystalline structure of the polyglobalide is disrupted when it is modified with the thiol-ene reaction, resulting in an amorphous material.

Claudino *et al.*[58] synthesized polyester by ring-opening co-polymerizing globalide and ε-caprolactone (Scheme 4.24). The globalide unsaturations were used to cross-link the polymers in the melt. The cross-linking was initiated with UV irradiation and a photo-initiator. A low cross-linking density was obtained under homopolymerization conditions (5–50% of the double bonds reacted). Much better cross-linking was achieved when thio-ene coupling was conducted with a stoichiometric amount of trimethylolpropane-trimercapto propionate (70–90% conversion of the double bonds). The resulting polymers exhibited low sol content and low T_g values.

Kolb and Meier[59] synthesized dicarboxylic acid methyl ester from undecenoate. Initially, they converted the methyl undecenoate into an unsaturated malonate diester derivative which was employed to synthesize polyester with 1,6-hexanediol. A variety of chemical moieties were grafted to

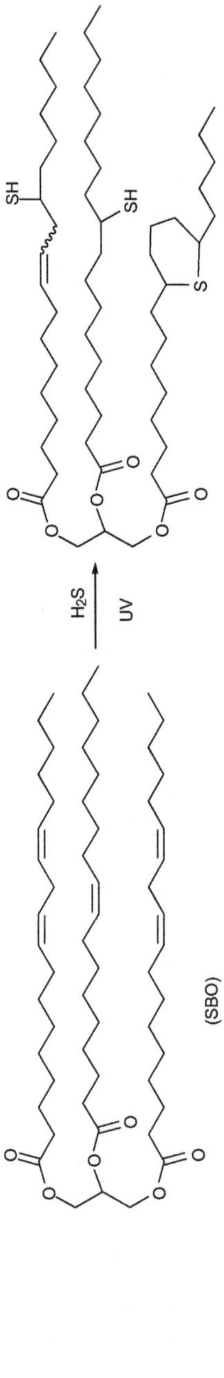

Scheme 4.22 Reaction of SBO with H_2S.[53]

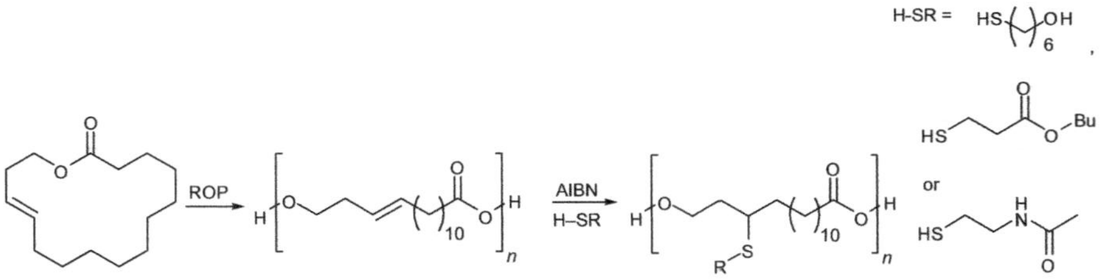

Scheme 4.23 Ring-opening polymerization (ROP) of globalide and the consecutive thiol-ene functionalization.[57] For simplicity, only one of two possible positions for the double bonds and one of three positions for the attachment of a sulfur atom are shown.

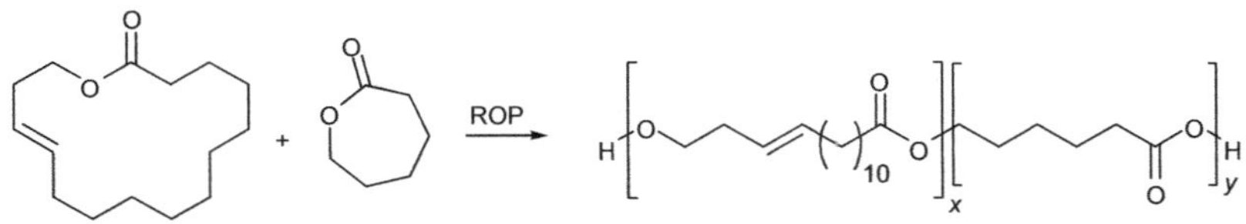

Scheme 4.24 The ROP between the globalide lactone and ε-caprolactone.[58] Only one of the two possible positions of the double bond is illustrated.

the polymer using the thiol-ene reaction (see Scheme 4.25). All five thiols modified the double bonds to sulfide with conversions above 99%, as determined by ^{1}H-NMR. The grafted polymers showed higher T_g values than the starting material.

4.4.3.2   Synthesis of Thiol-ene Polymers using Synthetically Introduced Thiol Groups

Goethals *et al.*[60] synthesized 10-undecenoylthiolactonamide from 10-undecenoic acid. Upon reaction with amine, the thiolactone structure generated a thiol capable of reaction with the double bonds (Scheme 4.26). A series of polymers was synthesized utilizing different amines. The polymers started degrading in air above 250 °C and the C–S bond was reported to be the most prone to oxidation. The T_g values ranged between 5 and 50 °C, and the elongation at break approached 1000%. Oxidation of the sulfide groups to sulfoxide and sulfone increased the elastic modulus, but decreased the elongation at break. Cross-linked polymers, with a soluble fraction of 5 to 23%, were prepared using diamines as cross-linkers.

van den Berg *et al.*[61] made polymers containing only bio-based carbons and hydrogens, achieved by a lengthy conversion of carboxylic groups into –CH$_2$SH groups (Scheme 4.27). The physical properties of the prepared polymer resemble those of low-density polyethylene (LDPE), with an elastic modulus of about 300 MPa and a melting temperature of 90 °C. Oxidation of the polythioether with hydrogen peroxide resulted in an intermediate sulfoxide-functionalized polymer, which was still soluble in common polar solvents. Further oxidation yielded a sulfone-functionalized polymer that was insoluble in any solvent tested. The elastic modulus of the polymer increased almost 1.5 times upon oxidation. The melting temperature increased to 175 °C, accompanied by the occurrence of multiple melting peaks. The onset of thermal decomposition was found to be around 350 °C.

The radical thiol-ene addition is well known to be anti-Markovnikov, *e.g.*, the sulfur atom is expected to attach to the 11th atom of the 10-undecenoic group. The authors report that some of the sulfur atoms attached to the 10th atom. This report suggests that similar partial Markovnikov attachment could occur undetected in similar reactions carried out with 10-undecenoic acid.

Firdaus *et al.*[55] used an indirect approach to introduce thiol groups to bio-based compounds with two olefinic bonds (undecenylic ether (UE) and D-limonene) per molecule (Schemes 4.10 and 4.28). They reacted the synthesized dithiols (di(11-thiol undecyl)ether (DTUE) and 5-((R)-1'-mercaptopropan-20-yl)-2-methylcyclohexanethiol (DTL)) with one of the starting olefinic materials (UE). The polymer formed from DTUE was insoluble in

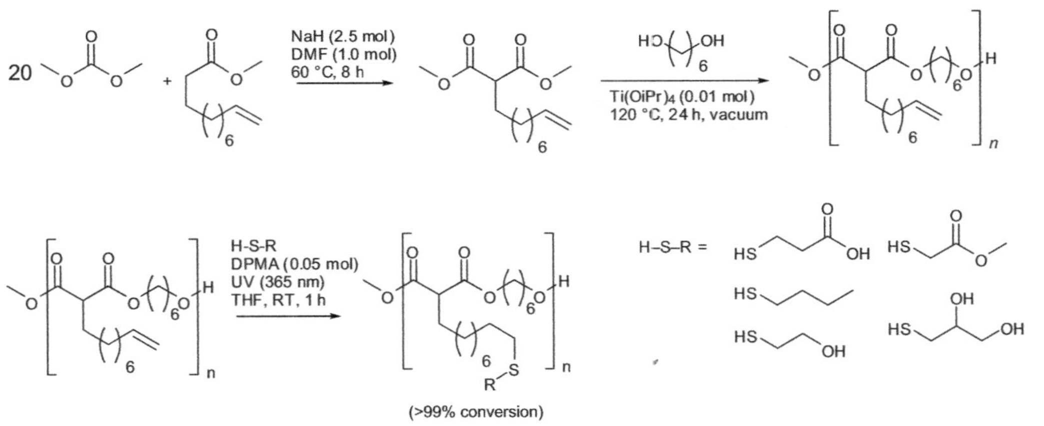

Scheme 4.25 The synthesis of bio-based polyester with dangling unsaturation, and grafting different groups to it, using the thiol-ene reaction.[59]

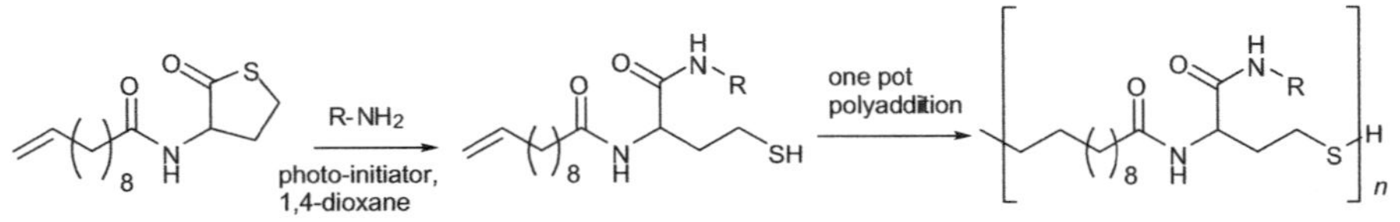

Scheme 4.26 10-Undecenoylthiolactonamide, upon reaction with a primary amine, generates a thiol, which reacts with the double bond to yield a polymer.[60]

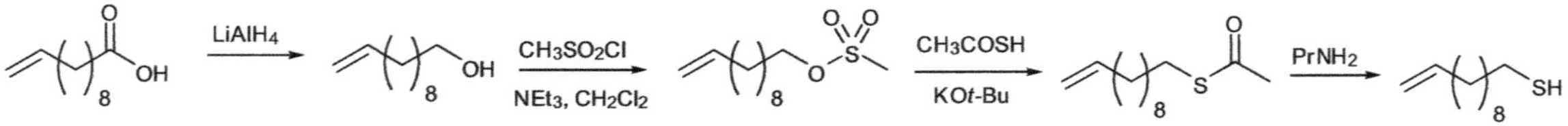

Scheme 4.27 The synthesis of bio-based 10-undecene-1-thiol.[61]

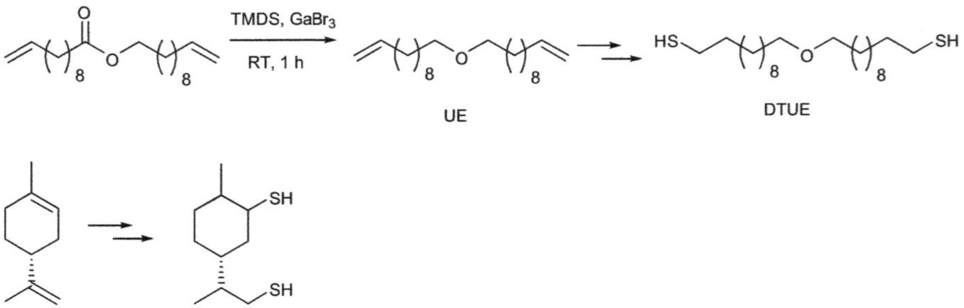

Scheme 4.28 The synthesis of dithiols (DTUE and DTL) from bio-based di-unsaturated materials.[55] DTUE and DTL were later reacted with undecenylic ether (UE) to form polymers *via* a thiol-ene reaction. TMDS: tetramethyldisiloxane.

THF and hexafluoroisopropanol; the one formed from DTL was soluble in THF. The melting temperature of the polymers was increased by 40–50 °C by oxidizing the sulfide groups to sulfone groups.

4.4.3.3 Synthesis of Polymers from Polyfunctional Enes and Thiols

Wu *et al.*[54] reported the kinetic study of the formation of bio-based thiols by reacting SBO with an excess of polyfunctional thiol (Scheme 4.29). Difficulties were reported with the miscibility of the thiols with SBO. Since the formed bio-based thiols had better miscibility with SBO than with the starting thiols, further reaction with SBO molecules led to oligomeric products with lower thiol functionalities.

Echeverri *et al.*[62] reacted SBO with glycerine to obtain a mixture of mono-, di- and triglycerides. The mixture was reacted with maleic anhydride to obtain mixed malonate–FA glycerides (see Scheme 4.30). The resulting product was reacted with stoichiometric amounts of tri- and tetrafunctional thiols, with DMPA as an initiator, and under UV irradiation. Within 12 h, the percentage of the soluble fraction dropped to 11% and the spectroscopy data indicated all the thiol groups and malonate double bonds had reacted, however, some FA double bonds remained unreacted. The T_g values of the polymers were ~ 7–8 °C, and the thermal degradation in air (thermogravimetric analysis, 5%) started at ~ 260 °C.

4.5 Bio-based Products of H-Phosphonate-ene Reactions

The reaction of H-phosphonates with double bonds is similar to the thiol-ene reaction, with the exception that *cis–trans* isomerization of double bonds has not been reported, which implies that the formation of the P–C bond may be irreversible.[63] UV light alone cannot initiate the H-phosphonate-ene reaction, however a γ-radiation-initiated reaction has been reported.[64] On the other hand, the phosphor-centered radicals are probably more reactive, as there are reports of these radicals abstracting allylic hydrogens the same way that thiyl radicals are reported to abstract bis-allylic hydrogens.[65] We are not aware of a report where the H-phosphonate-ene reaction has been successfully carried out to completion with compounds having bis-allylic hydrogens (linoleic acid, linolenic acid, their esters, or vegetable oils). Our attempts showed the reaction of H-phosphonates with methyl linoleate to be sluggish: the best achieved conversion was 30% in 48 h. In addition, the reaction was accompanied by transesterification of the different alkoxy groups, especially at higher reaction temperatures.[66] Nevertheless, there are reports where esters of oleic acid have been successfully modified using the general reaction of Scheme 4.31.

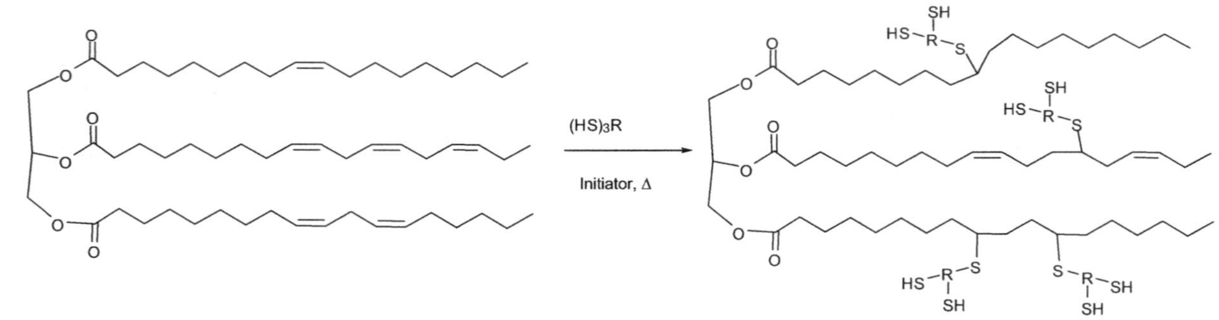

Scheme 4.29 Formation of bio-based polyfunctional thiols.[54] In addition, di- and tetrafunctional thiols were used as starting materials.

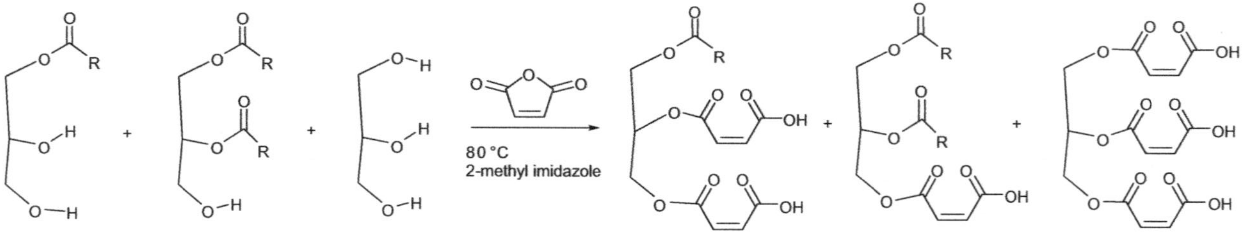

Scheme 4.30 Synthesis of maleated triglycerides.[62] R = FA chains (predominantly linoleic and oleic).

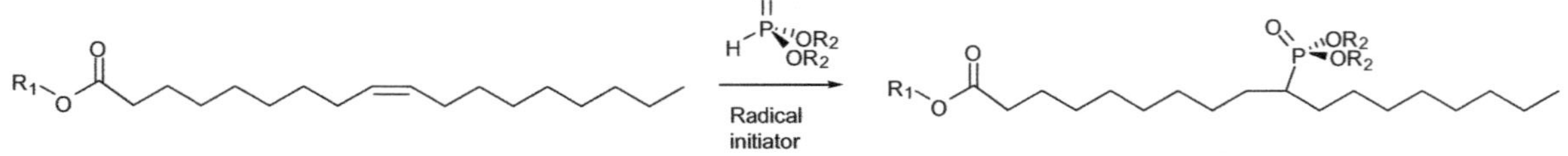

Scheme 4.31 General reaction scheme of the H-phosphonate-ene reaction, used for the synthesis of bio-based phosphonates. The 9-regioisomer is not shown.

Several bio-based phosphonates were disclosed by Hamilton and Williams.[67] They successfully reacted the double bonds of oleic acid and oleyl alcohol with di-*n*-butyl phosphite, using a di-*tert*-butyl peroxide initiator. In parallel to the H-phosphonate-ene addition reaction, they reported a transesterification reaction.

The synthesis of several alkyl 9(10)-(dialkyl phosphono) stearates was reported by Sasin *et al.*[68] The compounds were reported to have low Pour Points (−50 to −65 °C)[69] and to be good plasticizers for PVC.[70] Similar compounds were later evaluated as lubricant additives, and they showed good anti-wear properties.[71] Mod *et al.*[64] reacted such *N,N*-disubstituted amides of oleic acid with dialkyl H-phosphonate, (in their experiments, R_1 in Scheme 4.31 was a secondary N-group) and demonstrated that the resulting phosphonostearylamide derivatives had anti-bacterial properties against pathogenic microorganisms.

4.6 Reactions Related to the H-Phosphonate-ene Reaction

The reaction of H-phosphonates with double bonds has been carried out by Moreno *et al.*[72] in a non-radical fashion. First, they converted the double bonds in methyl oleate and high-oleic sunflower oils to enone groups and, afterwards, they reacted the enone groups using a $BF_3 \cdot EtO_2$-catalyzed phospha-Michael addition (see Scheme 4.32). It is noteworthy that they used a bis-phosphine oxide compound, so it acted as a cross-linking agent. Data showed that bio-based thermoset polymers have significantly improved flame-retardant properties when phosphorus is incorporated into their structure.

Another reaction worth mentioning is the Abramov reaction between aldehydes and H-phosphonates using a base catalyst (Scheme 4.33). It has been carried out by Cermak *et al.*[73,74] with aldehydes made from ricinoleic acid (the main component of castor oil) and lesquerolic acid (the main component of lesquerella oil). It has been well documented in the literature that α-hydroxy phosphonates and their corresponding phosphonic acids have potentially interesting biological activities.[75,76] Ricinoleic acid is a naturally occurring hydroxy FA used in a wide variety of applications, including medicine. Lesquerolic acid is a less-studied homolog of ricinoleic acid. It is of interest partially since the lesquerella plant is less toxic that the castor plant. α-Hydroxy phosphonates and their corresponding phosphonic acids are widely recognized as important structural moieties. They have been shown to exhibit a wide variety of biological activities as enzyme inhibitors of renin, enzyme 5-enolpyruvylshikimate-3-phosphate synthase, human immunodeficiency virus (HIV) protease, and farnesyl protein transferase.[77] They have also been utilized as pesticides, anti-biotics, and anti-cancer and anti-viral agents.[78]

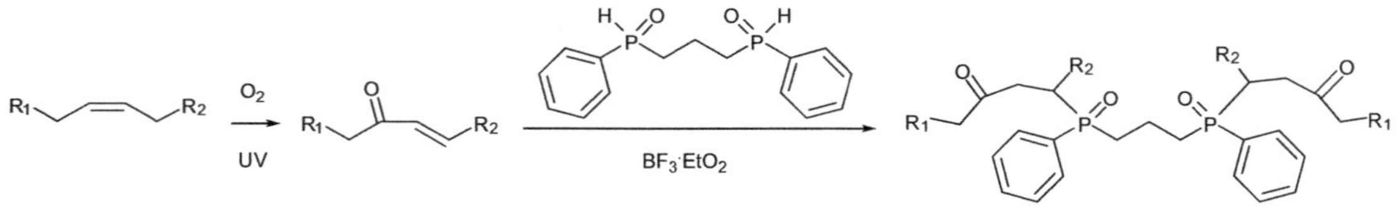

Scheme 4.32 Use of phospha-Michael addition to make bio-based polymers with improved flame-retardant properties.[72]

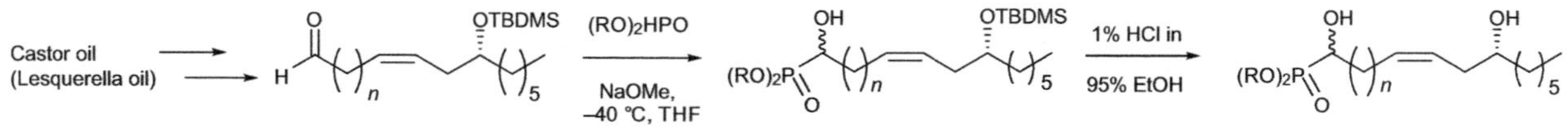

Scheme 4.33 Use of the Abramov reaction to form castor- ($n = 7$) and lesquerella- ($n = 9$) based α-hydroxy phosphonates.[73,74] The reaction was also carried out with FAs, where the C=C bond had been saturated. TBDMS: *tert*-butyldimethylsilyl.

4.7 Conclusions

The thiol-ene reaction has been demonstrated to have the potential to functionalize FA derivatives with a wide array of functional groups. The reaction is versatile and gives high yields when used with monounsaturated acids. Despite that, there are still unresolved problems with the reaction; especially the thiol-ene reaction with polyunsaturated FAs having bis-allylic protons. Literature examples with such FAs are less common, and the reported yields are lower, or the reaction conditions are harsher. The literature review in this chapter shows that there are still some discrepancies and gaps in the knowledge of the kinetics of the reaction.

The H-phosphonate-ene reaction also has promise for obtaining valuable compounds but, at the current stage, the reaction is more limited in scope. One of the problems is the lower reactivity, which is reflected in the fact that there are no published isolated products of H-phosphonate-ene reactions with polyunsaturated FAs. Another problem is the lack of availability of multifunctional H-phosphonate compounds. There is also less knowledge about the kinetics of H-phosphonate-ene reactions with double bonds compared to reactions based on thiol-ene kinetics.

References

1. J. Berkowitz, G. Barney Ellison and D. Gutman, *J. Phys. Chem.*, 1994, **98**, 2744.
2. (a) E. T. Denisov, *Zh. Fiz. Khim.*, 1993, **67**, 2416; (b) E. T. Denisov, T. G. Denisova and T. S. Pokidova,*Handbook of Free-Radical Initiators*, 2003, Wiley-Interscience, Hoboken.
3. D. Griller and D. D. M. Wayner, *Pure. Appl. Chem.*, 1989, **61**, 717.
4. T. L. Brown, H. E. LeMay Jr. and B. E. Bursten, *Chemistry: The Central Science*, Prentice Hall, Upper Saddle River, 8th edn, 2000.
5. C. Sivertz, W. Andrews, W. Elsdon and K. Graham, *J. Polym. Sci.*, 1956, **19**, 587.
6. C. Walling and W. Helmreich, *J. Am. Chem. Soc.*, 1958, **81**, 1144.
7. C. Sivertz, *J. Phys. Chem.*, 1959, **63**, 34.
8. N. P. Neureiter and F. G. Bordwell, *J. Am. Chem. Soc.*, 1960, **83**, 5354.
9. H. Kircher, *J. Am. Oil Chem. Soc.*, 1964, **41**, 351.
10. A. W. Schwab, L. E. Gast and J. C. Cowan, *J. Am. Oil Chem. Soc.*, 1968, **45**, 461.
11. A. W. Schwab and L. E. Gast, *J. Am. Oil Chem. Soc.*, 1970, **47**, 371.
12. A. W. Schwab, W. K. Rohwedder, L. W. Tjarks and L. E. Gast, *J. Am. Oil Chem. Soc.*, 1973, **50**, 364.
13. (a) A. W. Schwab, L. E. Gast and W. K. Rohwedder,*J. Am. Oil Chem. Soc.*, 1975, **52**, 236; (b) A. W. Schwab, L. E. Gast and W. K. Rohwedder,*J. Am. Oil Chem. Soc.*, 1976, **53**, 762; (c) A. W. Schwab, W. K. Rohwedder and L. E. Gast, *J. Am. Oil Chem. Soc.*, 1978, **55**, 860.

14. C. Ferreri, C. Costantino, L. Perrotta, L. Landi, Q. G. Mulazzani and C. Chatgilialoglu, *J. Am. Chem. Soc.*, 2001, **123**, 4459.
15. (a) C. Chatgilialoglu, C. Ferreri, M. Ballestri, Q. G. Mulazzani and L. Landi, *J. Am. Chem. Soc.*, 2000, **122**, 4593; (b) C. Chatgilialoglu, A. Altieri and H. Fischer, *J. Am. Chem. Soc.*, 2002, **124**, 12816.
16. C. Chatgilialoglu, A. Samadi, M. Guerra and H. Fischer, *ChemPhysChem.*, 2005, **6**, 286.
17. C. Ferreri, A. Samadi, F. Sassatelli, L. Landi and C. Chatgilialoglu, *J. Am. Chem. Soc.*, 2004, **126**, 1063.
18. C. Chatgilialoglu, *Helv. Chim. Acta*, 2006, **89**, 2387.
19. C. Chatgilialoglu, C. Ferreri, M. Melchiorre, A. Sansone and A. Torreggiani, *Chem. Rev.*, 2014, **114**, 255.
20. S. Adhikari, H. Sprinz and O. Brede, *Res. Chem. Intermed.*, 2001, **27**, 549.
21. H. Sprinz, J. Schwinn, S. Naumov and O. Brede, *Biochim. Biophys. Acta*, 2000, **1483**, 91.
22. C. Tronche, F. N. Martinez, J. H. Horner and M. Newcomb, *Tetrahedron Lett.*, 1996, **37**, 5845.
23. M. V. Encinas, P. J. Wagner and J. C. Scaiano, *J. Am. Chem. Soc.*, 1980, **102**, 1357.
24. M. Newcomb, A. G. Glenn and M. B. Manek, *J. Org. Chem.*, 1989, **54**, 4603.
25. (a) N. B. Cramer, T. Davies, A. K. O'Brien and C. N. Bowman, *Macromolecules*, 2003, **36**, 4631; (b) N. B. Cramer, S. K. Reddy, A. K. O'Brien and C. N. Bowman, *Macromolecules*, 2003, **36**, 7964.
26. C. Schoneich, U. Dillinger, F. von Bruchhausen and K.-D. Asmus, *Arch. Biochem. Biophys.*, 1992, **292**, 456.
27. U. Biermann, W. Butte, R. Koch, P. A. Fokou, O. Türünç, M. A. R. Meier and J. O. Metzger, *Chem.–Eur. J.*, 2012, **18**, 8201.
28. M. Claudino, M. Johansson and M. Jonsson, *Eur. Polym. J.*, 2010, **46**, 2321.
29. T. Y. Lee, T. M. Roper, E. S. Jonsson, C. A. Guymon and C. E. Hoyle, *Macromolecules*, 2004, **37**, 3606.
30. T. M. Roper, C. A. Guymon, E. S. Jonsson and C. E. Hoyle, *J. Polym. Sci., Part A: Polym. Chem.*, 2004, **42**, 6283.
31. L. T. Nguyen, M. T. Gokmen and F. E. Du Prez, *Polym. Chem.*, 2013, **4**, 5527.
32. G. B. Bantchev, J. A. Kenar, G. Biresaw and M. G. Han, *J. Agric. Food Chem.*, 2009, **57**, 1282.
33. J. Samuelsson, M. Jonsson, T. Brinck and M. Johansson, *J. Polym. Sci., Part A: Polym. Chem.*, 2004, **42**, 6346.
34. M. Desroches, S. Caillol, V. Lapinte, R. Auvergne and B. Boutevin, *Macromolecules*, 2011, **44**, 2489.
35. P. G. Gadd, H. M. Gillespie, F. W. Heywood and E. G. McKenna, *EU Pat.*, 0 713 867, 1996.
36. G. B. Bantchev, G. Biresaw, A. Mohamed and J. Moser, *Thermochim. Acta*, 2011, **513**, 94.

37. G. Biresaw, G. B. Bantchev and S. C. Cermak, *Tribol. Lett.*, 2011, **43**, 17.
38. R. E. Murray, G. B. Bantchev, R. O. Dunn, K. L. Ascherl and K. M. Doll, *ACS Sustainable Chem. Eng.*, 2013, **1**, 562.
39. B. K. Sharma, A. Adhvaryu and S. Z. Erhan, *J. Agric. Food Chem.*, 2006, **54**, 9866.
40. (a) B. K. Sharma, A. Adhvaryu and S. Z. Erhan, *Tribol. Int.*, 2009, **42**, 353; (b) S. Z. Erhan, A. Adhvaryu and B. K. Sharma, *US Pat.*, 7 279 448, 2007.
41. L. M. de Espinosa, A. Gevers, B. Woldt, M. Graß and M. A. R. Meier, *Green Chem.*, 2014, **16**, 1883.
42. C. Lluch, J. C. Ronda, M. Galia, G. Lligadas and V. Cádiz, *Biomacromolecules*, 2010, **11**, 1646.
43. O. Türünç and M. A. R. Meier, *Macromol. Rapid Commun.*, 2010, **31**, 1822.
44. R. J. Gonzalez-Paz, C. Lluch, G. Lligadas, J. C. Ronda, M. Galia and V. Cádiz, *J. Polym. Sci., Part A: Polym. Chem.*, 2011, **49**, 2407.
45. M. Desroches, S. Caillol, R. Auvergne, B. Boutevin and G. David, *Polym. Chem.*, 2012, **3**, 450.
46. M. Desroches, S. Caillol, R. Auvergne and B. Boutevin, *Eur. J. Lipid Sci. Technol.*, 2012, **114**, 84.
47. M. Stemmelen, C. Travelet, V. Lapinte, R. Borsali and J.-J. Robin, *Polym. Chem.*, 2013, **4**, 1445.
48. C. Lluch, G. Lligadas, J. C. Ronda, M. Galià and V. Cádiz, *Macromol. Rapid Commun.*, 2011, **32**, 1343.
49. C. Lluch, G. Lligadas, J. C. Ronda, M. Galià and V. Cádiz, *Macromol. Biosci.*, 2013, **13**, 614.
50. A. S. More, T. Lebarbé, L. Maisonneuve, B. Gadenne, C. Alfos and H. Cramail, *Eur. Polym. J.*, 2013, **49**, 823.
51. M. Stemmelen, F. Pessel, V. Lapinte, S. Caillol, J.-P. Habas and J.-J. Robin, *J. Polym. Sci., Part A: Polym. Chem.*, 2011, **49**, 2434.
52. O. Türünç, M. Firdaus, G. Klein and M. A. R. Meier, *Green Chem.*, 2012, **14**, 2577.
53. J. D. Byers, M. D. Refvik and C. W. Brown, *US Pat.*, 7 781 484 B2, 2010.
54. J. F. Wu, S. Fernando, D. Weerasinghe, Z. Chen and D. C. Webster, *ChemSusChem*, 2011, **4**, 1135.
55. M. Firdaus, M. A. R. Meier, U. Biermann and J. O. Metzger, *Eur. J. Lipid Sci. Technol.*, 2014, **116**, 31.
56. F. D. Gunstone, M. G. Hussain and D. M. Smith, *Chem. Phys. Lipids*, 1974, **13**, 71.
57. Z. Ates, P. D. Thorntona and A. Heise, *Polym. Chem.*, 2011, **2**, 309.
58. M. Claudino, I. van der Meulen, S. Trey, M. Jonsson, A. Heise and M. Johansson, *Polym. Chem.*, 2012, **50**, 16.
59. N. Kolb and M. A. R. Meier, *Eur. Polym. J.*, 2013, **49**, 843.
60. F. Goethals, S. Martens, P. Espeel, O. van den Berg and F. E. Du Prez, *Macromolecules*, 2014, **47**, 61.
61. O. van den Berg, T. Dispinar, B. Hommez and F. E. Du Prez, *Eur. Polym. J.*, 2013, **49**, 804.

62. D. A. Echeverri, V. Cádiz, J. C. Ronda and L. A. Rios, *Eur. Polym. J.*, 2012, **48**, 2040.

63. G. B. Bantchev, G. Biresaw, K. E. Vermillion and M. Appell, *Spectrochim. Acta, Part A*, 2013, **110**, 81.

64. R. R. Mod, J. A. Harris, J. C. Arthur Jr., F. C. Magne, G. Sumrell and A. F. Novak, *J. Am. Oil Chem. Soc.*, 1972, **49**, 634.

65. A. R. Stiles, W. E. Vaughan and F. F. Rust, *J. Am. Chem. Soc.*, 1958, **80**, 714.

66. G. B. Bantchev and G. Biresaw, *J. Am. Oil Chem. Soc.*, submitted.

67. L. A. Hamilton and R. H. Williams, *US Pat.*, 2 957 931, 1960.

68. R. Sasin, W. F. Olszewski, J. R. Russell and D. Swern, *J. Am. Chem. Soc.*, 1959, **81**, 6275.

69. D. Swern, W. E. Palm, R. Sasin and L. P. Witnauer, *J. Chem. Eng. Data*, 1960, **5**, 486.

70. D. Swern, W. E. Palm, R. Sasin and L. P. Witnauer, *J. Chem. Eng. Data*, 1960, **5**, 484.

71. G. Biresaw and G. B. Bantchev, *J. Am. Oil Chem. Soc.*, 2013, **90**, 891.

72. M. Moreno, G. Lligadas, J. C. Ronda, M. Galia and V. Cádiz, *J. Polym. Sci., Part A: Polym. Chem.*, 2013, **51**, 1808.

73. D. M. Cermak, S. C. Cermak, A. B. Deppe and A. L. Durham, *Ind. Crops Prod.*, 2012, **37**, 394.

74. J. S. P. Cusimano, M. M. Hart, D. M. Cermak, S. C. Cermak and A. L. Durham, *Ind. Crops Prod.*, 2014, **53**, 23.

75. D. M. Cermak, Y. Du and D. F. Wiemer, *J. Org. Chem.*, 1999, **64**, 388.

76. D. L. Pompliano, E. Rands, M. D. Schaber, S. D. Mosser, N. J. Anthony and J. B. Gibbs, *Biochemistry*, 1999, **31**, 3800.

77. D. F. Wiemer, *Tetrahedron*, 1997, **53**, 16609.

78. Q. Wu, J. Zhou, Z. Yao, F. Xu and Q. Shen, *J. Org. Chem.*, 2010, **75**, 7498.

CHAPTER 5

Plant-oil-based Polymeric Materials and their Applications

FEI LIU AND JIN ZHU

Ningbo Key Laboratory of Polymer Materials, Ningbo Institute of Material Technology and Engineering, Chinese Academy of Sciences, Ningbo, 315201, People's Republic of China
Email: jzhu@nimte.ac.cn

5.1 Introduction

The world is close to using up all of its non-renewable petroleum-based resources, which cost nature millions of years to produce. It was not until recent decades that people realized that nature produces even larger amounts of renewable and sustainable bio-based resources every day, which can be used for the synthesis of polymeric materials. Additionally, with increasing concerns about environmental protection and climate change, the use of renewable resources has become urgent. Plant oils, among other renewable resources, are interesting candidates, due to their universal availability, low price, low toxicity, bio-degradability and ease of chemical modification. These natural properties provide a versatile platform to build a better future for the coming generations. The most interesting plant oils are soybean oil and castor oil, yet other types of plant oils like tung oil, linseed oil and canola oil are also attractive. Much effort has been devoted to the field of the synthesis of polymeric materials from plant oils, with comparable properties, in order to substitute or partially substitute their

RSC Green Chemistry No. 29
Green Materials from Plant Oils
Edited by Zengshe Liu and George Kraus
© The Royal Society of Chemistry 2015
Published by the Royal Society of Chemistry, www.rsc.org

petroleum-based counterparts. Many reviews have been published in this area covering the use of plant oils for the synthesis of polymeric materials, focusing on various topics, such as different polymerization strategies[1–3] including radical polymerization, cationic polymerization, condensation polymerization and olefin metathesis and so on, or different polymeric materials[1,4,5] including polyester, polyamide, polyurethane, and bio-composites, or different applications[6,7] such as paints, coatings, plasticizers and adhesives as well as bio-medical applications. In this chapter, we will focus on the synthesis of polymeric materials based on different types of plant oils including soybean oil, castor oil, tung oils, linseed oil, and canola oil. Special attention will be paid to the first three plant oils. Their applications in coatings, adhesives, plasticizers and modifiers will also be reviewed.

5.2 Soybean Oil: Polymer Synthesis and Application

5.2.1 Use of Soybean Oil Directly for Polymer Synthesis

Soybean oil has an average number of double bond of 4.6 per triglyceride,[8] which makes it a semi-drying oil. The structure of a soybean oil triglyceride is shown in Figure 5.1, which illustrates that it mainly contains two kinds of fatty acids, oleic acid and linoleic acid. The unsaturation in the fatty acid chains makes them ideal monomers for polymerization. However, due to the relatively high molecular weights of the multiple-chain structures of soybean oil and the non-conjugated double bonds in the fatty acid chains, its reactivity is rather low.[9,10] For example, Acar *et al.*[11] exposed soybean oil to air under daylight at room temperature to get oxidized soybean oil *via* an autooxidation process. A cold-water-soluble soybean oil polymer was then obtained by reaction with diethanol amine, the resulting hydroxylated soybean oil polymer had a molecular weight of only 3800 to 5900 g mol^{-1}.

Therefore, homopolymerization of soybean oil and other semi-drying oils is not favored due to steric hindrance of the bulky oil moieties, co-monomers such as styrene or divinyl benzene are necessary for the preparation of polymers from soybean oil directly.[12] However, recently, Liu *et al.*[10,13] reported that cationic polymerization in supercritical carbon dioxide (scCO$_2$) media using boron trifluoride diethyl etherate (BF$_3 \cdot$OEt$_2$) as an initiator

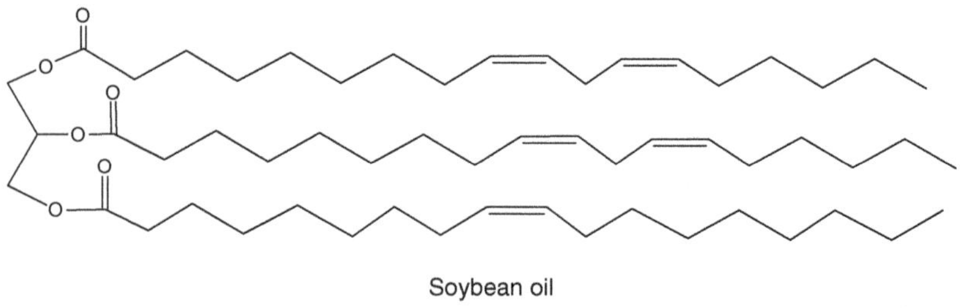

Soybean oil

Figure 5.1 Structure of a soybean oil triglyceride.

resulted in a high-molecular-weight poly soybean oil, ranging from 21 842 to 118 300 g mol^{-1}. This high-molecular-weight product is probably formed by both polymerization and an intermolecular Diels–Alder reaction. Additionally, the scCO$_2$ serves as a green reaction medium, making it a promising method for the polymerization of renewable starting materials.

In addition to reactions with its double bonds, soybean oil can undergo reactions with other functional groups. Wu *et al.*[14] generated thiol-functionalized oligomers *via* a thermal free-radical-initiated thiol-ene reaction between the soybean oil double bonds and the thiol functional groups. It was found that long reaction times and the use of a nitrogen reaction atmosphere were favorable for fast consumption of the soybean oil, and the reaction produced high-molecular-weight products. The synthesized soy-thiol oligomers can be used for renewable thiol-ene UV-curable materials and thiourethane thermal-cure materials.

Soybean oil fatty acids are not only monomers for the synthesis of polymeric materials, but also building blocks for the synthesis of more sophisticated monomers.[3] Recently, Chernykh *et al.*[15] synthesized a novel vinyl ether monomer from soybean oil by a base-catalyzed transesterification of 2-(vinyloxy) ethanol with soybean oil. The cationic polymerization of this monomer resulted in a polymer with a very low T_g value of −90 °C, with the expectation for the production of thermoplastic elastomers.

Co-polymerization of soybean oil with monomers containing heteroatoms such as silicon, boron and phosphorous would result in polymers with additional useful properties such as flame retarding. Sacristán *et al.*[16] prepared a co-polymer from soybean oil, styrene, divinylbenzene and *p*-trimethylsilylstyrene by cationic polymerization using BF$_3 \cdot$OEt$_2$ as an initiator. The obtained thermosets had limited oxygen index (LOI) values from 22.6 to 29.7, with T_g values ranging from 50 to 62 °C. This result indicated that these materials are useful alternatives for current non-renewable-based thermosets and that the flame-retardant properties of soybean-oil-based thermosets can be improved by adding covalently bonded silicon to the polymer. They also synthesized a boron-containing soybean-oil-based co-polymer with the same strategy,[17] where 4-vinylphenyl boronic acid was used to introduce the boron to the polymer. The LOI of the thermosets was in the range of 23.7 to 25.6, which is similar to those of their silicon counterparts. Ronda *et al.*,[18] on the other hand, synthesized a phosphorus-containing soybean oil co-polymer from dimethyl-*p*-vinylbenzylphosphonate. The resulting thermosets with just 1% of phosphorus had an LOI of about 24.0, indicating an improvement in the flame-retardant properties of the soybean-oil-based co-polymers.

5.2.2 Use of Soybean Oil after Chemical Modification for Polymer Synthesis

The double bonds on the fatty acid chains in soybean oil possess low activity, which drives people to perform modifications of soybean oil in order to increase the reaction activity as well as functionality diversity, so that soybean oil can be involved in more polymerization reactions for the synthesis of various polymeric materials.[1] Fortunately, the double bond is an excellent starting point, it can be chemically modified to generate many functional groups such as epoxide, acryl and hydroxyl groups and so on.

5.2.2.1 Epoxidized Soybean Oil

Epoxidation of soybean oil is usually achieved by treating soybean oil with hydrogen peroxide and acetic or formic acid in the presence of strong mineral acids such as H_2SO_4 and H_3PO_4 (ref. 19). The use of strong mineral acids can initiate undesirable oxirane ring-opening reactions with water, and this side-reaction would consume part of the epoxy groups, resulting in a final conversions of double bonds to epoxy groups of about 90%. Therefore, Vlcek *et al.*[19] optimized the epoxidation of soybean oil *via* a chemo-enzyme catalyst. A lipase, *Candida antarctica* (N435), immobilized on acrylic resin was used as a catalyst for the *in situ* epoxidation of soybean oil in the presence of hydrogen peroxide (Figure 5.2). The kinetic study showed a 100% conversion of double bonds to epoxides in 4 h. Additionally, by using the enzyme, a high conversion was achieved without adding free acids and the neutral pH of the reaction mixture was maintained. Furthermore, this epoxidized soybean oil (ESO) was then used to synthesize acrylated epoxide soybean oil (AESO),[20] which showed similar results compared

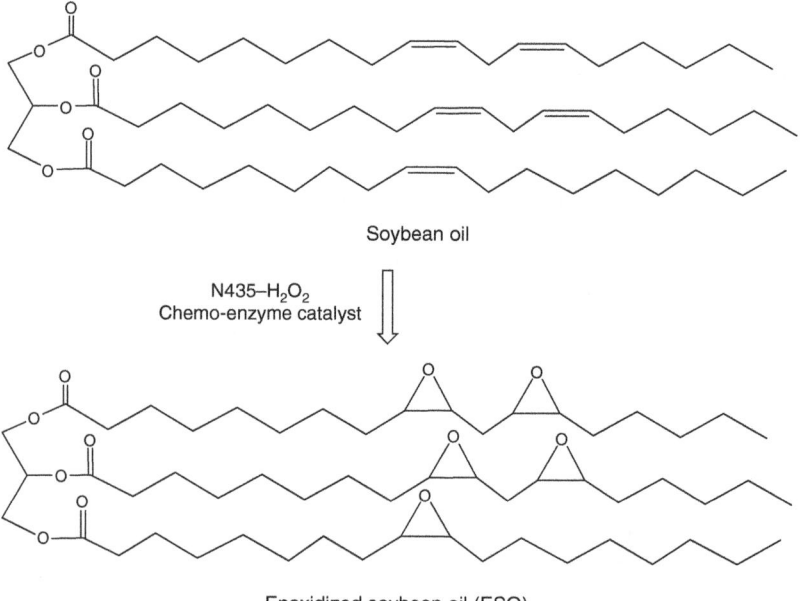

Figure 5.2 Synthesis of ESO from soybean oil *via* a chemo-enzyme catalyst.

with traditional ESO made through a chemical process. Thus it provides people with an alternative and more effective method for the synthesis of ESO.

With the presence of the epoxide functional groups in the fatty acid chains, ESO can undergo ring-opening polymerization (ROP) reactions. Liu *et al.*[21] carried out a ROP of ESO in methylene chloride using $BF_3 \cdot OEt_2$ as a catalyst. The product was a highly cross-linked polymer with T_g ranging from -16 to -48 °C. It was noted that the reaction was conducted in a toxic solvent, therefore the authors turned their attention to a green medium, subcritical CO_2 (ref. 22). The ROP of ESO catalyzed by the same catalyst was conducted under mild conditions such as room temperature and a subcritical CO_2 pressure of 65.5 bar. The formed product has a similar structure and properties, with T_g ranging from -11.9 to -24.1 °C. On the other hand, because of the concerns regarding the toxicity of the $BF_3 \cdot OEt_2$ catalyst, the application of these soy-based polymers in food and medicinal areas is limited. Therefore, they conducted the ROP of ESO in ethyl acetate catalyzed by the super acid $HSbF_6 \cdot 6H_2O$ (ref. 23). The results indicated that ESO was effectively polymerized by the catalyst and formed polymers with relatively high cross-link densities and with T_g values ranging from -13 to -21 °C. All these soybean-oil-based polymers synthesized *via* ROP of ESO can be functionalized to hydrogels by hydrolysis and the products can be used in personal and health care areas.[21–23]

In addition to the ROP reaction, ESO can be thermally cured by many systems such as maleinated linseed oil[24] and pyridine catalysts[25] or UV cured with different types of photo-initiators[26,27] for the synthesis of epoxy resins. However, there are some limitations of these systems such as long curing times, high curing temperatures or poor thermophysical properties. Recently, in order to overcome some of these problems, Tan *et al.*[28] reported that ESO can be thermally cured with a methylhexahydrophthalic anhydride (MHHPA) curing agent in the presence of a 2-ethyl-methylimidazole (EMI) catalyst. With the increase of EMI concentration, the rate of polyesterification, degree of conversion and cross-link density of the thermally cured ESO were higher and the T_g and storage modulus of cured ESO also increased. One possible explanation could be associated with the fact that a zwitterion was formed from MHHPA and EMI during the pre-mixed reaction and can initiate the polymerization of ESO with MHHPA. Another catalyst reported by Tan *et al.*[29] that can be used for ESO curing with MHHPA is tetraethylammonium bromide (TEAB). It is proposed that the ring opening of the MHHPA curing agent by the TEAB catalyst involves an S_N2 reaction. The triethylamine formed as a result of the dequaternization reaction of the TEAB catalyst, serves as a nucleophile and attacks the MHHPA curing agent to yield zwitterions, which are generated during the pre-mixed reaction. The zwitterions formed eventually react with the epoxy rings on the ESO backbone chains and subsequently generate alkoxide intermediates, which will cleave another MHHPA curing agent to yield carboxylate anions and then react with other ESO resins to yield reaction intermediate product and propagate the cycle. Apparently, the zwitterions played a very important role for the curing of ESO with MHHPA by different catalysts.

In addition to MHHPA, ESO can be cured with other anhydrides such as maleic anhydride (MA).[30] However, MA is a solid at room temperature, thus heating to a moderate temperature is necessary in order to obtain a homogenous mixture with ESO. For this reason, liquid cyclic anhydrides like MHHPA or nadic methyl anhydride (NMA) are more favorable.[31] NMA can be used as a cross-linking agent for the curing of ESO, and the cured product has improved mechanical and thermal properties with the addition of protein-based fillers such as ovalbumin.[31] With only 15 wt% of ovalbumin, the flexural modulus improved more than 150% compared with the unfilled material.

Because of its excellent performance in the synthesis of epoxy resins, ESO can be used to partially replace the synthetic epoxy pre-polymer based on the diglycidyl ether of bisphenol A (DGEBA). Ruseckaite *et al.*[32] combined DGEBA with 40 wt% of ESO in a resin using methyltetrahydrophthalic anhydride as a cross-linking agent and 1-methyl imidazole as an initiator. The product had an optimum set of properties; the Young's modulus in the glassy state was 93% of that of the neat DGEBA resin, the T_g value decreased by only about 11 °C and the impact strength increased by about 38% without loss of transparency. Another interesting strategy for using ESO in the synthesis of epoxy resins is to get free epoxidized fatty acids by hydrolysis of ESO.[33] The glycidyl esters of epoxidized fatty acids derived from soybean oil have a higher oxirane content, more reactivity and lower viscosity than ESO, thus leading to higher T_g values.

ESO-based elastomers are another interesting type of polymer because of their bio-degradability and bio-compatibility. Considering the structure, toxicity and operability, the design for an ideal cross-linked elastomer should meet the following principles:[34] (1) the raw materials are of low toxicity and can be metabolized *in vivo*; (2) no toxic additives are used, including the initiator, catalyst and solvent; (3) the cross-linked backbone used is bio-degradable; and (4) the preparation is easy with good repeatability. Based on these rules, Altuna *et al.*[35] synthesized a cross-linked smart material capable of stress relaxation and self-healing without the addition of any extrinsic catalyst from ESO with an aqueous citric acid without addition of any other catalyst or solvent. This was achieved by molecular rearrangements produced by the thermally activated transesterification reaction of β-hydroxyester groups generated in the polymerization reaction. Wang *et al.*[36] synthesized a new series of soybean-oil-based elastomers poly(ESO-*co*-decamethylene diamie) (PESD) by the ROP of ESO with decamethylene diamine (DDA). The glycerol center of ESO was broken by ammonolysis which resulted in uncross-linked elastomers with low T_g values ranging from −30 to −17 °C. And a cross-linked bio-elastomer was obtained from PESD cross-linked with succinic anhydride. The final bio-elastomer possessed good damping properties, low water absorption and a low degradation rate in phosphate-buffered solution. In this process, ESO was employed to react with a diamine by combining the ROP with ammonolysis in which the ROP contributes to chain growth and the ammonolysis breaks the glycerol center. Liu *et al.*[34] reported another series of bio-degradable and bio-compatible elastomers based on phosphoester cross-linked vegetable oils (PVOs) by a simple cross-linking reaction between phosphorylated castor oil (PCO) and ESO without any solvent or initiator at 37 °C. The PVOs can return to their original dimensions even after 1000 repeats of a cyclic compression test, accompanied by rapid restoration to their original states.

5.2.2.2 Acrylated Epoxidized Soybean Oil

Further modification of ESO generates another useful building block, acrylated epoxidized soybean oil (AESO), which is synthesized from ESO with acrylic acid (Figure 5.3). This modification not only increases the reactivity of the soybean oil, but also introduces more polymerizable functionalities such as acrylate and hydroxyl groups. There are also some unreacted epoxy groups which can be used for further modification such as with maleic acid (Figure 5.3), producing maleated acrylated epoxidized soybean oil (MAESO).[37] Therefore, AESO is another versatile chemical derived from soybean oil for the synthesis of various polymers and composites.

In addition to acrylic acid, different types of acrylates can be used for the synthesis of AESO. Rengasamy *et al.*[38] used 2-hydroxyethyl acrylate (HEA) for the synthesis of a new type of AESO, which possesses low viscosity and a high degree of acrylation. The low viscosity is due to the ether linkages in this AESO because they a have relatively increased freedom of rotation compared with ester linkages, and the esters are away from the fatty acid chains and may participate less in hydrogen bonding. Additionally, the use of an excess amount of HEA during the synthesis of AESO suppressed the formation of ether linkages between epoxy groups, which also decreases the viscosity because of less cross-linking. Because of its low viscosity, the demand for reactive diluents for the curing process also decreased.

When AESO is used for producing UV-curable coatings, different types of acrylates are involved. Wu *et al.*[39] incorporated hyperbranched acrylates (HBAs). The use of HBAs greatly increased the mechanical and thermal properties of coatings prepared from AESO, such as coating hardness, adhesion, impact resistance, tensile modulus and toughness as well as T_g. This is because of the low viscosity and high functionality, thus high reactivity, of HBAs, resulting in more cross-linking in the coating network. Kim *et al.*[40] synthesized poly(ethylene glycol) (PEG) diacrylate and poly(caprolactone) (PCL) diacrylate and cross-linked them with AESO to form a polymer network by UV-initiated free-radical polymerization. The tensile strengths and elongation-at-break ranged from 0.5 to 11 MPa and 7 to 200% respectively, depending on the composition of the reaction mixture. Jang *et al.*[41] used 2,5-furan diacrylate as a difunctional stiffener for the synthesis of AESO-based cross-linked polymer networks *via* UV photo-polymerization. The tensile strength was 7.2–14 MPa, and the elongation-at-break was 4.6–10%.

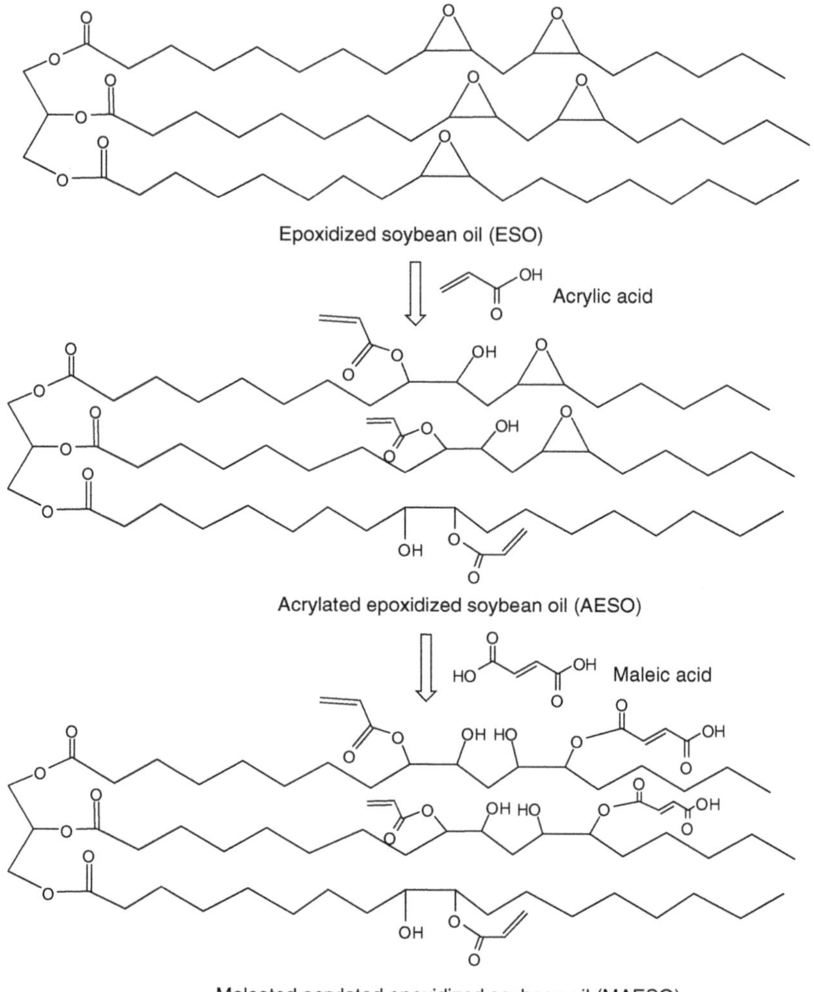

Epoxidized soybean oil (ESO)

Acrylic acid

Acrylated epoxidized soybean oil (AESO)

Maleic acid

Maleated acrylated epoxidized soybean oil (MAESO)

Figure 5.3 Synthesis and further modification of AESO.

These fully bio-based polymers derived from vegetable oil and sugar can be used as environmentally friendly renewable materials for bio-medical and other applications. With similar concerns, Ma *et al.*[42] synthesized a rosin-based diacrylate which is also a bio-based derivative, and serves as a rigid monomer. AESO is co-polymerized with the rigid monomer and the cured AESO has enhanced mechanical and thermal properties, especially the T_g, tensile strength and tensile modulus, compared with divinylbezene. This is another example of a fully bio-based thermosetting resin with promising mechanical properties. Janes *et al.*[43] used the advantages of photo-polymerization to produce chemically stable fibers containing over 50 wt% AESO with an average diameter of 30 μm without applied heat or solvents.

Dipentarythritol pentaacrylate (DPPA), pentaerythritol tetrakis(3-mercapto-propionate) (PETT) and Irgacure 2100 as the photo-initiator were mixed with AESO and the photo-polymerization was conducted during the fiber-formation process. This process can be viewed as replacing the thermal energy needed for melting processing with light energy.

A bio-based polyurethane (PU) acrylate was blended with AESO[44] in order to utilize the excellent properties of vegetable-oil-based PUs. Vegetable-oil-based PUs have excellent chemical and physical properties such as enhanced hydrolytic tendencies, high tensile strength and elongation and thermal stability, combined by other advantages such as low toxicity, inherent bio-degradability and high purity. Functional vegetable oils have been used as constituent component materials to prepare polymer network materials. PU and AESO are compatible due to the chemical cross-linking in these network blends. With an increase in AESO content, the T_g of the networks decreased from 40 to -4.8 °C, the tensile strength increased from 1.7 to 9.8 MPa and the elongation-at-break decreased from 470 to 70%.

AESO is also involved in the synthesis of composites. Albayrak *et al.*[45] synthesized nanocomposites from AESO combined with styrene monomers and montmorillonite (MMT) clay by using an *in situ* free-radical polymer-ization reaction. The nanocomposites had improved thermal and dynamic mechanical properties compared with a neat AESO-based-polymer matrix. Skrifvars *et al.*[46] on the other hand, mixed AESO with natural fibers without using any co-monomer such as styrene to produce structural composites with a high content of renewable material. The composites were prepared by spray impregnation followed by compression molding at elevated tempera-tures. The resin can be reinforced with up to 70 wt% of fibers without sac-rificing its processability. The tensile modulus ranged between 5.8 and 9.7 GPa.

5.2.2.3 Soybean-oil-based Polyols

Soybean oil has no hydroxyl functional groups in its fatty acid chains. However, the double bonds provide a starting point for introducing hydroxyl groups to the soybean oil. Polyols are mostly used for the synthesis of PUs. In order to get PUs with good properties, the polyols need to contain sufficient amounts of hydroxyl groups, which in turn requires a considerable number of double bonds in the vegetable oil. Soybean oil has an average number of double bonds of 4.6 per triglyceride, thus is suitable for conversion to polyols for the synthesis of PUs.

Typically, soybean-oil-based polyols are synthesized by epoxidation fol-lowed by oxirane ring opening.[5] Different catalyst systems are involved in the second step. Recently, Bailosky *et al.*[47] developed a new catalyst for the synthesis of soybean-oil-based polyols from ESO. Zinc triflate was a more efficient catalyst in catalyzing the ESO hydroxyl reaction than methane-sulfonic acid or $BF_3 \cdot OEt_2$ and gave the expected polyether polyols. Only 0.01% zinc triflate was found to give over 99% conversion with ESO and *n*-butanol in 3 h at 115 °C. Miao *et al.*,[48] on the other hand, synthesized a new polyol by reaction of ESO with isopropanolamine through simultaneous ring-opening and amidation reactions between both ester groups and epoxy groups in ESO with the amino group of isopropanolamine. The polyol has a high hydroxyl number of 317.0 mg KOH g^{-1}.

Another interesting strategy is to get polyols directly from soybean oil in a single-step reaction. Desroches *et al.*[49] established a method involving a single thiol-ene coupling reaction. Efficient thiol addition onto vegetable oils leads to bio-based polyols. The most important feature of this reaction is probably the number of double bonds per chain in the vegetable oil, which strongly influences the thiol grafting yield. This one-step route to produce fatty polyols represents a significant advance compared to the traditional epoxidation approach which occurs in two steps. Sun *et al.*[50] also developed a single-step reaction for the synthesis of polyols from soybean oil. The use of OsO_4 as the catalyst and 4-methylmorpholine-*N*-oxide as the oxidant resulted in polyols with excellent yields with hydroxyl numbers of up to 467.7 mg KOH g^{-1}. Advantages of this scheme include that two hydroxyl groups can be readily added to one double bond in a single step, and a wide range of hydroxyl numbers can be obtained by changing the concentration of the catalyst, and the reaction can be performed at room temperature.

PUs are one of the most interesting classes of copolymers with properties varying from rubbery materials to glassy thermoplastics and from linear polymers to thermosetting plastics. Recently, Lu *et al.*[51] synthesized novel surfactant-free core–shell hybrid latexes by seeded emulsion polymerization of 10–60 wt% vinyl monomers in the presence of a soybean-oil-based waterborne PU dispersions as seed particles. The soybean-oil-based waterborne PU, synthesized by reacting isophorone diisocyanate with methoxylated soybean oil polyols and dimethylol propionic acid, formed the latex shell and served as a polymeric high-molecular-weight emulsifier, whilst the vinyl polymers formed the core. The core–shell hybrid latex films showed a significant increase in thermal stability and mechanical properties when compared with the pure PU films, and exhibited a change in mechanical behavior from elastomeric polymers to tough and hard plastics, due to grafting and cross-linking in the hybrid latexes. The T_g values as well as the Young's moduli and tensile strengths were enhanced significantly by an increase in the cross-link density of both the shell and the core. Apparently, the cross-link density is one of the most important factors that affect the properties of the PU, and it is clearly related to the number of hydroxyl group in the polyols. Wang *et al.*[52] prepared novel polyurethane acrylates (PUAs) by the reaction of soybean polyols with isophorone diisocyanate and hydroxyethylacrylate. The cross-link density of PU and PUAs correlated with the hydroxyl number of the polyols. The T_g and initial decomposition temperature of the PU or PUAs were higher with larger hydroxyl numbers. Acrylation of PU to PUAs improved its thermal stability and damping properties. The tensile strengths of PUAs decreased with increasing hydroxyl number.

PU foam can be prepared from soybean-oil-based polyols. Yang *et al.*[53] prepared a series of PU rigid foams by mixing polyols with toluene diisocyanate using an isocyanate index of 1.1. These rigid foams possessed high thermal stability. When the mass fraction of soybean-oil-based polyol was increased up to 60%, the foam had the highest compressive strength of 292.34 kPa and a density of 425.29 kg m^{-3}, and the cells of the foam were small and uniform. Gu *et al.*[54] reported an *in situ* reaction of a methylene diphenyl diisocyanate (MDI) polyuria pre-polymer and a soybean-oil-based polyol for the synthesis of PU foam. The hydroxyl value of the soybean-oil-based polyol exerted important effects on the cell morphologies and foam properties. With high hydroxyl values, it exhibited superior tensile strength, high tensile elongation, high degradability and high T_g compared with low hydroxyl values.

Recently, Miao *et al.*[55] developed an interesting shape-memory PU from a series of different structural soybean-oil-based polyols with 1,6-diisocyanatohexane. It was found that they preserved the triglyceride structure and were fixed in a temporary shape at -20 °C and could completely regain their permanent shapes at 37 °C. Since several double bonds are present in vegetable oils, the functionality of vegetable oil polyols is usually higher than 2. Therefore, PU from vegetable oils consists of mostly cross-linked networks that can endow the polymers with shape-memory effects. Physical interactions such as vitrification and crystallization, might be able to fix a temporary shape, but the cross-linking structure helps to restore the original shape. Glycerol cross-linking and branch-linking were the two major structural patterns in the PU networks, and it was found that glycerol cross-linking was critical to the shape-recovery effect.

5.2.2.4 Other Modifications

Functional groups other than hydroxyl, acrylate and epoxide can also be introduced to the fatty acid chains of soybean oil by reacting with different reagents.

For example, an acrylated soybean oil can be obtained by a novel efficient one-step method by reacting soybean oil with acrylic acid under the catalysis of $BF_3 \cdot OEt_2$ (ref. 56). The number of acrylate groups could reach 3.09 per triglyceride molecule and the conversion of the double bonds was up to 75.7%.

Biswas *et al.*[57] developed an environmentally friendly water-based pathway to form the azide derivatives of soybean oil and fatty esters by first forming epoxides and then the azidization of these epoxides. The azidization reaction was carried out at high yields in water with a small amount of an ionic liquid as the catalyst. In addition, new monomers were prepared by introducing azide groups[57,58] into vegetable oils including castor, canola, corn, soybean and linseed. Polymerization of these azidated oils with alkynated soybean oil[58] under thermal click chemistry conditions (without using a solvent or a catalyst) yielded fully cross-linked elastomers of almost the same density.[59]

Isocyanated soybean oil (ISO) was synthesized by a two-step reaction.[60] Firstly, the triglyceride was brominated at the allylic positions by a reaction with N-bromosuccinimide, and in the second step these brominated species were reacted with AgNCO to convert them to isocyanate-containing triglycerides. About 60–70% of the bromine atoms were replaced by NCO groups while the double bonds remained intact. ISO can be used as chain extender for low-molecular-weight unsaturated polyesters (UPEs).[61] The results show that UPEs can be chain extended with ISO to shorten the polyesterification time substantially without alteration of the styrene solubility or gel time of the polyesters.

Echeverri et al.[62] focused on the photo-initiated cross-linking of maleated soybean oil (MSO) glycerides with tri- and tetrafunctional thiols without using any other reactive diluents, such as styrene, for the synthesis of elastomeric networks. The materials exhibited properties that resembled elastomers with T_g values below room temperature, and flexural moduli in the range 240–340 MPa.

Another interesting derivative of soybean oil is conjugated soybean oil (CSO). The conversion of vegetable oils to their conjugated counterparts is known to produce better drying oils which can prove to be valuable substitutes for the more expensive tung and castor oils. CSO has exceptional drying properties and the resulting coatings exhibit good solvent resistance. Catalysts based on homogeneous rhodium and ruthenium as well as platinum have been developed for the isomerization of soybean oil and other vegetable oils under mild conditions.[63] CSO can be polymerized via free-radical co-polymerization with many other monomers including divinylbenzene[12,64] or dicyclopentadiene.[64] Recently, CSO has been used to produce new type of co-polymer with acrylonitrile and dicyclopentadiene initiated by azobisisobutyronitrile (AIBN). The resulting thermosets were transparent and yellow.[65] Additionally, polylactic acid (PLA) was reported by Gramlich et al.[66] to be compatible for reaction with CSO. PLA modified with the reactive end-functional group N-2-hydroxyethylmaleimide (HEMI) was first synthesized and then reactively blended with CSO. The two components underwent a Diels–Alder reaction with high conversion, coupling the two immiscible components. Blends of HEMI-PLA with 5 wt% CSO resulted in a greater than 17-fold increase in elongation-at-break compared to the PLA homopolymer. Valverde et al.[67] synthesized a range of bio-based rubbery thermosets by the cationic co-polymerization of CSO and styrene with 1,5-hexadiene or isoprene as a flexible cross-linker. The moduli of the products were in the range 0.064 to 0.414 MPa and 0.243 to 0.63 MPa when using 1,5-hexadiene and isoprene as cross-linkers, respectively. On the other hand, the failure strain of the products was in the range 66 to 189% and 78 to 138% when 1,5-hexadiene and isoprene were used as cross-linker, respectively.

Further modification of ESO can generate different derivatives by using different chemicals. Öztürk et al.[68] synthesized radically polymerizable triglyceride-based monomers by the reaction of ESO with 4-vinyl benzene sulfonic acid (4VBSA). The products were the 1-(4-vinylbenzene sulfonyl)oxy-2-alkonols of epoxidized soybean oil (SESO). SESO was then free-radically polymerized and co-polymerized with styrene. Acyclic diene metathesis (ADMET) polymerization of SESO with a 2nd generation Hoveyda–Grubbs catalyst produced a thermoset polymer. The resulting polymers were rigid and non-hygroscopic at room temperature, but were thermally unstable beyond 67 °C. The hydrolytic stability of the polymer was also quite low and a fast hydrolysis was observed at 60 °C. These new polymers can be used in applications where heat- and moisture-triggered decomposition of a rigid plastic sample is needed. Additionally, Luo *et al.*[69] produced a novel class of thermosetting resins based on allylated and transesterified ESO (ATESO). The ATESO was prepared from ESO by oxirane ring-opening and then transesterification with allyl alcohol. A family of rubbery to glassy resins was prepared by radical co-polymerization of ATESO with MA. The T_g values of the resins were higher than 130 °C, the tensile moduli were up to 1.4 GPa and the tensile strength was up to 37 MPa.

By reviewing all the modifications applied to soybean oil, the ideal modification should allow direct polymerization to polymer products at low cost and with high bio-mass content. Thus more and more novel modifications of soybean oil are expected and are promising for the future development of soybean-oil-based polymeric materials.

5.2.3 Applications of Soybean Oil and its Derivatives

5.2.3.1 Plasticizers

Long fatty acid chains in soybean oil triglyceride provide it with good flexibility suitable for its use as plasticizer. ESO is a non-toxic plasticizer for non-degradable polymers such as poly(vinyl chloride) (PVC), phenolic resins and chlorinated rubbers, and is used to improve the stability and flexibility of these polymers. It is also an effective plasticizer for bio-degradable polymers such as poly(hydroxybutyrate-hydroxyvalerate) (PHBV) or PLA, because of its excellent plasticizer permanence and efficiency. Recently, Zhao *et al.*[70] used ESO as a novel plasticizer for another bio-degradable polymer, poly(butylene succinate) (PBS). With only 5 wt% of ESO, the elongation-at-break reached a maximum of 15 times that of pure PBS, while the tensile strength and modulus for the blends were lower than those for pure PBS. Additionally, the blends exhibited lower T_g, crystallization temperature (T_c) and melting temperature (T_m) values. ESO had very limited compatibility with PBS and phase separation was observed when more ESO was added.

A group of more effective plasticizers for PVC other than ESO is epoxidized fatty acid esters (EFAEs), specifically octyl epoxyoleate.[71] Traditional PVC plasticizers based on phthalates are widely used but are being challenged due to migration phenomena which result in toxicity problems. EFAEs are low-toxicity and low-migration plasticizers, and also have a double function inside the plastisol, acting as internal lubricants as well as stabilizers.

The epoxidized fatty acids can capture acid groups through catalytic degradation of their epoxide groups, thus the final product is stabilized. The tensile strength was reduced from 21 to 10 MPa when the amount of plasticizer increased from 23.1 to 44.4%, while the elongation-at-break increased from about 150 to 240%.

5.2.3.2 Toughening Agents

Soybean-oil-based polymers can also be used as toughening agents for different polymeric systems with poor toughness. Epoxy resins have poor toughness and low impact strength, and are brittle after curing. By adding a certain amount of flexibility while maintaining the strength of the epoxy resin, the toughness of the cured product increases. Chen *et al.*[72] used ESO as the flexible segment for the toughening of epoxy resins. Anhydride was used as the curing agent, opening both the epoxy group and ESO, allowing cross-linking with each other to form a network structure. The combination of commercial epoxy resin with 20 wt% ESO resulted in a bio-resin with the optimum properties, a high T_g of 130.5 °C, good thermal stability, a high tensile strength of 74.89 MPa and an impact resistance of 48.86 kJ m^{-2}. PLA is another brittle polymer with low impact strength, strain-at-break, and tensile toughness. Many approaches have been taken to improving these properties; the effect of the polymer stereochemistry, processing history and the addition of plasticizers have all been studied. In addition, PLA has been blended with many materials such as polyethylene, PCL, poly(hydroxyalkanoates), PBS and others. The challenge remains to find a completely renewable and bio-degradable toughening agent that enhances PLA's properties as effectively as existing non-renewable and non-bio-degradable blending partners. Robertson *et al.*[73] melt-blended PLA with polymerized soybean oil (PSO) in order to increase the toughness of PLA in an all-renewable blend. A critical interparticle distance of around 1 mm was found, below which increases in the tensile toughness were observed. As a result, the PLA-PSO blends had a tensile toughness and a strain-at-break four and six times greater than those of unmodified PLA, respectively.

In addition, bio-composites synthesized from PLA and starch also need a toughening agent, not only for enhancing their mechanical properties but also for increasing compatibility between these two components. Xiong *et al.*[74] melt-blended PLA-starch composites with ESO as a toughening agent as well as a reactive compatibilizer. The starch granules were grafted with MA to enhance their reactivity with ESO. The ready reactions between the epoxy groups on ESO and the MA groups on MA-grafted starch and the end carboxylic acid group of PLA brought blending components together and formed a compatible compound. The elongation-at-break was 140% for the blend compared with 5% for neat PLA, and the impact strength was 42 kJ m^{-2} which is much larger than for pure PLA (only 2.4 kJ m^{-2}). This worked showed that ESO can be used as a bio-based reactive plasticizer for PLA and starch compounds, and the effect was enhanced by chemically grafting the starch granules with MA. Similar to the above research, Kiangkitiwan et al.[75] grafted MA to soybean oil using dicumyl peroxide (DCP) as an initiator to get a maleated soybean oil (MSO). It was then employed for the surface modification of cassava starch powder, followed by mixing with PLA resulting in PLA-MSO-g-starch. Compared with PLA-starch, PLA-MSO-g-starch composites had improved impact strength and tensile strength. With a similar strategy, Brandelero et al.[76] applied soybean oil to starch and poly(butylene adipate terephthalate) (PBAT) films containing a large amount of starch, and the films exhibited good mechanical properties, and increasing humidity improved the performance of the films. The soybean oil acted as a good compatibilizer between starch and PBAT. The compatibilizer effect is related to the increase in groups that are characteristic of ester bonds and groups that are bonded to the glucose ring. Therefore, lipids are chemically associated with the polymers of the blends, increasing the interactions among the polymeric fractions and improving the polymeric mixture by increasing the interfacial adhesion of the polymers.

5.2.3.3 Coatings

Soybean oil is a semi-drying oil. It will form a thin, transparent and flexible film when exposed to oxygen in air because of the double bonds in the fatty acid chains. This process can be accelerated with a catalyst or with the modification of soybean oil to polyols or ESO.[6] Xia et al.[77] prepared a coating from soybean-oil-based cationic polyurethane dispersions (PUDs) which had anti-bacterial properties. The coatings were synthesized from methoxylated soybean oil polyols (MSOLs) with a hydroxyl number of 169 mg KOH g^{-1} with five amino polyols. These PUs as either dispersions or films, exhibited good anti-bacterial properties particularly towards the Gram-positive bacterium *Listeria monocytogenes*, due to the quaternary phosphonium or ammonium salts in the polymers. Bakhshi et al.[78] prepared another anti-bacterial PU coating based on ESO. It was subject to reaction with aniline using an ionic liquid as a green catalyst to generate a phenylamine-containing polyol intermediate, which was then methylated by reaction with methyl iodide to produce a polyol with pendant dimethylphenylammounium iodide groups, followed by reaction with isophorone diisocyanate to produce cross-linked PU coatings. The T_g values were in the range 50–82 °C, the initial moduli 13–299 MPa, the tensile strengths 4.5–13.8 MPa and the elongation-at-break 16–109%. The coating showed good adherence to aluminum and PVC substrates and promising anti-bacterial properties against both Gram-positive and Gram-negative bacteria with bacterial reduction in the range 83–100%. Alam et al.[79] synthesized a better coating system compared to the traditional soybean oil system. Vinylether functional monomers (VEFM) containing fatty acid pendant groups from soybean oil were produced by base-catalyzed transesterification, and a carbocationic polymerization process was developed for VEFM, allowing the synthesis of high-molecular-weight polymers without consuming any of the vinyl groups in the fatty acids portion of the monomers. Compared to soybean oil, which possesses on average 4.6 vinyl groups per molecule, VEFM possesses tens to thousands of vinyl groups per molecule depending on the molecular weight of the polymer produced. Therefore as a result, the coatings based on VEFM were shown to possess much higher cross-link densities at a given degree of functional group conversion compared to analogs based on conventional soybean oil. For coating systems based on cure by auto-oxidation, it was demonstrated that the use of VEFM can reduce the drying time by a factor of 4 to 6.5. In epoxy-amine cure systems, the use of epoxidized VEFM as opposed to ESO reduced the cure time at 120 °C by more than one order of magnitude. For cationic photo-cure systems, the use of VEFM substantially increased the cure rate and ultimately the function group conversion during photo-cure. In addition, the T_g and cross-link density of the cured films were significantly higher, and the use of acrylated epoxidized VEFM in place of AESO in free-radical photo-cure systems resulted in faster cure. All of these results can be attributed to the higher number of functional groups per molecule, which also results in the gel point being reached at a much lower conversion, and the attainment of a higher cross-link density at a given degree of functional group conversion. Waterborne PU urea dispersions as coatings were recently reported by Xia *et al.*[80] A series of soybean-oil-based amide diol isosorbide waterborne PU urea (PUU) dispersions has been successfully prepared, with the amounts of isosorbide ranging from 0 to 20 wt% of the total diol content. Young's modulus increased from 2.3 to 63 MPa and the ultimate tensile strength increased from 0.7 to 8.2 MPa, when the isosorbide amount increased from 1 to 20 wt%. The thermal stability decreased slightly with the incorporation of isosorbide. This work provides a new way of utilizing renewable materials such as isosorbide and a soybean-oil-based amide diol for the preparation of high-performance PUU coatings.

5.2.3.4   Adhesives

The use of soybean oil in the development of adhesives has attracted much attention. Adhesives can be synthesized from soybean oil by various chemical pathways, and the obtained products exhibit properties comparable to their petroleum-based counterparts.[6] Lu *et al.*[81] produced aqueous PU dispersions (PUDs), binary colloidal systems where the PU particles were dispersed in a continuous aqueous medium, as a substitute for traditional PUs. The PUDs had notably excellent adhesion properties and low-temperature film-forming abilities which made them very useful as adhesives, coatings and paints. The cationic PUDs exhibited very high adhesion to various ionic substrates especially anionic substrates such as leather and glass, leading to their wide application as adhesives and coagulants. These vegetable-oil-based aqueous cationic PUDs were synthesized from methoxylated soybean oil polyols (MSOLs). With an increase in –OH functionality of the MSOLs, the resulting PU films exhibited tensile stress–strain behavior ranging from elastomeric polymer to ductile plastics, and possessed Young's moduli ranging from 33.6 to 554.0 MPa, and elongation-at-break values ranging from 235 to 291%. This work provides a new way of utilizing renewable resources to prepare environmentally friendly bio-based polymers with high performance for coating and adhesive applications. Among all these adhesives, pressure-sensitive adhesives (PSAs) are one of the most widespread groups of adhesives. Ahn *et al.*[82] produced thermally stable transparent PSAs from epoxidized and dihydroxyl soybean oil (DSO) with peel strengths comparable to current PSAs. The main challenge in commercializing these oleo-based PSAs is reducing the curing time. They designed and synthesized a fast-curing co-polymer from ESO and DSO for PSA applications without using petrochemicals. Most current flexible petroleum-based plastics and PSAs have fairly low thermal stabilities, while low T_g, high T_m and low coefficient of thermal expansion (CTE) are favorable for PSAs. The PSAs synthesized from ESO and DSO have low T_g values of -34.29 °C, high T_m values of above 250 °C, and low CTE values of 11.5 ppm K^{-1}, with transparency similar to glass.

5.2.3.5 Other Applications

Soybean oil and its derivatives have many other applications. For example, ESO not only increases the toughness of PLA, but also improves the hydrolytic stability of poly(D-lactide) (PDLA). Fu *et al.*[83] synthesized a multi-arm star polymer ESO-PLA by ROP of DL-lactide using multifunctional ESO as an initiator. The results revealed that linear poly(DL-lactide) (PDLLA) films underwent water erosion more readily than the star-shaped ESO-PLA, and the decrease in molecular weight and weight loss ratio of the star-shaped ESO-PLA was lower than that of linear PDLLA. There are two reasons for this, one is that a strong intermolecular force may be present in the star-shaped polymer because of its long-arm chain undergoing twining behavior, preventing water erosion of the polymer. The other is that ESO can minimize the amount of trapped water which slows down the permeation of water into the polymer, resulting from the hydrophobic nature of ESO. Ren *et al.*[84] used isocyanate and AESO as a coupling agent to modify kenaf fibers and Lin *et al.*[85] expanded applications to bio-medical fields. They utilized soybean oil as a co-delivery system for DNA and subunit vaccines. Liposome-polymer transfection complexes (LPTCs) were formed by two hydrophilic polymers, polyethyleneimine (PEI) and PEG, with soybean oil. The soybean oil was used to form the liposome structure *via* sonication. Soybean oil may allow for the addition of immunostimulatory components such as the saponin adjuvant Quil A. Immunostimulatory agents are typically hydrophobic in nature and have immunogenicity, thus the addition of soybean oil through polar interactions improved the adjuvant effect of the vaccine. Additionally, Abdekhodaie *et al.*[86] used hydrolyzed polymers of soybean oil (HPSO) and of epoxidized soybean oil (HPESO) as drug-delivery systems and pharmaceutical excipients. HPSO and HPESO polymers were surface active and able to increase the wetting of solid tablets of the hydrophobic drugs ibuprofen and nifedipine, with HPSO being more effective than HPESO. The results suggested that these novel soybean-oil-based amphiphilic polymers have great potential for drug delivery and pharmaceutical formulations. Zhang *et al.*[87] applied ESO as an additive to wheat gluten (WG) to modify the properties of renewable and bio-degradable natural polymer materials. The combination of the plasticization and cross-linking effects derived from ESO resulted in good retention of mechanical strength for the plasticized WG-ESO materials as compared to those without 10 wt% mobile ESO additives. Palacios-Jaimes *et al.*[88] modified polypropylene (PP) membranes with AESO for the treatment of waste water. AESO was grafted onto the surface of microporous PP membranes by using a UV radiation method, which increased the hydrophilicity of the membrane surface and the modified membrane possessed filtration capabilities and removed 52% of the color and reduced 95% of the turbidity from industrial residual water. It is interesting that modifying PP membranes using vegetable oils as monomers can produce sustainable membranes, with filtration capabilities. These membranes, after reuse, could have the possibility of bio-degradation. The method is easy, low cost and environmentally friendly.

5.3 Castor Oil: Polymer Synthesis and Applications

5.3.1 Use of Castor Oil Directly for Polymer Synthesis

In contrast to soybean oil, castor oil is a non-food oil, with double bonds and hydroxyl groups evenly distributed throughout its triglyceride fatty acid chains. The average number of hydroxyl groups per triglyceride is about 2.7 and about 90% of the fatty acids in castor oil are ricinoleic acid (Figure 5.4).[5]

With this inherently available functionality, castor oil can be used as a polyol for the direct synthesis of cross-linked PU resins with various isocyanate compounds such as hexamethylene diisocyanate (HDI), tolyene-2,4-diisocyanate (TDI), and isophorone diisocyanate (IPDI) and so on. Unfortunately, the resulting PUs are soft due to the low cross-linking density resulting from the low hydroxyl number of castor oil. To overcome this problem, Teramoto *et al.*[89] used glycerol, an abundant bio-based triol with a

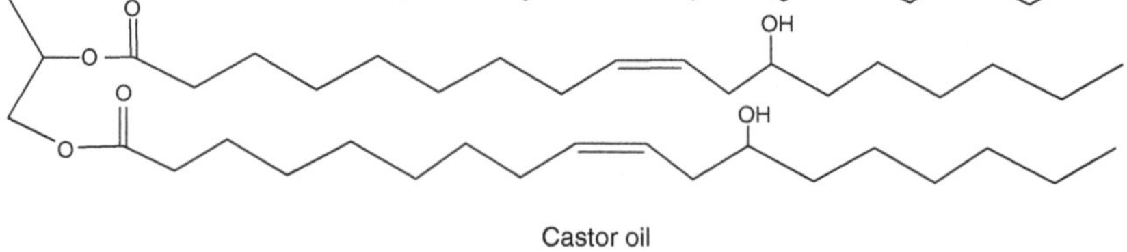

Castor oil

Figure 5.4 Structure of a castor oil triglyceride.

relatively high hydroxyl number, as a cross-linker for the synthesis of PU elastomers from castor oil with TDI-terminated poly(ethylene adipate) (PEA). The cured materials had a very high elongation-at-break of over 400%, with both tensile strength and tensile modulus two times higher than those without glycerol.

Another option, other than incorporation of additional triols, is to increase the hydroxyl number of castor oil by chemical modification. De *et al.*[90] synthesized a castor-oil-based hyperbranched polyol with a hydroxyl value of 420 mg KOH g^{-1}, which is almost two times higher than that of castor oil (157 mg KOH g^{-1}). Consequently, an outstandingly tough, highly elastic, biodegradable, and thermostable hyperbranched epoxy was synthesized by a simple polycondensation reaction between this polyol and bis(hydroxy methyl)propionic acid. The epoxy had a high tensile strength of 42 MPa.

Moreover, with the addition of nanoparticles such as silica or titanium oxide as a filler, the castor-oil-based PUs have enhanced mechanical properties. Xia *et al.*[91] reported that by adding only 2 wt% of silica nanoparticles, the Young's modulus and tensile strength increased from 32.3 to 116 MPa and 15.1 to 20.0 MPa respectively, for a waterborne castor-oil-based PU. This PU was chemically bonded to the silica nanoparticles and the cross-link density increased substantially, resulting in improved thermal stability and enhanced mechanical properties. Ristić *et al.*[92,93] chose titanium oxide nanoparticles as the fillers for a PU synthesized from castor oil and TDI. When 2 wt% of titanium oxide nanoparticles was used, the elongation-at-break increased by 3 times while the T_g decreased by 10 °C. This was due to changes in the segmental mobility influenced by interactions between the nanoparticles and the polymer chains.

Additional interesting properties were obtained when iron oxide was used as a filler. Mussatti *et al.*[94] prepared a series of polymeric composites with different iron oxide volume fractions through a casting process followed by compression molding at room temperature. By increasing the iron oxide content, not only did the tensile strength and Young's modulus values increase, but also the electrical conductivity values of the composites were increased. However, the thermal conductivity did not change significantly with the addition of iron oxide, both the compounds and PU were thermal insulators.

In addition to these inorganic nanoparticles, their organic counterparts can also be employed as fillers for the preparation of castor oil-PU composites. Recently, a new nanocomposite consisting of a castor-oil-based PU matrix filled with acetylated cellulose nanocrystals (ACNs) was developed by Lin *et al.*[95] When increasing the amount of ACN from 0 to 25%, the nanocomposite's tensile strength and Young's modulus increased from 2.79 to 10.41 MPa and from 0.98 to 42.61 MPa, respectively. The maximum value of elongation-at-break of the nanocomposites was achieved with 10 wt% of ACN, and was more than two times higher than that of the PU itself. A three-dimensional ACN network and strong interfacial interactions between the filler and matrix were responsible for the enhanced mechanical performance. Thakur *et al.*[96] took advantage of graphene oxide (GO) and used it to prepare a castor-oil-modified hyperbranched PU-GO nano-composite. Similarly, with only 2 wt% of GO, a tremendous enhancement of mechanical properties was observed. The tensile strength increased from 7 to 16 MPa and the elongation-at-break increased from 695 to 810%. More-over these nanocomposites exhibited excellent shape-memory behavior which can be explained by an increase in stored energy because of the homogeneous distribution of GO in the polymer matrix. The shape recovery was also found to increase with increasing amounts of GO in the matrix.

In addition to PU, castor oil can be used to synthesize a number of different polymers. For example, Sathiskumar *et al.*[97] synthesized a new family of castor-oil-based bio-degradable polyesters by a catalyst-free melt con-densation reaction between castor oil and diacids with D-mannitol. The resulting polymers were bio-degradable soft materials with a hydrophilic surface. A star-shaped polyester polyol was synthesized by Ristić *et al.*[98] *via* polymerization of L-lactide with castor oil as the initiator. Saravari *et al.*[99] synthesized a urethane alkyd by interesterification of castor oil with jatropha oil, followed by reaction with TDI. The castor-jatropha-oil-based urethane alkyd had a lower molecular weight and viscosity, a slightly lower hardness and a much longer drying time than conventional and commercial urethane alkyds, with excellent resistance to water and acid.

5.3.2 Use of Castor Oil after Chemical Modification for Polymer Synthesis

With both double bonds and hydroxyl groups in the fatty acid chains of castor oil, chemical modifications are available. Hirayama *et al.*[100] elimin-ated the hydroxyl groups from the fatty acid chains to obtain dehydrated castor oil (DCO), producing a semi-drying oil, which can be used extensively in paints and varnishes. The dehydration process is performed at about 250 °C in the presence of an acid catalyst such as H_2SO_4 and activated earth under an inert atmosphere or vacuum. The hydroxyl group and an adjacent hydrogen atom from the C11 or C13 position of the ricinoleic acid portion of the molecule are removed as water. This process forms not only conjugated 9,11-diene moieties but also non-conjugated 9,12-diene moieties with a ratio of about 41 : 59. The average number of double bonds per triglyceride is 4.8. The DCO is then used for the preparation of a thermosetting resin with 1,1′-(methylene-di-4,1-phenylene)bis-maleimide in 1,3-dimethyl-2-imidazolidi-none. It is worth noting that almost all the conjugated diene moieties of DCO reacted due to the Diels–Alder (DA) reaction between the maleimide groups and the conjugated dienes.

Instead of elimination of the hydroxyl groups, new functional groups such as dangling double bonds can be incorporated to the fatty acid chains *via* reaction of hydroxyl groups with acrylic acid, resulting in acrylated castor oil (ACO). Kim *et al.*[101] used ACO for the synthesis of novel cross-linked thin polymer networks by UV photo-polymerization with PEG diacrylate or PCL diacrylate as the cross-linking agent. Recently, a simple, low-environmental-impact procedure for preparing ACO was developed by Dillman *et al.*[102] using a hybrid acrylate isocyanate monomer, 2-acryloyloxyethyl isocyanate. The photo-polymerization of neat ACO and the co-polymerization of ACO with common low-molecular-weight acrylate monomers were rapid and reached high conversions. The resulting co-polymers ranged from highly flexible, low-glass-transition materials to rigid, high-glass-transition materials depending on the functionality and secondary functional groups of the commercial monomers used. The ACO oligomer was compatible with a variety of acrylate monomers and produced transparent films regardless of the co-monomer used. ACO-based materials provide a promising route to introducing renewable materials into many acrylate-based coating applications.

Similar to that of soybean oil, epoxidation of double bonds in the fatty acid chains of castor oil resulted in epoxidized castor oil (ECO). Bechi *et al.*[103] synthesized two series of organic-inorganic hybrid films from ECO and the inorganic precursor titanium(IV) isopropoxide (TIP), combined with silicon precursors, either 3-aminopropyltriethoxysilane or tetraethoxysilane with different organic-to-inorganic proportions. The hardness and tensile strength of the films increased with increased concentration of inorganic precursor. The combination of the silicon-rich inorganic precursors with TIP substantially improved the mechanical strength of the films. All of the films exhibited good adhesion to an aluminum surface.

Xia *et al.*[104] synthesized two castor-oil-based monomers. Norbornenyl-functionalized castor oil (NCO), which had about 0.8 norbornene rings per fatty acid chain, and norbornenyl-functionalized castor oil alcohol (NCA), which had about 1.8 norbornene rings per fatty acid chain. These two monomers can be used to conduct a ring-opening metathesis polymerization (ROMP) using a 2nd generation Grubbs catalyst, resulting in rubbery-to-rigid plastics with different cross-link densities. Higher cross-link densities gave better thermal properties, including increasing the T_g from -17.1 to $+65.4$ °C, and increasing the room temperature storage modulus, from 2.4 to 831.9 MPa, as well as enhanced mechanical properties, with a Young's modulus of 407 MPa and a tensile strength of 18 MPa at high cross-link densities.

Allauddin *et al.*[105] developed an alkoxysilane-functionalized castor oil (ASCO) for the synthesis of functional polyurethane-urea (PUU) coating films. Hydrolyzable $-Si-OCH_3$ groups were introduced into the castor oil backbone and were used subsequently for the development of polyurethane-urea – silica hybrid coatings. The ASCO was reacted with different ratios of IPDI to get an isocyanate-terminated hybrid PU pre-polymer which was cured under atmospheric moisture to get the desired coating films. The T_g values of the hybrid networks were found to be in the range 29 to 70 °C. The alkoxy silane-modified castor-oil-based coatings showed better mechanical and viscoelastic properties compared with the unmodified castor oil coatings.

5.3.3 New Monomers Derived from Castor Oil

Castor oil has been involved in the synthesis of monomers for a long time. The most common monomer derived from castor oil is 10-undecenoic acid (Figure 5.5), which is a platform for the synthesis of many new monomers such as 11-aminododecanoic acid, the monomer for the synthesis of nylon-11.[106] 10-Undecenoic acid or its methyl ester can be used directly for the synthesis of polymers. Recently, Bao *et al.*[107] synthesized hyperbranched polyesters with thioether linkages from methyl 10-undecenoate with 2-thioglycerol. The monomer was obtained in excellent yields through thiol-ene click chemistry in the presence of a photo-initiator under UV irradiation. Hyperbranched polyesters with high molecular weights and unusual crystalline properties were obtained by bulk polycondensation *via* a transesterification process. van den Berg *et al.*[108] synthesized polythioethers with molecular weights of up to 40 000 g mol^{-1} by UV- or thermal-initiated thiol-ene polyaddition polymerization of 10-undecene thiol, an AB-type monomer derived from 10-undecenoic acid. The thioether functionalities can be oxidized into sulfone linkages by hydrogen peroxide, resulting in polysulfones, which are valuable engineering plastics. The physical properties of the prepared polymer were similar to those of low-density polyethylene (LDPE), with an elastic modulus of about 300 MPa and T_m of 90 °C.

Dong *et al.*[109] used 10-hydroxycapric acid (HDA), a new monomer derived from 10-undecenoic acid, for the synthesis of a novel photo-cross-linkable and bio-degradable polyester with 3,4-dihydroxycinnamic acid, which was derived from lignin. The presence of HDA in the co-polymers enhanced the flexibility of the macromolecular chain, thus lowering the T_g and accelerating the photo-reactivity of the polyester. These co-polymers exhibited fairly good tensile properties, and after UV photo-cross-linking, the tensile

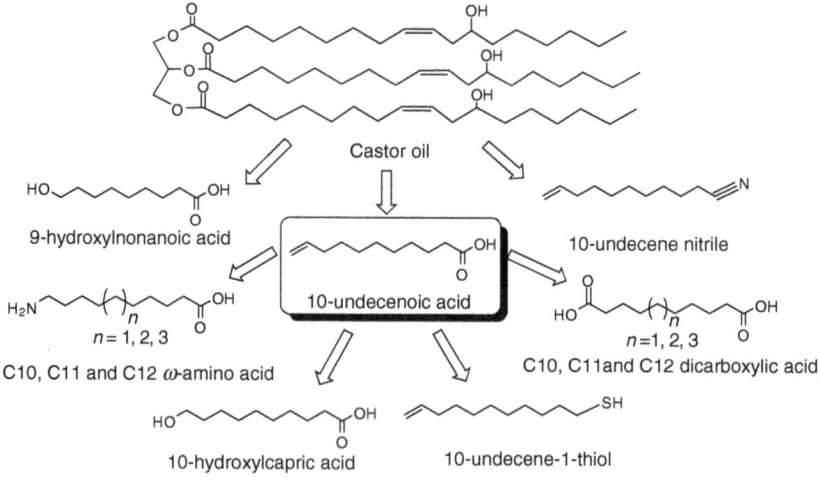

Figure 5.5   New monomers derived from castor oil.[106,108–112]

strength was further improved and the elongation-at-break decreased. Koh *et al.*[110] synthesized a series of new monomers from 10-undecenoic acid including ω-amino fatty acids and α,ω-dicarboxylic acid with 10 to 12 carbons in the aliphatic chain.

Another new monomer derived from castor oil is 9-hydroxynonanoic acid (HNME). The castor oil triglyceride was subjected to ozonolysis followed by methanolysis to produce HNME.[111] High-molecular-weight linear polyester can be prepared from HNME. The polymer was an analog of PCL with highly crystalline, longer hydrocarbon chains between ester groups, imparting intermediate properties between polyesters and polyethylene. It had higher T_m and T_g values and better thermal stability, but lower solubility in chlorinated solvents than PCL.

Miao *et al.*[112] performed a cross-metathesis reaction with 10-undecene nitrile, another monomer derived from castor oil. A C12 nitrile ester was synthesized with a high turnover number, from 10-undecene nitrile with acrylonitrile catalyzed by a ruthenium catalyst. This product has potential as a new bio-sourced intermediate for the production of polyamides.

5.3.4 Applications of Castor Oil and its Derivatives

Castor oil and its derivatives have wide applications including as coatings, toughening agents, plasticizers, adhesives, additives and so on. Thakur *et al.*[113] developed a hyperbranched PU as a coating. The castor-oil-based hyperbranched polyurethanes (HBPUs) were synthesized *via* an A2 + B3 approach. Castor oil or a monoglyceride of castor oil was used as the hydroxyl-containing B3 reactant and TDI was used as an A2 reactant, with 1,4-butane diol as a chain extender and PCL diol as a macroglycol. Both HBPUs behaved as dielectric materials thus could be used as advanced surface-coating materials. Gharibi *et al.*[114] synthesized a castor-oil-based coating with anti-corrosion properties. A simple and efficient synthetic methodology was developed for the combination of castor-oil-based PUs and polypyrrole moieties. The resulting materials were promising anti-corrosion coatings with inhibition efficiencies within the range 85–98%.

Castor oil was used for toughening PLA by Robertson *et al.*[115] Addition of 5 wt% castor oil to PLA significantly enhanced the overall tensile toughness with minimal reduction in the modulus and no plasticization of the PLA matrix. The binary PLA-castor oil blend with 5 wt% castor oil exhibited a tensile toughness 7 times greater than neat PLA, and the impact strength was 1.5 times greater than neat PLA. Xiong *et al.*[116] on the other hand, used castor oil as a plasticizer for a PLA-starch composite. The enrichment of starch with castor oil had a profound effect on the properties of the blend. In order to achieve this, HDI was grafted onto starch granules, and the ready reaction between the hydroxyl groups on castor oil and the isocyanate groups on the HDI-grafted starch, allowed the enrichment of starch with castor oil molecules. Differential scanning calorimetry analysis showed that the castor oil layer on starch had a positive effect on the crystallization of PLA in the ternary blend. The accumulation of castor oil on starch greatly improved the toughness and impact strength of PLA-starch blends.

Klinger *et al.*[117] used castor oil as an additive for polystyrene (PS). The aim was to use castor oil as a desirable alternative to phathalates, which are arguably the most commonly utilized plasticizers, and are suspected to be hazardous to human health. Tensile strength measurements indicated that the additive renders PS stronger at low loading, and plasticizes PS at high loadings. This is due to the fact that at low loading, the castor-oil-derived additive acted as an anti-plasticizer, leading to an increase in the tensile strength of the polymer, while at higher loadings, it acted as a plasticizer so the tensile strength of the polymer decreased with an increase in the amount of additive.

Silva *et al.*[118] prepared a solvent-less castor-oil-based PU adhesive. Its foam joints showed peeling strength values 75% higher than those of a solvent-based commercial adhesive. Its wood joints showed lap shear strength values 20% higher than those of a commercial solvent-based adhesive used for wood. Adhesives free of volatile organic compounds are less toxic and avoid environmental pollution, therefore alternative solvent-less adhesives would be an important contribution to more sustainable products and would present less risk to human health and to the environment from exposure to chemicals. The NCO groups of diisocyanate compounds react with the –OH groups of castor oil and become part of the network therefore they would not vaporize out of the adhesive. PU adhesives produced using only castor oil and TDI with or without catalysts are alternative solvent-free adhesives to the currently used commercial adhesives containing solvents. An increase in the NCO/OH molar ratio led to a decrease in the adhesive density because of bubble formation due to the release of CO_2 during the curing reaction, and an increase in the T_g and hardness values due to a higher adhesive cross-linking level.

Oprea[119] treated castor oil as a trifunctional cross-linker for the synthesis of PU from poly(1,4-butane diol) with aliphatic 1,6-hexamethylene diisocyanate. This resulted in a series of cross-linked PU elastomers with different hard segment structures. The increased hard segment molar ratio and dangling chains present in the triglyceride structures, which acted as plasticizers, increased T_g by 13 °C for PU with an increased content of castor oil. The PUs displayed relatively low T_g values of −70 to −57 °C, a tensile strength of maximum 9 MPa and elongation-at-break of 630%.

Castor oil is also useful in drug-delivery systems.[120] It was used to develop a novel microsphere with Pluronic F-68, in order to study the controlled release of 5-fluorouracil (5-FU) using a solvent-evaporation method. The insertion of double bonds into castor oil gives versatile chemical resistance, hardness, elongation and tensile strength properties and highly compatible poly(ether-urethane)s (PEUs). The drug release mainly depends on the amount of Pluronic present in the matrix, it acts as hydrophilic filler in the formulations, which helps to control the swelling and release of hydrophobic drugs from the microspheres. The formation and dissociation of hydrogen bonds results in the swelling and collapse of PEUs and this special property is used to control the delivery of drugs or other active molecules.

5.4 Tung Oil

5.4.1 Use of Tung Oil and its Derivatives for Polymer Synthesis

Tung oil has the highest average number of double bonds per triglyceride (7.5).[8] 84% of its fatty acid chains are from elaeostearic acid, which contains a conjugated triene structure. This highly unsaturated property makes it a drying oil that has been used for more than 1000 years in China.[121] It can be cationically polymerized with styrene, resulting in a co-polymer with good mechanical and damping properties as well as shape-memory effects.[122] The T_g values are close to room temperature and generally increase with the content of styrene. The Young's modulus increased from 4.89 to 13.92 MPa with the increase of styrene from 30 to 70 wt%. Moreover, these hard elastomers present shape-memory behavior with high recovery and fixity ratios, opening possibilities for practical applications that require a material response close to room temperature. Furthermore, a third co-monomer such as divinylbenzene (DVB) can be used to obtain a co-polymer based on tung oil with styrene.[123,124] The use of DVB as a co-monomer in the cationic polymerization of the triglyceride resulted in polymers with high thermal stability in an ample temperature range and a room-temperature Young's modulus close to 1 GPa. The co-polymer also displayed shape-memory effects with the switch temperatures in the range 25 to 40 °C. When the amount of DVB increased, the T_g and Young's modulus also increased, this can be associated with the increase of cross-linking density and the contribution of the rigid aromatic structure of the DVB co-monomer. Das *et al.*[125] blended an unsaturated polyester resin/styrene mix with tung oil. This novel material had improved impact strength, creep resistance, Young's modulus, and hardness. With only 1 wt% of tung oil, the impact strength, Young's modulus, and hardness increased by 15, 20 and 41%, respectively.

Huang *et al.*[126] prepared a UV-curable resin based on tung oil. The resin was synthesized *via* a Diels–Alder reaction of tung oil with MA, followed by a non-isocyanate polyurethane (NIPU) reaction and an acrylation modification with glycidyl methacrylate. The higher concentration of double bonds resulted in a faster curing rate but a lower final conversion of the double bonds. All the cured films had two T_g values, which were in the range 8–10 °C and 46–48 °C, respectively. Excellent thermal stability was observed with decomposition temperature as high as 350 °C, indicating that they can be used for UV coatings for substrates with high temperature resistance. Additionally, the introduction of structure by NIPU is helpful for enhancing adhesion of the cured films to substrates.

Luo *et al.*[127] synthesized a tung-oil-based thermoset, which was reinforced by tannin lipid. The polymer films were synthesized by oxidative co-polymerization of tannin linoleate/acetate mixed esters with tung oil, resulting in products ranging from soft rubbers to rigid thermosets. It was found that tannin incorporation into the formulations was essential for the final product to achieve ample mechanical strength. The film stiffness, T_g and cross-link density increased with greater tannin linoeate/acetate content because the tannin component provided rigidity through polyphenolic aromatic rings and unsaturated chains as cross-linking sites. The films had ambient modulus values between 0.12 and 1.6 GPa, with T_g values in the range 32 to 72 °C.

Huang *et al.*[128] developed two new monomers from tung oil, and epoxy resins were prepared accordingly. The new monomers were a 21-carbon dicarboxylic acid (C21DA) and a 22-carbon tricarboxylic acid (C22TA). They were converted to triglycidyl esters (DGEC21 and TGEC22, respectively) and cured with nadic methyl anhydride. Both triglycidyl esters were liquids at room temperature and had lower viscosity and higher reactivity than the commercial Bisphenol A (BPA) epoxy resin DER332. The resulting resins exhibited T_g values in the order DER332 (168 °C) > TGEC22 (131 °C) > DGEC21 (80 °C) > ESO (37 °C). Both cross-link density and molecular rigidity contributed to this order. Firstly, the cured TGEC22 had a significantly higher cross-link density than the cured DGEC21, which in turn had a higher cross-link density than that of the cured ESO, thus TGE22 had a higher T_g than that DGE21 and ESO. On the other hand, although the cured DER332 had a lower cross-link density than the cured TGEC22, the former had a much more rigid molecular structure than the latter, thus displayed the highest T_g.

Another new monomer derived from tung oil was described by Liu *et al.*[129] The monomer was prepared *via* the alcoholysis of tung oil with pentaerythritol (PER) to obtain the corresponding alcoholysis products (TO-PER), followed by a maleinization reaction to get the maleate half ester (TO-PER-MA). It was then co-polymerized with 33% styrene and cured subsequently, to give a polymer matrix with promising mechanical properties: the tensile strength and modulus were 35.9 and 1.94 GPa, whereas the flexural strength and modulus were 46.2 and 2.08 GPa.

The chemical modification of tung oil utilizes the conjugated triene in the fatty acid chains. New functional groups such as hydroxyl groups can be introduced. Ribeiro da Silva *et al.*[130] synthesized a tung-oil-based polyol in a two-step process. Firstly, the hydroxylated tung oil was formed by reaction of tung oil with hydrogen peroxide and formic acid, and then, this intermediate was reacted with dry triethanolamine to form the polyol, with an average hydroxyl number of 450 mg KOH g^{-1}. This tung-oil-based polyol was then used for the synthesis of PU foam, which was further reinforced by addition of rice husk ash (RHA), a residue from the rice processing industry, as a rigid filler because of its high silica content. The thermal stability of the composite was not affected by the RHA, while a relatively higher thermal conductivity was observed. Wood flour and microcrystalline cellulose can also be used as reinforcement fillers for tung-oil-based PUs. Mosiewicki *et al.*[131] found that with 10 wt% wood flour, the tensile modulus and tensile strength increased from 0.91 to 1.23 GPa and from 26 to 36 MPa, respectively, while incorporation of 10 wt% microcrystalline cellulose did not result in a significant change in the mechanical properties. This is probably due to the better particle dispersion of wood flour in the polymer matrix.

5.4.2 Applications of Tung Oil and its Derivatives

A tung-oil-based composite has been used as a sealing material for more than 1000 years in China.[121] Fang *et al.*[121] conducted a study on a piece of chu-nam putty discovered on an ancient ship in China, and revealed that the putty was prepared by mixing tung oil, lime and oakum. This special organic-inorganic hybrid composite invented by ancient Chinese people had excellent sealing performance including excellent waterproofing and bonding properties. The application of chu-nam putty in a wooden ship led to improvements in sailing technologies and ship safety issues.

Xiong *et al.*[132] used tung oil anhydride (TOA) as a plasticizer for PLA-starch composites. The ready reaction between the MA on TOA and the hydroxyl groups on starch resulted in an accumulation of TOA molecules on starch, which increased the compatibility of the PLA-starch blends. A layer was formed upon accumulation of TOA on starch, which had an effect on the thermal behavior of PLA in the ternary blend. The T_g and T_m values of the ternary blend were slightly lower than those of neat PLA, and increasing the amount of tung oil from 5 to 12 wt% in the ternary blend did not affect the T_g and T_m values of the blends. The enrichment of starch with TOA also improved the toughness and impact strength of the PLA-starch blends. The elongation-at-break reached a maximum value of 30% when 7 wt% of TOA was used in the blend, and the impact strength of this blend was almost two times of that of neat PLA.

Liu *et al.*[133] used tung oil as a reactive toughening agent for an unsaturated polyester resin terminated with dicyclopentadiene (DCPD-UPR) *via* an intermolecular Diels–Alder reaction occurring at the later stage of melt polycondensation. These polymers were further blended with a styrene co-monomer and cured *via* free-radical polymerization to give cross-linked thermosetting polymers. The thermal and mechanical properties of these bio-materials showed an enhanced toughness with increasing tung oil content. With 20% tung oil, the matrix obtained from DCPD-UPR-tung oil had maximum increases of 373% and 875% in impact strength and tensile failure strain, respectively, due to the synergistic effects of phase separation and cross-link density. The optimum amount of tung oil is 10% because a stiffness–toughness balance can be achieved for the polymer matrix at this point.

Ma *et al.*[134] used tung oil to modify rosins *via* a Diels–Alder addition reaction and the modified rosins (GTR) were further used in the formulation of glycerin esters with flexible characteristics. Increasing the amount of tung oil resulted in a decrease in the bromine value and molecular weight, as well as in the softening points and viscosities for GTRs. However, a slight increase in thermostability was observed due to the incorporation of flexible fatty chains into the rigid hydrophenanthrene units in the rosins. When applied in a PU adhesive as a tackifier, an increase in the amount of tung oil in GTRs led to an initial increase and then a decrease in both the miscibility of the GTR with PU and the T-peel strength of the adhesives. The elongation-at-break of the films increased monotonically, but their tensile strengths initially increased and then decreased with increasing amounts of tung oil in the GTRs.

5.5 Other Plant Oils

Corn oil is one of the cheapest commercially available vegetable oils and is mainly used in food and livestock feed.[8] It contains an average number of double bonds of 4.1 per triglyceride, which can be involved directly in the synthesis of polymers. Epoxidation of corn oil is also interesting. Sun *et al.*[135] reported an enzymatic epoxidation of corn oil using hydrogen peroxide as an oxygen donor and stearic acid as an active oxygen carrier in the presence of Novozym 435. Epoxidized corn oil is also useful in renewable, bio-degradable and non-toxic lubricants, polymer stabilizers, and intermediates. Epoxidation is also an interesting modification of other plant oils like linseed oil. Epoxidized linseed oil can be used as a bio-based plasticizer, a possible substitute for phthalate in PVC.[136] Moreover, a 99.5% bio-derived highly flexible transparent film with significant water resistance was obtained from epoxidized linseed oil with a bio-based diacid cross-linker.[137] Euphorbia oil on the other hand, is the only plant oil that possesses epoxy groups of its own. Therefore, direct polymerization of euphorbia oil by ROP was carried out by Liu *et al.*[138] in liquid CO_2, catalyzed by $BF_3 \cdot OEt_2$. The resulting polymers were cross-linked with T_g values in the range -15.0 to -22.7 °C. Palaskar *et al.*[139] synthesized a series of sunflower-oil-based polyols as PU precursors. While Alemdar *et al.*[140] developed a sunflower-oil-based macro-initiator, which was then styrenated using reversible addition–fragmentation chain transfer polymerization (RAFT) in the presence of phenacyl morpholine dithiocarbamate (PMDC) as a chain-transfer agent. The resulting material gave transparent films with good film properties. Kong *et al.*[141] synthesized two novel bio-based poly(ether ester) polyols with high functionality and low viscosity from canola oil by a simple, three-step reaction including epoxidation, hydroxylation and transesterification. These canola-oil-based polyols were used for the production of PUs with commercial petrochemical-derived diisocyanate. This work established the production of polyols and their corresponding PUs from vegetable oil starting materials with the glycerol backbone being removed explicitly during the polyol synthesis reaction.

5.6 Conclusions

From the point of view of sustainable development, the utilization of plant oils for polymer synthesis is promising, and much progress has been made in the last decade. This progress will undoubtedly continue for the coming decades. We have witnessed numerous efforts in developing all kinds of polymers such as polyurethanes, polyesters, polyamides, and epoxy resins as well as various composites based on different plant oil resources. Plant oils can be involved directly in the synthesis of polymeric materials, taking advantage of their own functionalities like double bonds, hydroxyl groups and epoxy groups. Further modification of plant oils allows us to increase the reactivity by introducing new functionalities. As a result, the modified plant oils find themselves involved in all kinds of polymerization methods including cationic polymerization, free-radical polymerization, ring-opening polymerization, metathesis polymerization, condensation polymerization, living polymerization and so on. The obtained polymeric materials thus have applications such as coatings, adhesives, plasticizers and additives *etc.*, with comparable properties to their petroleum-based counterparts, providing us with excellent candidates for the replacement of petroleum-based materials in many fields.

Nonetheless, there are challenges to achieving these goals. Although further efforts can be made to use plant oils directly for polymer synthesis, or to develop more and more novel chemical modification methods on plant oils, the utilization of plant oils or their derivatives has disadvantages due to their complicated triglyceride structures and multiple functionalities, which makes it difficult to control and tune the structures and properties of the polymer products precisely. Thus a more favorable process would be to use the plant oils as raw materials for developing traditional polymerizable monomers. 10-Undecanoic acid from castor oil is one of the best examples of this strategy. However, several criteria must be fulfilled for this strategy, including developing highly efficient methods for converting plant oils to monomers, which require not only mild conditions during the process but also have high yields. Bio-transformation seems a promising method to meet these criteria. The resulting monomers can be utilized for the synthesis of polymers by traditional methods. Consequently, fully bio-based polymeric materials could be developed.

References

1. Y. Xia and R. C. Larock, *Green Chem.*, 2010, **12**, 1893.
2. U. Biermann, U. Bornscheuer, M. A. R. Meier, J. O. Metzger and H. J. Schäfer, *Angew. Chem., Int. Ed.*, 2011, **50**, 3854.
3. L. Montero de Espinosa and M. A. R. Meier, *Eur. Polym. J.*, 2011, **47**, 837.
4. M. A. Mosiewicki and M. I. Aranguren, *Eur. Polym. J.*, 2013, **49**, 1243.
5. D. P. Pfister, Y. Xia and R. C. Larock, *ChemSusChem*, 2011, **4**, 703.
6. G. Lligadas, J. C. Ronda, M. Galià and V. Cádiz, *Mater. Today*, 2013, **16**, 337.
7. M. G. A. Vieira, M. A. da Silva, L. O. dos Santos and M. M. Beppu, *Eur. Polym. J.*, 2011, **47**, 254.
8. V. Sharma and P. P. Kundu, *Prog. Polym. Sci.*, 2006, **31**, 983.

9. Z. Petrović, *Polym. Rev.*, 2008, **48**, 109.

10. Z. Liu and S. Z. Erhan, *J. Polym. Environ.*, 2010, **18**, 243.

11. M. Acar, S. Çoban and B. Hazer, *J. Macromol. Sci., Part A: Pure Appl. Chem.*, 2013, **50**, 287.

12. F. Li, M. Hanson and R. Larock, *Polymer*, 2001, **42**, 1567.

13. Z. Liu, *BioEnergy Res.*, 2013, **6**, 1230.

14. J. F. Wu, S. Fernando, D. Weerasinghe, Z. Chen and D. C. Webster, *ChemSusChem*, 2011, **4**, 1135.

15. A. Chernykh, S. Alam, A. Jayasooriya, J. Bahr and B. J. Chisholm, *Green Chem.*, 2013, **15**, 1834.

16. M. Sacristán, J. C. Ronda, M. Galià and V. Cádiz, *Biomacromolecules*, 2009, **10**, 2678.

17. M. Sacristán, J. C. Ronda, M. Galià and V. Cádiz, *Polymer*, 2010, **51**, 6099.

18. J. C. Ronda, M. Galià, V. Ca and M. Sacristán, *J. Appl. Polym. Sci.*, 2011, **122**, 1649.

19. T. Vlcek and Z. S. Petrović, *J. Am. Oil Chem. Soc.*, 2006, **83**, 247.

20. P. Saithai, *eXPRESS Polym. Lett.*, 2013, **7**, 910.

21. Z. Liu and S. Z. Erhan, *J. Am. Oil Chem. Soc.*, 2009, **87**, 437.

22. Z. Liu, K. M. Doll and R. A. Holser, *Green Chem.*, 2009, **11**, 1774.

23. Z. Liu and A. Biswas, *Appl. Catal., A*, 2013, **453**, 370.

24. T. Takahashi, K. Hirayama, N. Teramoto and M. Shibata, *J. Appl. Polym. Sci.*, 2008, **108**, 1596.

25. A. P. Gupta, S. Ahmad and A. Dev, *Polym.-Plast. Technol. Eng.*, 2010, **49**, 657.

26. R. A. Ortiz, D. P. López, M. D. L. G. Cisneros, J. C. R. Valverde and J. V. Crivello, *Polymer*, 2005, **46**, 1535.

27. J. V. Crivello and K. D. Carlson, *Macromol. Rep.*, **33**, 251.

28. S. G. Tan and W. S. Chow, *J. Am. Oil Chem. Soc.*, 2010, **88**, 915.

29. S. G. Tan, *eXPRESS Polym. Lett.*, 2011, **5**, 480.

30. J. M. España, L. Sánchez-Nácher, T. Boronat, V. Fombuena and R. Balart, *J. Am. Oil Chem. Soc.*, 2012, **89**, 2067.

31. V. Fombuena, L. Sánchez-Nácher, M. D. Samper, D. Juárez and R. Balart, *J. Am. Oil Chem. Soc.*, 2012, **90**, 449.

32. R. A. Ruseckaite, P. M. Stefani, F. I. Altuna and L. H. Espo, *J. Appl. Polym. Sci.*, 2011, **120**, 789.

33. R. Wang, *eXPRESS Polym. Lett.*, 2013, **7**, 272.

34. Z. Liu, Y. Xu, L. Cao, C. Bao, H. Sun, L. Wang, K. Dai and L. Zhu, *Soft Matter*, 2012, **8**, 5888.

35. F. I. Altuna, V. Pettarin and R. J. J. Williams, *Green Chem.*, 2013, **15**, 3360.

36. Z. Wang, X. Zhang, R. Wang, H. Kang, B. Qiao, J. Ma, L. Zhang and H. Wang, *Macromolecules*, 2012, **45**, 9010.

37. S. N. Khot, J. J. Lascala, E. Can, S. S. Morye, G. I. Williams, G. R. Palmese, S. H. Küsefoğlu and R. P. Wool, *J. Appl. Polym. Sci.*, 2001, **82**, 703.

38. S. Rengasamy and V. Mannari, *Prog. Org. Coatings*, 2013, **76**, 78.

39. J. F. Wu, S. Fernando, K. Jagodzinski, D. Weerasinghe and Z. Chen, *Polym. Int.*, 2011, **60**, 571.

40. H.-M. Kim, H.-R. Kim and B. S. Kim, *J. Polym. Environ.*, 2010, **18**, 291.
41. N. R. Jang, H.-R. Kim, C. T. Hou and B. S. Kim, *Polym. Adv. Technol.*, 2013, **24**, 814.
42. Q. Ma, X. Liu, R. Zhang, J. Zhu and Y. Jiang, *Green Chem.*, 2013, **15**, 1300.
43. D. W. Janes, K. Shanmuganathan, D. Y. Chou and C. J. Ellison, *ACS Macro Lett.*, 2012, **1**, 1138.
44. S. Oprea, *J. Mater. Sci.*, 2009, **45**, 1315.
45. Ö. Albayrak, S. Şen, G. Çaylı and B. Ortaç, *J. Appl. Polym. Sci.*, 2013, **130**, 2031.
46. M. Skrifvars and P. Walkenstro, *J. Appl. Polym. Sci.*, 2009, **114**, 2502.
47. L. C. Bailosky, L. M. Bender, D. Bode, R. A. Choudhery, G. P. Craun, K. J. Gardner, C. R. Michalski, J. T. Rademacher, G. J. Stella and D. J. Telford, *Prog. Org. Coatings*, 2013, **76**, 1712.
48. S. Miao, S. Zhang, Z. Su and P. Wang, *J. Appl. Polym. Sci.*, 2013, **127**, 1929.
49. M. Desroches, S. Caillol, V. Lapinte and B. Boutevin, *Macromolecules*, 2011, **44**, 2489.
50. L. J. Sun, C. Yao, H. F. Zheng and J. Lin, *Chin. Chem. Lett.*, 2012, **23**, 919.
51. Y. Lu, Y. Xia and R. C. Larock, *Prog. Org. Coatings*, 2011, **71**, 336.
52. C. Wang, X. Chen, J. Chen, C. Liu and H. Xie, *J. Appl. Polym. Sci.*, 2011, **122**, 2449.
53. L.-T. Yang, C.-S. Zhao, C.-L. Dai, L.-Y. Fu and S.-Q. Lin, *J. Polym. Environ.*, 2011, **20**, 230.
54. R. Gu, S. Konar and M. Sain, *J. Am. Oil Chem. Soc.*, 2012, **89**, 2103.
55. S. Miao, N. Callow, P. Wang, Y. Liu, Z. Su and S. Zhang, *J. Am. Oil Chem. Soc.*, 2013, **90**, 1415.
56. P. Zhang and J. Zhang, *Green Chem.*, 2013, **15**, 641.
57. A. Biswas, B. K. Sharma, J. L. Willett, A. Advaryu, S. Z. Erhan and H. N. Cheng, *J. Agric. Food Chem.*, 2008, **56**, 5611.
58. J. Hong, Q. Luo and B. K. Shah, *Biomacromolecules*, 2010, **11**, 2960.
59. J. Hong, Q. Luo, X. Wan, Z. S. Petrović and B. K. Shah, *Biomacromolecules*, 2012, **13**, 261.
60. G. Cayli and S. Küsefoğlu, *J. Appl. Polym. Sci.*, 2008, **109**, 2948.
61. E. Taylan and S. H. Küsefoğlu, *J. Appl. Polym. Sci.*, 2011, **119**, 1102.
62. D. A. Echeverri, V. Cádiz, J. C. Ronda and L. A. Rios, *Eur. Polym. J.*, 2012, **48**, 2040.
63. R. C. Larock, X. Dong, S. Chung, C. K. Reddy and L. E. Ehlers, *J. Am. Oil Chem. Soc.*, 2001, **78**, 447.
64. M. Valverde, D. Andjelkovic, P. P. Kundu and R. C. Larock, *J. Appl. Polym. Sci.*, 2008, **107**, 423.
65. L. Yang, C. Dai, L. Ma and S. Lin, *J. Polym. Environ.*, 2010, **19**, 189.
66. W. M. Gramlich, M. L. Robertson and M. A. Hillmyer, *Macromolecules*, 2010, **43**, 2313.
67. M. Valverde, S. Yoon, S. Bhuyan, R. C. Larock, M. R. Kessler and S. Sundararajan, *Macromol. Mater. Eng.*, 2011, **296**, 444.

68. C. Öztürk, H. Mutlu, M. A. R. Meier and S. H. Küsefoğlu, *Eur. Polym. J.*, 2011, **47**, 1467.
69. Q. Luo, M. Lui, Y. Xu, M. Ionescu and Z. S. Petrović, *J. Appl. Polym. Sci.*, 2013, **127**, 432.
70. Y. Zhao, J. Qu, Y. Feng, Z. Wu, F. Chen and H. Tang, *Polym. Adv. Technol.*, 2012, **23**, 632.
71. O. Fenollar, D. Garcia-Sanoguera, L. Sanchez-Nacher, T. Boronat, J. López and R. Balart, *Polym.-Plast. Technol. Eng.*, 2013, **52**, 761.
72. Y. Chen, L. Yang, J. Wu, L. Ma, D. E. Finlow, S. Lin and K. Song, *J. Therm. Anal. Calorim.*, 2012, **113**, 939.
73. M. L. Robertson, K. Chang, W. M. Gramlich and M. A. Hillmyer, *Macromolecules*, 2010, **43**, 1807.
74. Z. Xiong, Y. Yang, J. Feng, X. Zhang, C. Zhang, Z. Tang and J. Zhu, *Carbohydr. Polym.*, 2013, **92**, 810.
75. N. Kiangkitiwan and K. Srikulkit, *Sci. World J.*, 2013, 860487.
76. R. P. H. Brandelero, M. V. Grossmann and F. Yamashita, *Carbohydr. Polym.*, 2012, **90**, 1452.
77. Y. Xia, Z. Zhang, M. R. Kessler, B. Brehm-Stecher and R. C. Larock, *ChemSusChem*, 2012, **5**, 2221.
78. H. Bakhshi, H. Yeganeh and S. Mehdipour-Ataei, *J. Biomed. Mater. Res., Part A*, 2013, **101**, 1599.
79. S. Alam and B. J. Chisholm, *J. Coatings Technol. Res.*, 2011, **8**, 671.
80. Y. Xia and R. C. Larock, *ChemSusChem*, 2011, **4**, 386.
81. Y. Lu and R. C. Larock, *Prog. Org. Coatings*, 2010, **69**, 31.
82. B. K. Ahn, S. Kraft, D. Wang and X. S. Sun, *Biomacromolecules*, 2011, **12**, 1839.
83. C. Fu, B. Zhang, C. Ruan, C. Hu, Y. Fu and Y. Wang, *Polym. Degrad. Stab.*, 2010, **95**, 485.
84. X. Ren and K. Li, *J. Appl. Polym. Sci.*, 2013, **128**, 1101.
85. Y.-L. Lin, L.-Y. Chen, C.-H. Chen, Y.-K. Liu, W.-T. Hsu, L.-P. Ho and K.-W. Liao, *J. Nanomater.*, 2012, **2012**, 427306.
86. M. J. Abdekhodaie, Z. Liu, S. Z. Erhan and X. Y. Wu, *Polym. Int.*, 2012, **61**, 1477.
87. X. Zhang, M. D. Do, L. Kurniawan and G. G. Qiao, *Carbohydr. Res.*, 2010, **345**, 2174.
88. M. L. Palacios-Jaimes, F. Cortes-Guzman, D. A. González-Mártínez and R. M. Gómez-Espinosa, *J. Appl. Polym. Sci.*, 2012, **124**, E147.
89. N. Teramoto, Y. Saitoh, A. Takahashi and M. Shibata, *J. Appl. Polym. Sci.*, 2010, **115**, 3199.
90. B. De, K. Gupta, M. Mandal and N. Karak, *ACS Sustain. Chem. Eng.*, 2013, **2**, 445.
91. Y. Xia and R. C. Larock, *Macromol. Rapid Commun.*, 2011, **32**, 1331.
92. I. S. Ristić, Z. D. Bjelović, B. Holló, Mészáros K. Szécsényi, J. Budinski-Simendić, N. Lazić and M. Kićanović, *J. Therm. Anal. Calorim.*, 2012, **111**, 1083.

93. I. S. Ristić, J. Budinski-Simendić, I. Krakovsky, H. Valentova, R. Radičević, S. Cakić and N. Nikolić, *Mater. Chem. Phys.*, 2012, **132**, 74.
94. E. Mussatti, C. Merlini, G. M. de Oliveira Barra, S. Güths, A. P. Novaes Oliveira and C. Siligardi, *Mater. Res.*, 2013, **16**, 65.
95. S. Lin, J. Huang, P. R. Chang, S. Wei, Y. Xu and Q. Zhang, *Carbohydr. Polym.*, 2013, **95**, 91.
96. S. Thakur and N. Karak, *RSC Adv.*, 2013, **3**, 9476.
97. P. S. Sathiskumar and G. Madras, *Polym. Degrad. Stab.*, 2011, **96**, 1695.
98. I. S. Ristić, M. Marinović-Cincović, S. M. Cakić, L. M. Tanasić and J. K. Budinski-Simendić, *Polym. Bull.*, 2013, **70**, 1723.
99. O. Saravari and S. Praditvatanakit, *Prog. Org. Coatings*, 2013, **76**, 698.
100. K. Hirayama, T. Irie, N. Teramoto and M. Shibata, *J. Appl. Polym. Sci.*, 2009, **114**, 1033.
101. H.-M. Kim, H.-R. Kim, C. T. Hou and B. S. Kim, *J. Am. Oil Chem. Soc.*, 2010, **87**, 1451.
102. B. F. Dillman, N. Y. Kang and J. L. P. Jessop, *Polymer.*, 2013, **54**, 1768.
103. D. M. Bechi, M. A. Luca, M. De Martinelli and S. Mitidieri, *Prog. Org. Coatings*, 2013, **76**, 736.
104. Y. Xia and R. C. Larock, *Polymer*, 2010, **51**, 2508.
105. S. Allauddin, R. Narayan and K. V. S. N. Raju, *ACS Sustain. Chem. Eng.*, 2013, **1**, 910.
106. M. Van der Steen and C. V. Stevens, *ChemSusChem*, 2009, **2**, 692.
107. Y. Bao, J. He and Y. Li, *Polym. Int.*, 2013, **62**, 1457.
108. O. van den Berg, T. Dispinar, B. Hommez and F. E. Du Prez, *Eur. Polym. J.*, 2013, **49**, 804.
109. W. Dong, J. Ren, L. Lin, D. Shi, Z. Ni and M. Chen, *Polym. Degrad. Stab.*, 2012, **97**, 578.
110. M. Koh, H. Kim, N. Shin, H. S. Kim, D. Yoo and Y. G. Kim, *Bull. Korean Chem. Soc.*, 2012, **33**, 1873.
111. Z. S. Petrović, J. Milić, Y. Xu and I. Cvetković, *Macromolecules*, 2010, **43**, 4120.
112. X. Miao, C. Fischmeister, P. H. Dixneuf, C. Bruneau, J.-L. Dubois and J.-L. Couturier, *Green Chem.*, 2012, **14**, 2179.
113. S. Thakur and N. Karak, *Prog. Org. Coatings*, 2013, **76**, 157.
114. R. Gharibi, M. Yousefi and H. Yeganeh, *Prog. Org. Coatings*, 2013, **76**, 1454.
115. M. L. Robertson, J. M. Paxton and M. A. Hillmyer, *ACS Appl. Mater. Interfaces*, 2011, **3**, 3402.
116. Z. Xiong, L. Zhang, S. Ma, Y. Yang, C. Zhang, Z. Tang and J. Zhu, *Carbohydr. Polym.*, 2013, **94**, 235.
117. M. Klinger, L. P. Tolbod and P. R. Ogilby, *J. Appl. Polym. Sci.*, 2010, **118**, 1643.
118. B. B. R. Silva, R. M. C. Santana and M. M. C. Forte, *Int. J. Adhes. Adhes.*, 2010, **30**, 559.

119. S. Oprea, *J. Am. Oil Chem. Soc.*, 2009, **87**, 313.
120. R. S. Harisha, K. M. Hosamani, R. S. Keri, N. Shelke, V. K. Wadi and T. M. Aminabhavi, *J. Chem. Sci.*, 2010, **122**, 209.
121. S. Fang, H. Zhang, B. Zhang, G. Wei, G. Li and Y. Zhou, *Thermochim. Acta*, 2013, **551**, 20.
122. C. Meiorin, M. I. Aranguren and M. A Mosiewicki, *Polym. Int.*, 2012, **61**, 735.
123. C. Meiorin, M. I. Aranguren and M. A. Mosiewicki, *J. Appl. Polym. Sci.*, 2012, **124**, 5071.
124. C. Meiorin, M. A. Mosiewicki and M. I. Aranguren, *Polym. Test.*, 2013, **32**, 249.
125. K. Das, D. Ray, C. Banerjee, N. R. Bandyopadhyay, A. K. Mohanty and M. Misra, *J. Appl. Polym. Sci.*, 2011, **119**, 2174.
126. Y. Huang, L. Pang, H. Wang, R. Zhong, Z. Zeng and J. Yang, *Prog. Org. Coatings*, 2013, **76**, 654.
127. C. Luo, W. J. Grigsby, N. R. Edmonds and J. Al-Hakkak, *Acta Biomater.*, 2013, **9**, 5226.
128. K. Huang, P. Zhang, J. Zhang, S. Li, M. Li, J. Xia and Y. Zhou, *Green Chem.*, 2013, **15**, 2466.
129. C. Liu, X. Yang, J. Cui, Y. Zhou, L. Hu, M. Zhang and H. Liu, *BioResources*, 2012, **7**, 447.
130. V. Ribeiro da Silva, M. A. Mosiewicki, M. I. Yoshida, M. Coelho da Silva, P. M. Stefani and N. E. Marcovich, *Polym. Test.*, 2013, **32**, 438.
131. M. A. Mosiewicki, U. Casado, N. E. Marcovich and M. I. Aranguren, *Polym. Eng. Sci.*, 2009, **49**, 685.
132. Z. Xiong, C. Li, S. Ma, J. Feng, Y. Yang, R. Zhang and J. Zhu, *Carbohydr. Polym.*, 2013, **95**, 77.
133. C. Liu, W. Lei, Z. Cai, J. Chen, L. Hu, Y. Dai and Y. Zhou, *Ind. Crops Prod.*, 2013, **49**, 412.
134. G. Ma, T. Zhang, J. Wu, C. Hou, L. Ling and B. Wang, *J. Appl. Polym. Sci.*, 2013, **130**, 1700.
135. S. Sun, G. Yang, Y. Bi and H. Liang, *J. Am. Oil Chem. Soc.*, 2011, **88**, 1567.
136. J. Lo, R. Balart, O. Fenollar and L. Sa, *J. Appl. Polym. Sci.*, 2012, **124**, 2550.
137. N. Supanchaiyamat, P. S. Shuttleworth, A. J. Hunt, J. H. Clark and A. S. Matharu, *Green Chem.*, 2012, **14**, 1759.
138. Z. Liu, S. N. Shah, R. L. Evangelista and T. Isbell, *Ind. Crops Prod.*, 2013, **41**, 10.
139. D. V. Palaskar, A. Boyer, E. Cloutet, J.-F. Le Meins, B. Gadenne, C. Alfos, C. Farcet and H. Cramail, *J. Polym. Sci., Part A: Polym. Chem.*, 2012, **50**, 1766.
140. N. Alemdar, A. T. Erciyes and N. Bicak, *J. Appl. Polym. Sci.*, 2012, **125**, 10.
141. X. Kong, G. Liu and J. M. Curtis, *Eur. Polym. J.*, 2012, **48**, 2097.

CHAPTER 6

Green Polyurethanes and Bio-fiber-based Products and Processes

RUIJUN GU*[a] AND MOHINI SAIN*[a,b]

[a] Centre for Biocomposites and Biomaterials Processing, Faculty of Forestry, University of Toronto, 33 Willcocks Street, Toronto, Ontario, M5S 3B3, Canada; [b] King Abdulaziz University, Abdullah Sulayman, Jeddah 22254, Kingdom of Saudi Arabia
*Email: ruijun.gu@utoronto.ca; m.sain@utoronto.ca

6.1 Introduction

Polyurethanes (PUs) are the most widely used polymers in coating,[1,2] foam,[3–5] building and construction,[6–11] transportation,[12–17] and elastomer[18] applications because of their chemical versatility and high durability, which have hastened the development of worldwide PU production and consumption.[19] The key raw materials used in PU manufacturing are isocyanates, polyols and other additives. The rebound in the automotive and construction industries in North America and Europe, and the rapid economic growth in the Asia-Pacific region is expected to drive the PU market,[20] specifically PU foams.[20,21] Polyols comprise the largest volume of PU production. According to a market report,[20] PU demand in 2012 was worth US$43.2 billion and is expected to reach US$66.4 billion by 2018. This will drive the PU market and affect the polyol market, which consisted of over 7.5 million tons in 2012, and is expected to be over 10.4 million tons by 2018.

RSC Green Chemistry No. 29
Green Materials from Plant Oils
Edited by Zengshe Liu and George Kraus
© The Royal Society of Chemistry 2015
Published by the Royal Society of Chemistry, www.rsc.org

Petroleum-based PU takes hundreds of years to break down in nature due to its insufficient carbon and nitrogen sources for microbial growth, even when exposed to extreme conditions.[22] The growing demand for bio-based PUs has manufacturers around the world increasing their commitment to using renewable and eco-friendly bio-materials in their products. Thus, a dynamic development is foreseen for bio-based PUs, which are chemically identical to their petrochemical counterparts, derived from bio-mass with increasing regulatory pressure for sustainable solutions. Bio-based PUs are spearheaded by bio-based polyols which are derived from feedstocks.[23,24] The use of bio-based polyols decreases the petrochemical content of PU formulations and increases their economic allocation (defined as economic partitioning the input or output flows of a process or a product system, *i.e.* between the product system under study and other product systems).[25] According to industrial estimates, natural-oil-derived polyols produce 36% less greenhouse gas (GHG) emissions, use 61% less non-renewable energy, and have 23% less total energy demand.[25,26] Bio-based PUs have significantly improved biodegradability compared to PU products made from petrochemicals.[27,28]

Traditionally, the isocyanates and their derivatives used for PU manufacture were prepared by well-known methods *via* nitrene intermediates, as described in Scheme 6.1.[29] Currently, there are no 100% sustainable PU products through non-isocyanate reactions. However, some laboratory-scale methods have reported bio-based diisocyanates from fatty acids *via* the phosgene and azide methods.[30] Hojabri and co-workers have prepared fatty-acid-derived diisocyanate, using a Curtius rearrangement, which was similar to traditional aliphatic diisocyanate.[31] So far, commercial technology for bio-based isocyanates is still under evaluation.

It is known that a few plant oils, such as cashew nut oil, are rich in phenolic compounds.[32] However, triglycerides are the main constituents of most vegetable or plant oils. According to a report by Twitchett,[29] aliphatic and aromatic isocyanates can be obtained by phosgenation of the corresponding primary amines, indicated in Scheme 6.2.

Though isocyanates are non-bio-degradable, green PU products are mostly manufactured from green polyols derived from plant and vegetable oils,[23,24]

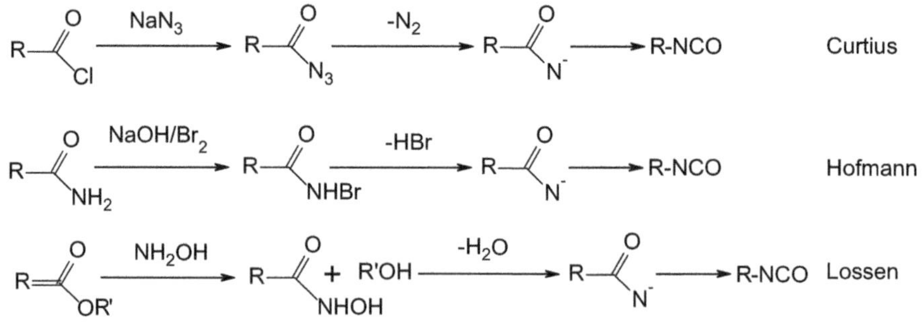

Scheme 6.1 Isocyanate syntheses on a laboratory scale.
Reproduced from ref. 29.

$$RNH_2 \xrightarrow{\text{COCl}_2} RNHCOCl \xrightarrow{\text{-HCl}} R\text{-NCO}$$

Scheme 6.2 Isocyanate preparation *via* the phosgene method. Adapted from ref. 29.

bark liquefaction[34,35] and lignin.[36] It is known that PU foams are the most dominant product, accounting for over 65% of the total PU demand in 2011.[19,20] PU foams are mainly employed in the construction, automotive and furniture industries and also for footwear, packaging *etc.*[20] The construction and automotive industries are expected to be the key growth market for green PUs as environmental and social concerns increase. As a lightweight and durable core material, green PU foams are used throughout the construction and automobile industries in spray foam insulation[37] and transportation seating systems.[38] According to *PU Magazine*, the drawbacks of manufacturing PU with bio-based polyols are the need for higher levels of additives, the installation of new mixing heads and the higher scrap rate experienced by several foam makers.

A number of groups have conducted research into bio-based isocyanates[30,31] and non-isocyanate routes for bio-based PU synthesis from vegetable oils, and cashew nut shell liquid-based cyclocarbonate oligomers.[39] So far, 100% renewable PU foams made with both bio-degradable isocyanates and polyols are still far off. Currently, bio-based PU products in the current market are prepared with petroleum-based isocyanates and renewable polyols. Because the OH groups of polyols are involved in a reaction with the NCO groups of isocyanates, some hydroxyl group containing bio-masses such as lignin, natural fibers, bark and their liquids could be used as renewable polyols. In this chapter, we will discuss renewable PU foams and their method of preparation using renewable polyols and/or the employment of biomass.

6.2 Bio-based PU Foams made with Bio-based Polyols

Polyols are the major component of PU products. Renewable polyols come from plant and vegetable oils,[23,24] and bio-mass liquefaction such as bark,[34,35] lignin[40] and nut shells.[33] Bio-based polyester polyols became available for making PUs with higher renewable contents and equal performance, following the commercialization of bio-based succinic acid.[41,42]

6.2.1 Bio-based Polyols Derived from Plant or Vegetable Oils

Castor oil is commercially unique among the naturally occurring oils composed of ricinoleic acid. It is known that a typical castor oil contains the hydroxyl-functionalized unsaturated C18 triesters of rincinoleic acid and glycerin[43] (at least 80%, see Figure 6.1). The unreacted hydroxyl groups are on C12 and its hydroxyl value is in the range of 150–180 mg KOH g^{-1} (ref. 43). Thus, castor oil can be used as a reactive monomer in PU foam preparations

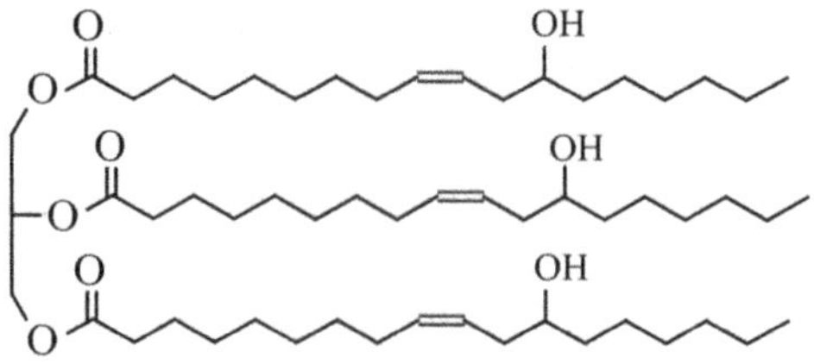

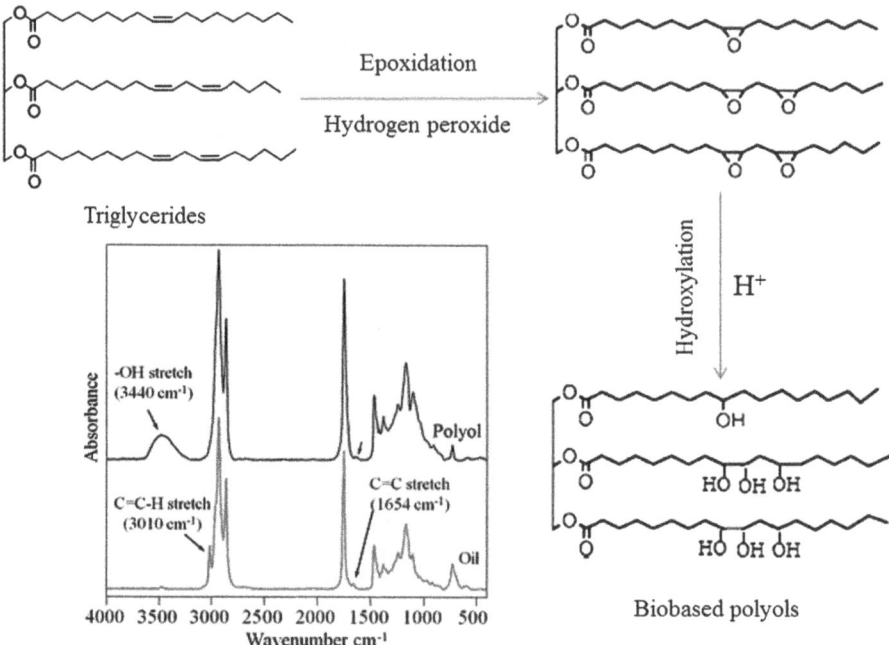

Figure 6.1 The major component of castor oil.

Figure 6.2 Preparation of vegetable-oil-based polyols.
Reproduced from ref. 5 with permission from Elsevier Ltd.

directly.[44] Icynene Inc. reports that it has produced environmentally friendly spray foam insulations with an *R*-value, a measure of resistance to heat flow through a given thickness of material, of 3.7 per inch using renewable castor oil for 30% replacement of petroleum-based polyols.[45]

Other vegetable oils such as soybean oil and canola oil are composed of hydroxyl-free and unsaturated triglycerides. Therefore, hydroxyl groups must be introduced onto the carbon chains of triglycerides through epoxidation and hydroxylation of the C–C bonds, as shown in Figure 6.2. Plant- or vegetable-oil-based polyols generally have low hydroxyl functionalities and high molecular weights, which make them more suitable for use in spray foams[4,6,7] and flexible foams,[5] rather than rigid foams. One breakthrough in natural-oil-based polyol technology came from Cargill and Biobased Technologies, which reported that they can be used in PU rigid foams as automotive parts.[12,14] Using proprietary manufacturing processes, Cargill and Biobased Technologies have produced soya-oil-based polyols with hydroxyl numbers in the range of 56–370 mg KOH g^{-1} which can produce foams with a wide range of physical properties for furniture and bedding applications, color pastes and automotive applications.[46] These bio-based polyols, both the BiOH® and Agrol® series, reportedly have a high renewable content, of more than 86%, and excellent compatibility with conventional polyols.[46,47]

Major chemical differences between bio-based and petroleum-based polyols were identified using FT-IR, as demonstrated in Figure 6.3. Bio-based PU foams using vegetable-oil-based polyols contain a typical spectroscopic fingerprint for triglycerides in the region of 2700–3000 cm^{-1}, whereas petroleum-based PU foams show a typical fingerprint of polyether polyol.[5] It was observed that bio-based PU foam had a different cell structure compared to petroleum-based foam with the same formulation (Figure 6.4), where the

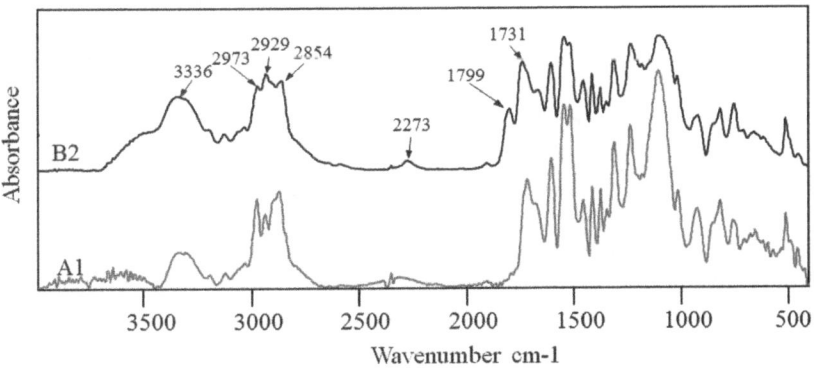

Figure 6.3 FT-IR spectra of PU foams (A1 = petroleum-based PU foam, B2 = soybean-oil-based PU foam).
Adapted from ref. 5 with permission from Elsevier Ltd.

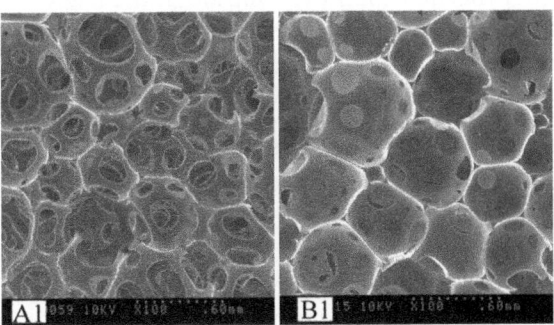

Figure 6.4 Cell structures of PU foams (A1 = petroleum-based PU foam, B1 = soybean-oil-based PU foam).
Adapted from ref. 5 with permission from Elsevier Ltd.

cells in bio-based foam were covered by round window membranes, compared with the porous cell structures in petroleum-based foam.[5] The round window membranes are typical cell structures in bio-based foams due to the presence of triglycerides. This cell morphology results in higher closed-cell content (closed-cell content is the percentage of cells with closed structures in whole foam cells; higher closed-cell content means a superior *R*-value) and dimensional stability of bio-based foams.

It is known that primary hydroxyl groups react faster with isocyanates than secondary hydroxyl groups.[48] According to a report from Biobased Technologies,[49] natural oil polyols are suitable for producing rigid PU foams as they enable rigid networks to form, due to their high content of primary hydroxyl groups (reported to be up to 70%).[47] In addition, their complete miscibility with conventional polyether polyols and hydrocarbon blowing agents makes natural oil polyols suitable for bio-based PU foam manufacturing. According to a study by Gautam and co-workers,[50] polyester PU is more susceptible to biodegradation by bacteria under controlled laboratory conditions, compared to polyether PU. It is known that bio-based polyols made from plant or vegetable oils are typical polyether polyols, due to their high levels of triglycerides. As a result of their higher renewable content, bio-based PUs will be broken down more easily, through their labile chemical moieties, by fungal attack.[51] Similar results were also reported by Mathur and Prasad,[52] who reported that bio-based PU foams made with castor oil had a high resistance to thermal degradation, and could be used for roof insulation at temperatures above 80 °C.[53] However, bio-based PU foams can still degrade under thermal conditions. Benes and co-workers reported that PU materials made with a castor-oil-based polyol were decomposed successfully at 180 °C by basic hydrolysis and the transesterification of the triglycerides of castor oil.[27] According to a report by Hou,[54] bio-based PU foams were degraded either under cyclic compressive loading conditions or aging conditions under ASTM (American Society for Testing and Materials) D3574, when they were exposed to 140 °C for 22 h with a relative humidity of 45%, or to 50 °C for 22 h at 95% relative humidity.

It is known that a higher isocyanate index produces polyisocyanurate-type foams with higher thermal stability.[55] Although the triglyceride structures in vegetable oils consist of unsaturated chains, which have poor oxidative stability,[56] unsaturated structures in triglycerides at 3010 cm^{-1} and 1654 cm^{-1} disappeared after epoxidation and hydroxylation, and the hydroxyl peak of bio-based polyols appeared at 3440 cm^{-1}, as indicated in Figure 6.2.[5] According to Gu's study,[5] soybean-based PU foams have higher glass-transition temperatures and inferior cryogenic properties compared to petroleum-based foams. However, it has lower thermal degradation in the degradation of urethane segments due to natural molecular chains with lower thermal stability than petroleum skeletons. Also, bio-foams had better thermal stability at a high-temperature level. Guo also reported that cyclopentane blown-rigid PU foams produced using soybean-oil-based polyols had high thermal stabilities.[57] The increase of thermal stability of bio-based PU foams, results in excellent dimensional stability. Naturally occurring oils contain triglycerides which can generate stable three-dimensional networks after reacting with the NCO groups of isocyanates. A very small shrinkage and expansion problem was observed for a bio-based PU foam prepared from a naturally occurring oil-based polyol in a water-blown system.[58,59] The dimensional changes of the bio-PU foam, which was made with a palm-oil-derived polyol and MDI (methylene diphenyl isocyanate) in the ratio of 100 : 140 parts and blown by 4.5 parts of water, were −0.09% and +0.012% at −15 °C and +70 °C at 95% relative humidity for 24 h, respectively. Moreover, when a petroleum-based polyol was used to replace the bio-based polyol, major shrinkage and expansion problems were observed in an identical formulation.[59]

6.2.2 Bio-based Polyols Derived from Bio-based Succinic Acid

Succinic acid is produced by sugar fermentation *via* carbon neutral fermentation.[60,61] Generally, bio-based butanediol can be made from carbon neutral succinic acid.[62,63] BioAmber has reported a commercial bio-based butanediol, which is chemically identical to petroleum-based 1,4-butanediol. Bio-based polyols can be produced *via* esterification of bio-based succinic acid with butanediol at approximately 170–200 °C (Scheme 6.3).[64] These bio-based polyols are used for bio-based PU foams instead of adipic acid and 1,4-butanediol. Succinic acid based polyester polyols have similar thermal and mechanical properties to adipic acid polyols.

6.2.3 Bio-based Polyols Derived from Bio-mass Liquefaction

Industrial bio-mass such as lignin, bark and nut shell is an economical material in nature for the production of PU foams, as a source of renewable precursors of polyols or bio-mass fillers. Due to bio-mass containing a high content of aromatic structures, polyols made from bio-mass are rich in phenolic components. Liquefaction of these materials is an appealing route to producing polyols for the preparation of bio-based PU foams. Waste bio-mass is inexpensive since it is a byproduct of the logging and pulping

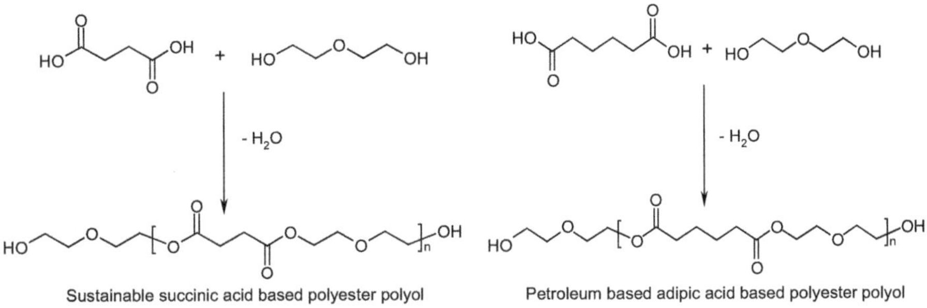

Sustainable succinic acid based polyester polyol   Petroleum based adipic acid based polyester polyol

Scheme 6.3 Polyol structures made from bio-based succinic acid and petroleum-based adipic acid.

industries. Bio-mass is also rich in extractive compounds which have high hydroxyl functionalities and can be extracted easily.

Although polyol structures vary greatly depending on the liquefaction conditions, their phenolic structures impart thermal stability and fire-resistant properties. Cheradame and co-workers have reported that lignin-based liquids can react with hexamethylebe diisocyanate under rather mild conditions.[36] Ionescu reported that cashew nut shell bio-mass liquefied with polyethylene glycol has a high content of phenolic components. Due to cashew nut shell liquid being rich in phenolic structures,[33] Mannish polyols were successfully obtained from them, producing rigid PU foams with good physical and mechanical properties. It was also reported that a high aromatic content in PU foam results in low fire-retardant properties.[33] Bark-based polyols for PU foams were liquefied with glycol and polyethylene glycol by Zhao and D'Souza.[34,35] Though some holocellulose was converted into levulinate and formic esters at the temperature of 130–160 °C, the lignin fraction can only be extracted above its soft point of 150–160 °C. In addition, extractives from lignin, bark or waste nut shell underwent condensation reactions with the liquefaction solvents.

6.3 Bio-based PU Foams with the Reinforcement of Bio-mass

6.3.1 Wood Fibers

It is known that microclay and wood fiber particles both have nucleation effects in foamed plastics.[65–67] Considering the hydroxyl groups in wood fiber, wood fiber is expected to be a reactive polysaccharide in PU foams. Gu and co-workers have reported that the presence of a small amount of wood fiber furnished PU foams with higher decomposition temperatures,[3,6] lower β-transition temperatures and higher glass-transition temperatures.[6] With the increase in the amount of water, the content of hard segments in bio-based PU foams increased, as indicated by the lower β-transition and higher glass-transition temperatures [Figure 6.5(a) and (a′)]. As a rigid polysaccharide, the employment of wood fiber in the foams also offered lower β-transition and higher glass-transition temperatures [Figure 6.5(b) and (b′)], which means that some hard structures were generated. It was understood that smaller fiber particles have a lower degree of fiber condensation, which leads to a lower cross-linked density. Unlike increasing the amount of wood fiber, smaller fiber particles resulted in PU foams with higher β-transition and glass transition temperatures [Figure 6.5(c) and (c′)].

Because wood fiber has the ability to react with the NCO groups of isocyanates, PU foams containing wood fibers have superior compressive and tensile strengths compared with neat foams[5,6] and microclay-reinforced foams with the increase of isocyanate index (Figure 6.6).[3] It was also reported

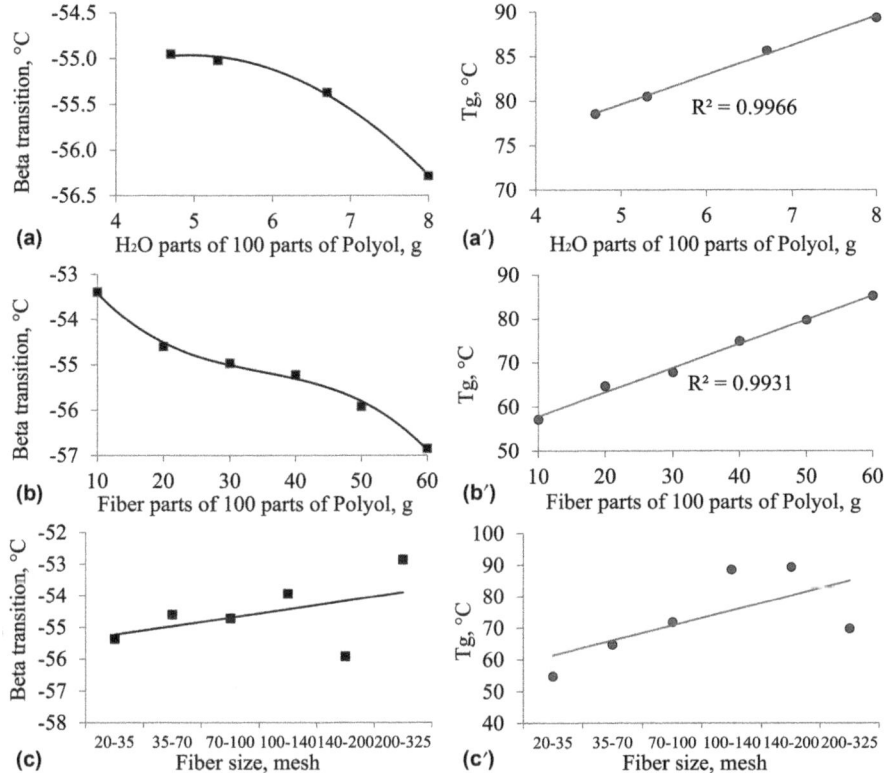

Figure 6.5 The effect of water and fibers on the DSC (differential scanning calorimetry) behavior of bio-based PU spray foams.
Reproduced from ref. 6 with permission from the authors.

that both wood fiber and microclay could improve foam tensile and compressive strengths with the increase of isocyanate index from 110 to 250 (ref. 3). Moreover, wood fiber has better reinforcement properties than microclay with higher isocyanate indexes, as shown in Figure 6.6(b) and (c). Therefore, wood fiber is a better reactive bio-mass in PU foam manufacturing than microclay, when considering increasing environmental concerns and economic considerations.

With the demands of decreasing carbon footprints, bio-mass materials are attracting interest in PU foam manufacturing, typically for building PU insulation. Gu and co-workers have done lots of work on wood-fiber-reinforced PU spray foam composites with the potential for use in building and construction applications.[4,6,7] With the exposure of hydroxyl groups on the fiber surface, the existence of wood fiber increased the foam density and the cell size[4,7] (Figure 6.7) but decreased the foam tensile strength at low isocyanate index.[3,4] The tensile strength of the PU spray foam decreased, while the compressive strength increased, as the fiber amount increased (Figure 6.8). Wood

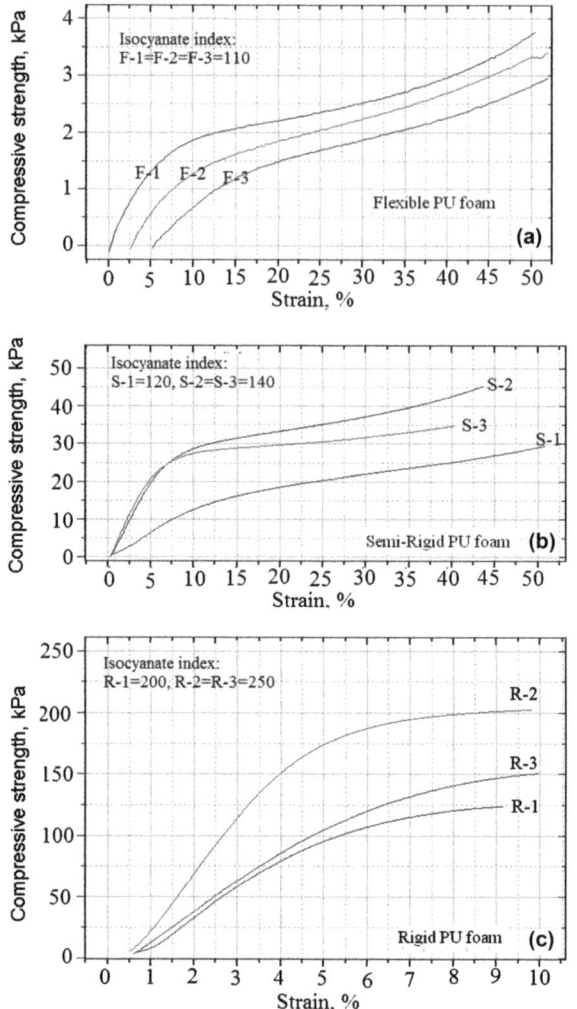

Figure 6.6 The typical compressive behavior of bio-based PU foams with different isocyanate indexes.
Adapted from ref. 3 with permission from Springer Science. 1-neat PU foam; 2-with 2.5 parts of wood fiber; 3-with 2.5 parts of microclay.

fibers act as polysaccharides. Smaller fiber particles mean lower hydroxyl functionalities. That is the reason that both the tensile and compressive strength decreased following the decrease in fiber size (Figure 6.9). It is known that PU spray foams have higher thermal stabilities when more water is used as the blowing agent.[6] Like water, the amount and size of wood fiber particles effects the thermal stability of PU foams. Gu has reported that larger wood fiber particles and a higher content of wood fiber produced PU spray foams with higher thermal stabilities, as indicated in Tables 6.1 and 6.2.

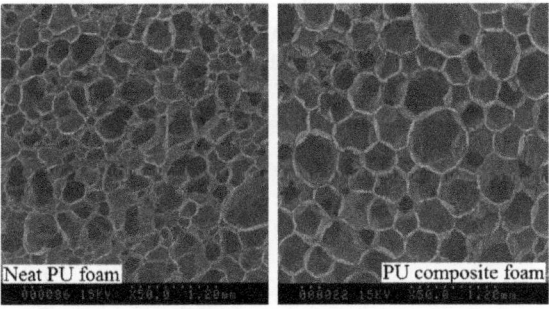

Figure 6.7 SEM images of bio-based PU foams.
Adapted from ref. 4 with permission from Elsevier Ltd.

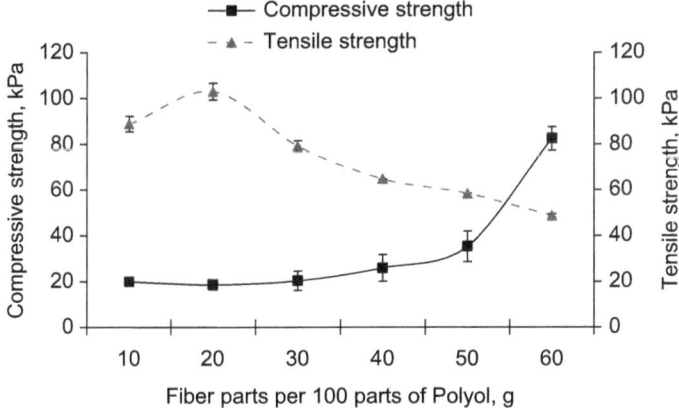

Figure 6.8 The effect of the amount of wood fiber on the mechanical properties of a PU foam.
Adapted from ref. 6 with permission from the authors.

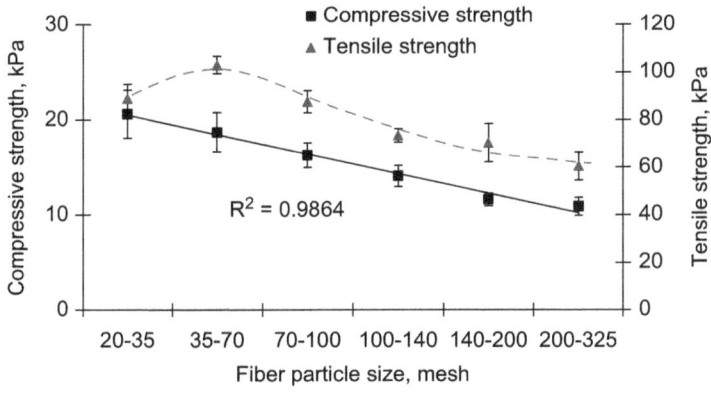

Figure 6.9 Effect of fiber size on PU foam mechanical properties.
Adapted from ref. 6 with permission from the authors.

Table 6.1 Effect of fiber content on cell thermal degradation. Adapted from ref. 6 with permission © Elsevier, 2011.

35–70 Mesh fiber/php[a]	$T_{d5}/^\circ C$	$T_{d50}/^\circ C$
0	254	429
10	258	429
20	256	439
30	260	437
40	262	437
50	265	468
60	274	467

[a]php: per hundred parts. Generally, 100 parts refer to polyol.

Table 6.2 Effect of fiber particle size on cell thermal degradation. Adapted from ref. 6 with permission © Elsevier, 2011.

Fiber particle size/20 php	$T_{d5}/^\circ C$	$T_{d50}/^\circ C$
20–35	255	443
35–70	256	439
70–100	255	426
100–140	256	427
140–200	256	429
200–325	251	410

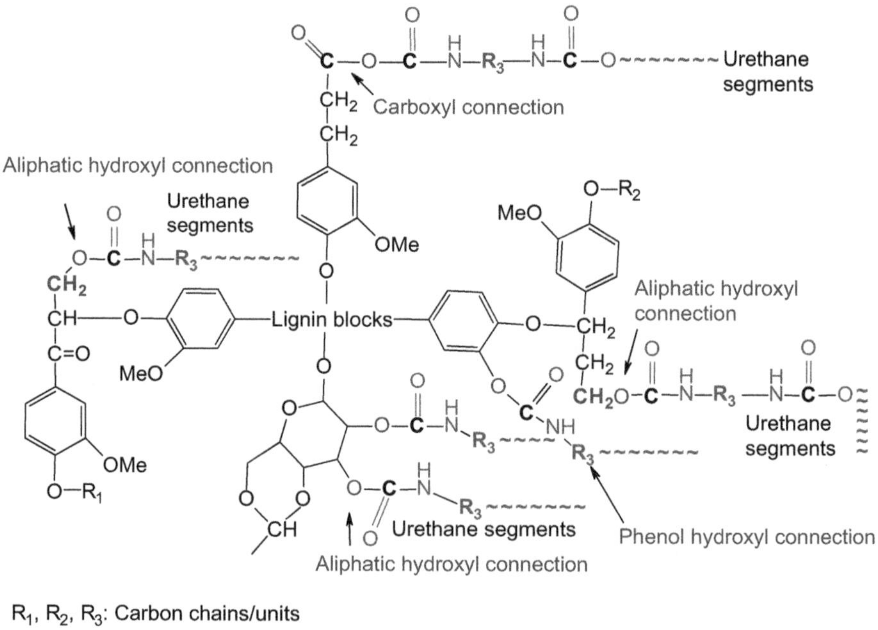

R_1, R_2, R_3: Carbon chains/units

Figure 6.10 Reaction mechanism of isocyanates with a lignin block.

6.3.2 Lignin

Lignin is the second richest bio-resource after cellulose. Although lignin undergoes significant structural changes when isolated from plant cell walls, it still has abundant phenol, carboxyl and hydroxyl groups, which can react with isocyanates to form urethane bonds (see Figure 6.10). Lignin acts as a network former due to its higher functionality compared to conventional vegetable-oil-based polyols. Due to lignin containing aromatic hydroxyl groups, the majority of lignin, either organic solvent lignin or kraft lignin, is chemically cross-linked with the isocyanates during the urethane reaction, not just physically trapped in the foams. That is the reason lignin is used as filler rather than a polyol precursor when replacing petroleum-based polyols.[12,68–71] Generally, the aromatic hydroxyl groups in lignin provide it with a higher reactivity than the aliphatic hydroxyl groups in wood fiber and polyols.[72]

According to a report by Yoshida and co-workers,[68] the effective contribution of lignin to the formation of cross-linked PU resin is apparent at low NCO/OH ratios (less than 1), where the aromatic hydroxyl groups in lignin compete with the aliphatic hydroxyl groups in polyols. Variable PUs, from soft to hard, can be prepared at low NCO/OH ratios (0.5–1.2) by combining the effects of increasing cross-link density and chain stiffness with increasing lignin content. Moreover, the introduction of lignin increased the glass-transition temperature due to the increase in cross-link density.[70] In addition, either flexible but weak, or tough PU materials can be obtained depending on the NCO/OH ratio with low lignin content.[68] However, the cross-link density of PU composites increased, as did their Young's moduli at various NCO/OH ratios with increasing lignin content.[68–70] In addition, a high lignin content resulted in hard and brittle PU materials regardless of the NCO/OH ratio used, due to the combined effect of the increased cross-link density caused by the high functionality of the lignin, and the increase in chain stiffness.[68,69] At high levels of lignin (over 30%) and low NCO/OH ratios (less than 1.5), the ultimate stress increased with the increase in lignin content (up to 35%).[68,70]

Different lignins from different separation methods have different reinforcement behaviors in PU foams.[71] Nonetheless, rigid PU foams had acceptable cell structures and compressive strengths with the addition of lignin in the range of 20–30%. There is no doubt that the addition of lignin to polyols can increase the viscosity of their blend.[71] As discussed above, the ratio of NCO to OH affected the reinforcement behavior of lignin in PU foams. PU foams had lower compressive strengths and foam densities, compared to neat foams, when they had a high level of lignin content and the ratio of NCO to OH was 1.1 and 1.3.[71] However, the presence of lignin can increase the foam density, compressive and impact strength and modulus of rigid PU foams when used with a low content (around 5%) at higher NCO/OH ratios (1.4).[12] Nevertheless, the effects of lignin on PU foams are similar to those seen by reinforcement with wood fibers, as mentioned above.

Lignin bio-mass reinforcement produced PU foams with bio-degradability. It is known that lignin, as a polyphenolic material, has an intrinsically high

Table 6.3 Flammability resistance of PU foams with lignin reinforcement. Reproduced from ref. 75 with permission © Springer, 2013.

Sample	LOI/ %	TTI/s	Time to PHRR/s	PHRR/ kW m^{-2}	THR/ MJ m^{-2}	Mean MLR/g s^{-1}	FPI/ m^2 s kW^{-1}	FGI/ kW s m^{-2}
			Cone calorimetry data of PU foams					
PU	20.0	7	62	401	31.7	0.046	0.017	6.468
PU/PFAPP[a]	23.5	10	53	208	29.9	0.043	0.048	3.924
PU/PFAPP/ PL10[b]	23.0	6	50	229	26.9	0.042	0.026	4.580
PU/PFAPP/ PL20[b]	23.5	27	84	193	23.4	0.036	0.140	2.297
PU/PFAPP/ PL30[b]	24.5	43	96	165	23.1	0.027	0.261	1.719

[a]PFAPP = Phenolic encapsulated ammonium polyphosphate.
[b]PL10, PL20 and PL30 represent 10, 20 and 30% replacement of polyol using lignin, which was modified with formaldehyde, triethylamine and ethylene glycol.

flammability.[33] According to Cateto's study, PU foams made with polyester polyol were highly resistant to bio-degradation, whilst foams made with lignin had higher weight loss by fungi to produce lignin peroxidase.[73] The presence of lignin in PU foams noticeably increases their fire resistance over lignin-free PU foams.[74,75] According to Xing's study, the flammability resistance of PU foams was improved following an increase in lignin concentration, by increasing the limiting oxygen index (LOI), peak heat release rate (PHRR), total heat release (THR), mass loss rate (MLR) and fire growth index (FGI) while increasing the time to ignition (TTI), the time to PHRR and the fire performance index (FPI), as shown in Table 6.3.[74] The flammability resistance of lignin was also found in the reinforcement of expanded poly(acetic acid) based materials.[76] Lewin also reported that brominated lignin can be used as a highly efficient fire retardant.[77] This is due to the fact that as brominated products are not attacked by wood fungi, it is a way to develop sustainable fire retardants through lignin modification.

6.3.3 Nanocellulose

Nanocellulose is isolated from cellulose source materials such as wood pulp,[78] agricultural feedstocks[79] and plants[80] using mechanical defibrillation, acid treatment or bacterial treatment.[81] Nanocrystalline cellulose (NCC) is obtained from native fibers by an acid hydrolysis giving highly crystalline and rigid particles.[82–84] However, NCC is shorter than the nanofibrils obtained through the homogenization route. Different nanocelluloses have different morphologies and fiber lengths when obtained *via* defibrillation from different sources, as shown in Figures 6.11 and 6.12.

ArboraNano, a Canadian forest nanoproducts network, has funded nanocellulose diversification in wood coatings, thermoplastics and PU automotive foams.[85] Nanocellulose with a high aspect ratio has more hydroxyl groups exposed to the fiber surface compared to traditional natural fibers. By mixing nanocellulose with polyols, it is possible to manufacture

Figure 6.11   Nanocelluloses made by different methods.

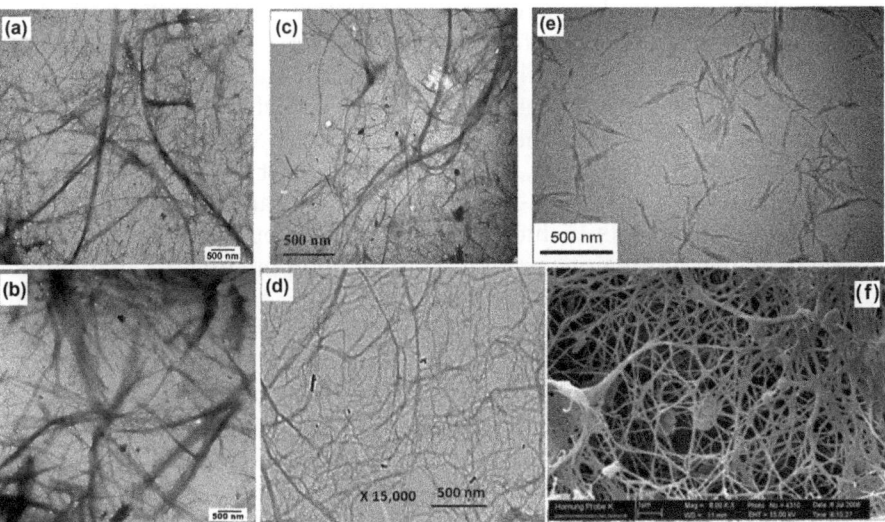

Figure 6.12   TEM images of nanocelluloses made by different methods and from different sources.

new, lightweight and strong PU products which can be utilized in many industrial and consumer applications.

Li and co-workers found that rigid PU foams with 0.75% cellulose whiskers had superior compressive and tensile strengths and compressive moduli compared to neat foams.[86] They also discovered that the presence of cellulose whiskers in PU foams helped with cell dispersion and produced a better cell size distribution. Like wood fibers, nanocellulose was shown to react with PU foams *via* FT-IR results.[86] However, there was a drawback in that freeze-dried nanocellulose must be first dispersed in dimethylformamide (DMF) by sonication, which is difficult to remove even under reduced

Table 6.4 Mechanical improvements of PU composites. Reproduced from ref. 88
with permission © Springer, 2013.

Material	Flexural strength/MPa	Flexural modulus/MPa	Impact strength/J m^{-1}
PU	3.03 (0.62)^b	104 (31)	82.8 (5.7)
PU + 0.125% BC^a	3.81 (0.22)	125 (80)	52.1 (2.4)
PU + 0.250% BC	6.00 (0.21)	135 (11)	48.3 (9.0)
PU + 0.375% BC	5.28 (0.30)	152 (21)	42.1 (3.1)

^{a}BC-Bacterial cellulose.
^{b}Standard Deviations in Parentheses.

pressure. In order to mix dried nanocellulose with polyols well, different
methods have been adapted to modify nanocellulose before employing it in
the foaming process. Though nanocellulose is fluffy after freeze-drying, it is
still hard to mix with polyols uniformly due to its high aspect ratio and
hydrogen bonding. In order to mix them well, hydrophobic nanocelluloses
were prepared using silane or hydrophobic polybutyl acrylate (PBA) latex to
improve the compatibility between nanocellulose and polyols, both in PU
resins and foams.[12,87] Zhang found that PUs with 1.5% silane-treated NCC
had high abrasion resistance and high hardness.[87] In addition, Seydibeyoglu
et al. succeeded in mixing bacterial nanocellulose with polyol using a high-
speed homogenizer, and obtained a PU composite with improved flexural
tensile strength and modulus, but decreased impact strength (Table 6.4).[88]

Currently, 5–30% fiber-glass-reinforced PU foams are used for a number of
automotive components including front and rear bumpers. In addition to
bumpers, which are mainly exterior, interior components can also be functionalized to provide multiple built-in load-bearing applications. One of the
major changes in recent years in the automotive industry has been to
introduce carbon neutral materials into parts manufacturing. Durable and
flexible PU foams are used in seating systems, whilst stiffer and lighter
reinforcing PU rigid foams, in conjunction with green materials, are considered to be among the viable options for components such as bumpers.
Faruk has reported the preparation of green bumpers made from rigid PU
foam with reinforcement by lignin and nanofibers.[12] NCC and nanocellulose, with or without carbon fibers, are considered to be materials for the
future development of lightweight parts. The primary function of these PU
foams will be to reduce the part weight by at least 10% compared to that of
existing foam parts reinforced with fiber glass. The development of PU foam
components reinforced with nanocellulose and hybrid components containing nanocellulose and other bio-mass will take advantage of their
hydroxyl reactivity and bio-degradability. Unlike wood fiber and lignin,
nanocellulose will be capable of carrying higher loads and providing greater
flexibility to automotive manufacturers. One of the leading automotive part
suppliers, Woodbridge Foam Corporation, has pioneered the development
of bio-based PU foams for use in seat cushions and others applications to
reduce their environmental footprint, partnering with Cargill and Ford.[16]

6.4   Conclusions

Generally speaking, both bio-based polyols and the renewable materials used in PU product manufacturing are non-toxic and potentially economically and environmentally viable. The supply of bio-mass continues to increase with the return to a steady increase in global oil demand. Billions tons of bio-mass is produced in the forest and agricultural industries every year. These materials can be converted to bio-based polyols, pyrolysis oils and sustainable fillers for the replacement of petroleum-based polyols and inorganic fillers in PU manufacturing through chemical, biological and liquefaction methods. PU products have reduced weights and result in a reduction in equipment abrasion. They also provide a more user-friendly and safer operational environment by reducing the amount of toxic isocyanate used in foam processing, and avoiding splinter damage from glass-fiber-reinforced parts.

However, nanocellulose has not been extensively used in PU products. In the current markets, much more petroleum based PU products have been substituting with bio-based alternatives without compromising the integrity while improving bio-degradability. However, more efforts still need developing and commercializing bio-based isocyanates to achieve 100% green PU products.

References

1. D. K. Chattopadhyay and K. V. S. N. Raju, *Prog. Polym. Sci.*, 2007, **32**, 352.
2. H. V. Patel, J. P. Raval and P. S. Patel, *Int. J. Chem. Tech. Res.*, 2010, **2**, 532.
3. R. Gu and M. M. Sain, *J. Polym. Environ.*, 2013, **21**, 30.
4. R. Gu, M. Sain and S. Konar, *Ind. Crop. Prod.*, 2013, **42**, 273.
5. R. Gu, S. Konar and M. M. Sain, *J. Am. Oil Chem. Soc.*, 2012, **89**, 2103.
6. R. Gu, M. Khazabi and M. Sain, *BioResources*, 2011, **6**, 3775.
7. M. Khazabi, R. Gu and M. Sain, *BioResources*, 2011, **6**, 3757.
8. O. Loebel, presented at UTECH2012, Maastricht, 2012.
9. M. T. Bomberg and M. K. Kumaran, Use of field-applied polyurethane foams in building, *Construction Technology Updates*, no. 32, Institute for Research In Construction, NRC, 1999.
10. *Green Building Insulation: The Environmental Benefits*, Honeywell, 2008, http://www.greenguard.org/Libraries/GG_Documents/GreenBuildingInsulation TheEnvironmentalBenefits_1.sflb.ashx (accessed 15th August 2014).
11. A. Kruger and C. Seville, *Green Building: Principles and Practices in Residential Construction*, Cengage Learning, 2012.
12. O. Faruk, M. Sain, R. Farnood, Y. Pan and H. Xiao, *J. Polym. Environ.*, 2014, **22**, 279.
13. R. Eller, *Polyurethane in Automotive Interiors: Opportunities and Challenges*, Robert Eller Associates, 2003.
14. R. Stanciu and H. Khalil, presented at the Bio Based Chemical Symposium Mini-Conference, Edmonton, 2010.
15. R. Gu and M. Sain, presented at the 4th International Conference on Polymeric Materials in Automotive Processing, Bratislava, 2011.

16. E. C. Lee, C. M. Flanigan, K. A. Williams, D. F. Mielewski and C. Perry, presented at the Processing Global Plastics Environmental Conference, Atlanta, 2006.
17. C. M. Flanigan, C. Perry and D. F. Mielewski, presented at the Processing Global Plastics Environmental Conference, Orlando, 2008.
18. O. Faruk and M. Sain, *J. Biobased Mater. Bioenergy*, 2013, **7**, 309.
19. *MDI, TDI and Polyurethane Market by Type, Applications, Prices, Regulations Trends & Global Forecasts 2011–2016*, MarketsandMarkets Report, 2011, #CH 1596.
20. V. Mondhe, *Polyols and Polyurethanes Market For Furniture, Construction, Electronics & Appliances, Automotive, Footwear and Packaging - Global Industry Analysis, Size, Share, Growth, Trends and Forecast, 2012–2018*, 2013.
21. *Foamed Plastics (Polyurethane)-A Global Strategic Business Report*, Global Industry Analysis Inc., 2011, #MCP-2069.
22. M. Urgun-Demirtas, D. Singh and K. Pagilla, *Polym. Degrad. Stabil.*, 2007, **92**, 1599.
23. Z. S. Petrović, *Polym. Rev.*, 2008, **48**, 109.
24. V. Sharma and P. P. Kundu, *Prog. Polym. Sci.*, 2008, **33**, 1199.
25. *Life cycle impact of soybean production and soy industrial products*, Omni Tech International prepared for the United Soybean Board, 2010, 20100201 GEN 422. http://www.biodiesel.org/reports/20100201_gen-422.pdf.
26. L. M. Sherman *Plast. Technol.*, 2007, (1), http://www.ptonline.com/articles/polyurethanes-bio-based-materials-capture-attention (accessed 15th August 2014).
27. H. Benes, R. Cerna, A. Durackova and P. Latalova, *J. Polym. Environ.*, 2012, **20**, 175.
28. A. Loredo-Treviño, G. Gutiérrez-Sánchez, R. Rodríguez-Herrera and C. N. Aguilar, *J. Polym. Environ.*, 2012, **20**, 258.
29. H. J. Twitchett, *Chem. Soc. Rev.*, 1974, **3**, 209.
30. A. E. Rheineck and S. Shulman, *Eur. J. Lipid Sci. Technol.*, 1968, **70**, 75.
31. L. Hojabri, X. Kong and S. S. Narine, *Biomacromolecules*, 2009, **10**, 884.
32. G. Çayli and S. Küsefoğlu, *J. App. Polym. Sci.*, 2008, **109**, 2948.
33. M. Ionescu, X. Wan, N. Bilic and Z. S. Petrović, *J. Polym. Environ.*, 2012, **20**, 647.
34. J. D'Souza and N. Yan, *ACS Sustain. Chem. Eng.*, 2013, **1**, 534.
35. Y. Zhao, N. Yan and M. Feng, *J. Appl. Polym. Sci.*, 2012, **123**, 2849.
36. H. Cheradame, M. Detoisien, A. Gandini and F. Pla, *Br. Polym. J.*, 1989, **21**, 269.
37. M. T. Bomberg and M. K. Kumaran, *Use of Field-applied Polyurethane Foams in Building*, National Research Council of Canada, 1999.
38. G. R. Blair, J. I. Reynolds and M. D. Weierstall, *A Review: Automotive Cushioning through the Ages*, Molded Polyurethane Foam Industry Panel, Ontario, 2008, available at http://www.moldedfoam-ip.com/linkedpdf/Technical%20Info%20-%20Automotive%20Cushioning%20Through%20the%20Ages.pdf.

39. A. R. Mahendran, N. Aust and G. Wuzella, *J. Polym. Environ.*, 2012, **20**, 926.

40. M. Zhang, Y. Zhou, X. Yang and L. Hu, *Adv. Mat. Res.*, 2011, **250–253**, 974.

41. Bio-based polyurethanes that work. *Myriant polyol technical bulletin 2013*, http://www.myriant.com/pdf/myriant-polyol-technical-bulletin-2013.pdf (accessed 9th March 2014).

42. Enabling sustainable polyurethanes, Reverdia, The Netherlands, http://www.reverdia.com/wp-content/uploads/Biosuccinium-PU-brochure-PAGES-v16-web.pdf (accessed 9th March 2014).

43. D. K. Chowdhury and B. K. Mukherji, *J. Am. Oil. Chem. Soc*, 1956, **22**, 189.

44. O. S. Ogunfeyitimi, A. O. Okewale and P. K. Igbokwe, *Int. J. Multidiscip. Sci. Eng.*, 2012, **3**, 10.

45. Icynene Inc. LD-R-50 Spray Foam Insulation, www.icynene.com.

46. Cargill BiOH polyol, www.bioh.com.

47. Biobased Technologies, Technical data, www.agrolinside.com/ (accessed 10th March 2014).

48. M. Ionescu, *Chemistry and Technology of Polyols for Polyurethanes*, Rapra Technology Ltd, Shawbury, 2005, p. 13.

49. N. Luo, J. Qian, J. Cupps, Y. Wang, B. Zhou, L. Armbruster and P. Frenkel, *Natural Oil Polyol of High Reactivity for Rigid Polyurethanes*, Biobased Technologies, Arkansas, http://www.agrolinside.com/img/pdf/Natural%20Oil%20Polyol%20of%20High%20Reactivity%20for%20Polyurethanes.pdf (accessed 15th August 2014).

50. R. Gautam, A. S. Bassi, E. K. Yanful and E. Cullen, *Int. Biodeterior. Biodegrad.*, 2007, **60**, 245.

51. S. Oprea and F. Doroftei, *Int. Biodeterior. Biodegrad.*, 2011, **65**, 533.

52. G. Mathur and R. Prasad, *Appl. Biochem. Biotechnol.*, 2012, **167**, 1595.

53. G. T. Cardoso, S. C. Neto and F. Vecchia, *Front. Archit. Res.*, 2012, **1**, 348.

54. C. Hou, K. Czubernat, S. Y. Jin, W. Altenhof, E. Maeva, I. Seviaryna, S. Bandyopadhyay-Ghosh, M. Sain and R. Gu, *Int. J. Fatigue*, 2014, **59**, 76.

55. I. Javni, W. Zhang and Z. D. Petrović, *J. Polym. Environ.*, 2004, **12**, 123.

56. F. Shahidi, *Bailey's Industrial Oil and Fat Products*, John Wiley & Sons, Hoboken, 6th edn, 2005, vol. 1, ch.1, p. 15.

57. A. Guo, I. Javni and Z. Petrović, *J. Appl. Polym. Sci.*, 2000, **77**, 467.

58. N. Nodelman, L. Hall, J. Qian, N. Shackelford and W. Parker, *Polyurethane Rigid Foams from Agrol® Polyols*, Biobased Technologies, Arkansas, http://www.agrolinside.com/img/pdf/BioBased%20Technologies%20PU2011%20Paper.pdf (accessed 15th August 2014).

59. K. H. Badri, Biobased polyurethane from palm kernel oil-based polyol, in *Polyurethane*, ed. F. Zafar and E. Sharmin, *InTech*, 2012, ch. 20, p. 447.

60. R. Datta, Process for the production of succinic acid by anaerobic fermentation, *US Pat.*, 5 143 833, 1992.

61. H. Song and S. Y. Lee, *Enzyme Microb. Technol.*, 2006, **39**, 352.

62. S. H. Chung, M. S. Kim, H. J. Eom and K. Y. Lee, Catalytic hydrogenation of bio-based succinic acid for the production of 1,4-butanediol through

the indirect pathway, presented at the 2013 AiChE Annual Meeting, San Francisco, 2013.

63. B. Ahn, S. H. Kim, Y. H. Kim and J. S. Yang, *J. Appl. Polym. Sci.*, 2001, **82**, 2808.
64. C. Hess, *Bioplast. Mag.*, 2012, **7**, 24.
65. L. Urbanczyk, C. Calberg, C. Detrembleur, C. Jérôme and M. Alexandre, *Polymer*, 2010, **51**, 3520.
66. Y. H. Lee, K. H. Wang, C. B. Park and M. Sain, *J. App. Polym. Sci.*, 2006, **103**, 2129.
67. D. Rodrigue, S. Souici and E. Twite-Kabamba, *J. Vinyl Addit. Technol.*, 2006, **12**, 19.
68. H. Yoshida, R. Morck and K. P. Kringstad, *J. Appl. Polym. Sci.*, 1987, **34**, 1187.
69. H. Yoshida, R. Morck and K. P. Kringstad, *J. App. Polym. Sci.*, 1990, **40**, 1819.
70. A. Reimann, R. Morck, H. Yoshida, H. Hatakeyama and K. P. Kringstad, *J. Appl. Polym. Sci.*, 1990, **41**, 39.
71. X. Pan and D. C. Webster, *ChemSusChem*, 2012, **5**, 419.
72. J. Laine and P. Stenius, *Cellulose*, 1994, **1**, 145.
73. C. A. Cateto, M. F. Barreiro, C. Ottati, M. Lopretti, A. E. Rodrigues and M. N. Belgacem, *J. Cell. Plast.*, 2014, **50**, 81.
74. W. G. Glasser and R. H. Leitheiser, *Polym. Bull.*, 1984, **12**, 1.
75. W. Xing, H. Yuan, P. Zhang, H. Yang, L. Song and Y. Hu, *J. Polym. Res.*, 2013, **20**, 234.
76. C. Reti, M. Casetta, S. Duquesne, S. Bourbigot and R. Delobel, *Polym. Adv. Technol.*, 2008, **19**, 628.
77. M. Lewin, *J. Fire Sci.*, 1997, **15**, 29.
78. S. Janardhnan and M. Sain, *J. Polym. Environ.*, 2011, **19**, 615.
79. A. Alemdar and M. Sain, *Bioresour. Technol.*, 2008, **99**, 1664.
80. B. Wang, M. Sain and K. Oksman, *Appl. Compos. Mater.*, 2007, **14**, 89.
81. R. Gu, B. V. Kokta, K. Frankenfeld and K. Schlufter, *BioResources*, 2010, **5**, 2195.
82. S. Elazzouzi-Hafraoui, Y. Nishiyama, J.-L. Putaux, L. Heux, F. Dubreuil and C. Rochas, *Biomacromolecules*, 2008, **9**, 57.
83. B. Huang, L. Tang, D. Dai, W. Ou, T. Li and X. Chen, Preparation of Nanocellulose with Cation–Exchange Resin Catalysed Hydrolysis, in *Biomaterials Science and Engineering*, ed. R. Pignatello, InTech, 2011, ch. 6.
84. H. Dong, K. E. Strawhecker, J. F. Snyder, J. A. Orlicki, R. S. Reiner and A. W. Rudie, *Carbohydr. Polym.*, 2012, **87**, 2488.
85. ArboraNano Research highlights, www.arboranano.ca.
86. Y. Li, H. Ren and A. J. Ragauskas, *Nano-Micro Lett.*, 2010, **2**, 89.
87. H. Zhang, Y. She, S. Song, H. Chen and J. Pu, *BioResources*, 2012, **7**, 5190.
88. M. O. Seydibeyoglu, M. Misra, A. Mohanty, J. J. Blaker, K.-Y. Lee, A. Bismarck and M. Kazemizadeh, *J. Mater. Sci.*, 2013, **48**, 2167.

CHAPTER 7

Production of Low-cost Polyesters by Microwaving Heating of Carboxylic Acids and Polyol Blends

BRENT TISSERAT*[a] AND ZENGSHE LIU[b]

[a] USDA, ARS, National Center for Agricultural Utilization Research, Functional Foods Research Unit, 1815 N. University Street, Peoria, IL 61604, USA[†]; [b] USDA, ARS, National Center for Agricultural Utilization Research, Bio-Oils Research Unit, 1815 N. University Street, Peoria, IL 61604, USA[†]
*Email: brent.tisserat@ars.usda.gov

7.1 Status of the Synthetic and Bio-plastic Industries

7.1.1 Scope of the Current Synthetic Plastic Industry

Plastics are an important part of our society. Their versatility allows them to be employed in a myriad of products.[1] The mass production of plastics derived from petroleum feedstocks commenced in the 1940s.[2] In 1941, fewer than a million tons of plastic were produced. Today, over 260 million tons of plastic are used globally per annum and account for 8% of the World's oil

---

[†]Mention of trade names or commercial products in this publication is solely for the purpose of providing specific information and does not imply recommendation or endorsement by the U.S. Department of Agriculture. USDA is an equal opportunity provider and employer.

RSC Green Chemistry No. 29
Green Materials from Plant Oils
Edited by Zengshe Liu and George Kraus
© The Royal Society of Chemistry 2015
Published by the Royal Society of Chemistry, www.rsc.org

production.[2] The plastics industry revenue has grown an average of 3.5% annually over the last 30 years. The five largest plastics produced are poly-olefins [polypropylene (PP) and polyethylene (PE)], polyvinyl chloride (PVC), polystyrene (PS), expanded polystyrene (EPS) or Styrofoam, and polyethylene terephthalate (PET), which account for 70% of the total global demand, *i.e.*, 200 million tons.[1,3] The USA plastics industry ranks third in revenue for all USA industries, representing close to US$375 billion in yearly shipments, and employs over 1 million people.

The environmental problems created by plastics are enormous.[4] Over 10% of waste generation is plastic, although some plastics are recycled, much is disposed of in landfill. Plastic debris pollution has now accumulated in all natural habitats. Plastics represent 60 to 80% of the total marine debris, generated by improper waste disposal.[5] Most marine debris is generated from land-based pollution, *i.e.*, cigarettes/cigarette filters, food wrappers/containers, (plastic) bags and beverage bottles. Ninety-six percent of the plastic found in the North pacific consists of small pieces of plastic derived from manufactured plastic products and pre-production plastic pellets (*i.e.*, industrial pellets, virgin pellets, plastic resin beads, or nurdles).[6] Aside from the mega- and macroplastic pollution problems there is evidence that plastics are fragmenting in the environment to become minute indigestible ingredients consumed by an ever wider range of organisms.[4,6] To enhance the performance of plastics, a wide range of chemicals (plasticizers, flame retardants, stabilizers, anti-oxidants, anti-microbials *etc.*) are added during manufacturing. These chemicals are often toxic and may cause health problems to a wide range of wildlife.[4]

However, plastics do have some environmental advantages over other materials: (1) less energy consumption is used in their manufacturing (it takes 30% less energy to make polystyrene than paperboard containers); (2) plastics are lighter than other materials thus require less fuel to transport them; and (3) many thermoplastics can be recycled.[1]

7.1.2   Scope of the Current Bio-plastic Industry

Bio-plastics can be defined as "plastics that contain bio-based content, are bio-degradable, or both".[7] Given this broad definition, inclusion of any or-ganic compounds in a plastic (such as wood-plastic composites) or deriving a plastic from an organic feed source qualifies the product to be termed a "bio-plastic". Bio-plastics have the potential to reduce the long-term pol-lution problems currently associated with petroleum-based plastics.[8] Much interest exists in replacing petroleum-based (or petro-based) plastics with bio-plastics.[8–12] Bio-plastic and bio-polymer usage is hindered by their price, performance and availability.[7,10] Currently, less than 1% of total global plastic usage is bio-plastics. Nevertheless, bio-plastic production is steadily increasing worldwide. For example, the current USA bio-plastic market, worth US$490 million in 2010 and 26.7% of the global total, and is expected to increase to 32.9% of the global total by 2015. The bio-plastic market is evolving, currently the top five bio-plastics in 2010 based on production were: bio-based polyethylene (bio-PE), bio-degradable starch blends, poly-lactic acid (PLA), polyhydroxyalkanoate (PHA), and bio-degradable poly-esters.[7] By 2015, starch blends will be replaced by bio-based polyethylene terephthalate (bio-PET). Unfortunately, both bio-PE and bio-PET are not bio-degradable or compostable. Factors that hinder the growth of the bio-plastic industry include: confusing terminology (*i.e.*, not all bio-based plastics are bio-degradable); lack of infrastructure to develop or reprocess bio-plastics (*i.e.*, slow acceptance for food waste diversion programs and lack of adequate composting and industrial bio-degradation infrastructure); lack of funding to develop new bio-plastics (*i.e.*, few public offerings as well as few govern-mental grants); and finally and most importantly, the limited availability of bio-based feedstocks to manufacture bio-plastics (*i.e.*, the bio-based chem-ical supplies are limited and their costs are high).[7]

Nevertheless, bio-plastics are expected to increase in use by 40% from 2010 to 2015.[3] Factors promoting the use of bio-plastics include: the un-certain and fluctuating cost of petroleum prices; regulatory and legislative actions aimed at the reduction of plastic waste in order to minimize its detrimental environmental effects; and the productive utilization of bypro-ducts generated from the food and agricultural processing industries as feedstock materials.[7,13] Undoubtedly, the most significant reason for re-placement of petroleum-based plastics with bio-based plastics is the in-creasing evidence that petroleum plastics may be responsible for a myriad of health and environmental problems.[7] Consumer acknowledgement of the environmental problems associated with petroleum-based plastics was clearly recognized in the 2011 Cone/Echo Global Corporate Responsibility Survey where 94% of the respondents indicated that they buy "green" products over "non-green" products.[7]

7.2   Alternative Bio-plastics and their Potential Uses

7.2.1   Epoxidized Vegetable Oil Polymers

Vegetable oils that have high contents of unsaturated fatty acids can be converted into epoxy fatty acids by conventional epoxidation, catalytic acidic ion–exchange resins, catalyst epoxidation, or by using chemo-enzymatic epoxidation.[14–17] Epoxidized vegetable oils (EVOs) can be used as a raw material for the synthesis of several chemicals including alcohols (polyols), glycols, olefinic compounds, lubricants, plasticizers, and stabilizers for polymers.[14] The most common EVOs are derived from oil palms and soy-bean due to their high cultivation acreage. EVOs may also be potential re-placements for petroleum-based epoxy, polyester and vinyl esters since they are obtained from renewable feedstocks, pose fewer environmental health risks in their utilization, and are bio-degradable. The use of EVOs, especially epoxidized soybean oil (ESBO), in the production of stable polyesters has been well documented over the last 25 years.[14–21]

7.2.2 Carboxylic Acid Glycol Polyesters

The production of diacid glycol polyesters employed in food products has been well recognized since the 1970s.[22,23] Bade[22] generated citric acid esters from citric acid and partial fatty acids to stabilize fatty emulsions and act as synergistic agents for anti-oxidants. Since the 1980s, several studies have been conducted to obtain polyester-type plastics from polyfunctional acids and glycols (*e.g.*, dicarboxylic or tricarboxylic acid and glycerol) through polycondensation.[24–45] These polyesters were generated using various procedures and methods but primarily heating is required to promote esterification, usually followed by a subsequent lower heating process in order to promote curing and polymerization.[27,28,30–33,36,37,39–46] Glycerol-based polymers have been reported to have potential bio-medical applications because they are benign and bio-degradable.[24] Citric acid glycerol polyesters or citrate glycerides, see Table 7.1, exhibited high rates of degradation in several different environments. Interest in the development of bio-degradable polyesters for drug delivery or temporary surgical prosthesis has been expressed.[25,34–36]

Bio-degradable polymers have uses as bio-medical materials. Absorption of degradable scaffolds by host tissues may overcome the long-term bio-compatibility problems associated with persistent implants, as well as eliminate the high cost of patient morbidity associated with follow-up surgery.[35,47] These new polymers are synthesized by simple catalyst-free polyesterification processes using glycol and diol monomers.[35] These polymers can be further exploited for bio-functionality by attaching biomolecules such as proteins. Such materials will have applications in tissue engineering.[35] However, to date no commercial utilization of a diacid glycol polymer has

Table 7.1 Response of citrate glyceride polyesters to different solvent environments. Percentage weight remaining from polyesters of 1:1 molar ratio citric acid to glycerol in various solvents after a 10-day treatment.

Treatment	Foam[a]/%	Powder[b]/%
Water	85	75
0.1 M NaOH	74	72
0.5 M NaOH	12	19
1.0 M NaOH	4	6
0.1 M HCl	75	75
0.5 M HCl	63	69
1.0 M HCl	57	62
Chloroform	102	94
Dichloromethane	94	82
Ethanol	96	78
Methanol	94	80
Dimethyl sulfoxide	112	131

[a]Foam sample consists of 150 mg, microwave construction.[28]
[b]Powder is 150 mg obtained by grinding the foam with a Wiley mill through a 1 mm screen.

occurred. Nevertheless, the low cost of this polyester and its rapid bio-degradation properties have promoted continued interest in its development.[27–28,33,46] There are certain problems associated with these polyesters: they are highly hydrophilic (water sensitivity) and exhibit high brittleness. Both of these conditions have limited their use.

7.3   Ingredients and Cost of Bio-plastics

7.3.1   The Cost of Current Bio-plastics

Polylactic acid is the one of the most popular bio-plastics in today's market, being manufactured from fermented plant starch (usually corn).[10,48] PLA is often considered an alternative to petroleum-based plastics but it has some major disadvantages such as poor mechanical performance, especially at high temperatures and in wet environments, poor bio-degradability and a relatively high price *versus* petroleum-based polyolefins.[48] In order to improve the mechanical and thermal properties of PLA, commercial additives are included with PLA during processing.[10] PLA could be employed as a substitute for PS.[49] The price of PLA has declined considerably over the years from US$2.00 per lb in 2005[48] to around US$1.00 per lb in 2007.[10] The price of other "green" thermoplastics is considerably more than petroleum-based plastics. For comparative purposes the current costs of petroleum-based plastics are shown in Table 7.2.

For example, bio-based PE is about 50% more expensive than petroleum PE (US$1.50–1.80 per lb *versus* US$1.04–1.23 per lb).[50] Similarly, bio-based

Table 7.2   Prices of petroleum-based plastics.

Plastic	Cost/US per lb[a]
Thermoplastic:	
High Density Polyethylene (HDPE)	1.04–1.23
Low Density Polyethylene (LDPE)	1.15–1.14
Linear Low Density Polyethylene (LLDPE)	0.97–1.28
Polystyrene (PS)	1.27–1.77
Polyvinyl Chloride (PVC)	0.94–1.12
Polypropylene (PP)	1.15–1.60
Acylonitrile Butadiene Styrene (ABS)	1.15–1.70
Acrylic	1.25–1.30
Polyethylene terephthalate (PET)	97–1.12
Thermoset:	
Epoxy	1.00–2.97
Melamine molding compound	1.35–1.75
Phenolic	0.75–0.85
Polyester unsaturated	1.71–2.65
Polyurethane isocyanates	1.05–1.20
Urea molding compound	0.95–1.10
Vinyl ester	2.04–2.37

[a]Prices presented were obtained from a current plastic resin price chart accessed 02/02/14.[54]

PET is 30–40% more expensive than petroleum PET (US$1.50–1.80 per lb *versus* US$1.04–1.23 per lb).[51] Despite the higher costs of bio-based plastics, there is a driving force in the packaging industry to replace petroleum-based plastics with a renewable source to achieve a "green" rating approval.[49,51,52]

7.3.2   Cost and Sources of Citrate Glycerides

The cost of the "green" feedstock ingredients is often the deciding factor in the commercial viability of a bio-plastic.[49,51,53,54] If the cost of the feedstock chemicals to manufacture a bio-plastic exceeds the cost of purchasing the commercial plastic polymer it is to replace, it is essentially economically infeasible to utilize this bio-plastic. Our line of interest is to explore the possibility of utilizing low-cost and "green" feedstocks to manufacture biodegradable bio-plastics. To this end the citric acid glycerol polyesters satisfy these two parameters. The cost of carboxylic acids and polyols, see Table 7.3, varies considerably.

Clearly, citric acid and glycerol are considerably less expensive than other polyfunctional acids or epoxidized oils (Table 7.3). In fact, citric acid glycerol polyesters (citrate glycerides) are probably the least expensive of all natural polyester polymers (Table 7.3). In terms of their ingredient costs, a high-purity citric acid source costs around US$0.35–0.86 per lb and high purity glycerol source costs US$0.30–0.43 per lb. Therefore, an equal molar polyester of these two compounds would cost ~US$0.33 per lb excluding preparation costs. The ingredient costs for the preparation of other polyfunctional acid glycol polyesters are so prohibitively expensive they cannot compete in price with current synthetic or natural thermoplastic polymers. This study is confined to the exclusive discussion of these citrate glycerol polyesters.

Glycerol is relatively non-toxic, as recognized by the Food and Drug Administration and is employed in a multitude of food, pharmaceutical, and industrial products.[55] Glycerol production has mushroomed from 200 000 MT in 2004 to 2.8 million MT in 2011 and is projected to increase to 3.0 million MT in 2014.[56,57] About two-thirds of all glycerol is produced by the bio-diesel industry,[56] as a byproduct.[57] Bio-diesel production consumes 10% of the 200 million MT of natural oil and fats globally produced.

Table 7.3   Cost of ingredients used to manufacture carboxylic acid glycols.

Ingredient	Cost/US$ per lb	Purity	Source of quote[a]
Citric acid	0.35–0.86	>99%	ref. 72
Adipic acid	0.77–0.82	99%	ref. 72
Azelaic acid	8.18–22.00	>99%	ref. 72
Suberic acid	20.00–45.00	>99%	ref. 72
Sebacic acid	1.82–2.50	>99%	ref. 72
Glycerol	0.30–0.43	95%	ref. 72
ESBO	1.50–1.62	>99%	ref. 73

[a]Prices presented were accessed on 02/02/14 and sold in ton or more quantities, free on board.

Bio-diesel production generates about 10% (w/w) glycerol as its main by-product. For every gallon of bio-diesel produced, approximately 1.05 pounds of glycerol is generated. By 2016, 37 billion gallons of bio-diesel will be produced which translates into the production of 4 billion gallons (38.8 billion pounds) of crude glycerol.[57] Refined glycerol sells for about US$0.30 per lb, compared to the US$0.70 it sold for prior to the expansion of the USA bio-diesel market. The price of crude glycerol is about US$0.05 per lb.[24,57] Crude glycerol is currently being used as a feedstock for production of other value-added chemicals and animal feeds.[57,58] Utilization of crude glycerol to produce citrate glycerides has been demonstrated and will go further towards decreasing the cost of manufacturing this polyester.[24]

Citric acid has wide applications in numerous fields and can be produced from a variety of low-cost sugar substrates by fermentation.[59–61] Citric acid is inexpensive, bio-degradable and non-toxic.[62] Citric acid is one of most widely produced organic chemical commodities. Global production of citric acid was 1.5 million tons in 2005 and steadily increased to 1.8 million tons in 2010.[63] Because of its ease of manufacturing by fermentation it is considerably less expensive than other tri- or diacids which are more complex to manufacture and employ higher priced ingredients (Table 7.3). Clearly, coupling these two low-cost ingredients to produce low-priced polyesters is a way of competing with current bio- and petroleum-based plastics.

7.4 Preparation of Carboxylic acid Glycerol Polyesters

Several novel carboxylic acid glycerol polyesters have been produced with unique thermal and mechanical properties.[24,26,35,41,43,44,64] Generally, the procedure followed to produced polyesters, see Scheme 7.1, involves mixing the carboxylic acid and glycerol in various monomeric ratios (usually 1:1 M), and heating close to the melting point of the carboxylic acid until the polycondensation reaction has occurred sufficiently to allow for

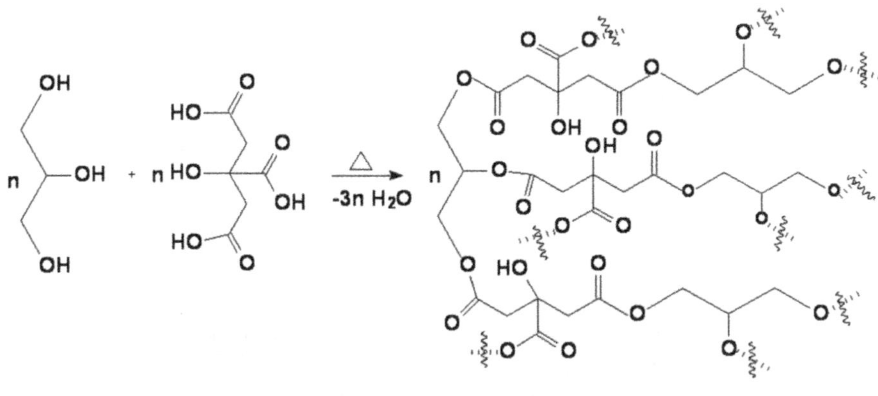

Scheme 7.1   Esterification of glycerol and citric acid to produce citrate glycerides. The citrate glyceride structure presented is theoretical.[27]

esterification.[24,26–33,35,39–41,46,64] Following esterification, a thermal curing is administered to eliminate moisture, induce solidification and complete the esterification process.[24,26–33,35,39–41,46,64]

7.5   Mechanical Properties of Carboxylic acid Glycerol Polyesters

For the purposes of this study, polyesters consisting of only two ingredients were studied in detail, citric acid and glycerol. Tensile test bars were prepared from equimolar amounts of glycerol and citric acid by microwaving heating, curing for up to 72 h at 100 °C, grinding the polyester and then compression molding. Microwave heating was conducted in silicone cupcake muffin molds (70 mm diameter×30 mm height; 80 mm³ capacity) containing 12 g of ingredients to achieve a temperature of ∼210 °C requiring around 1 min at 1200 Watts. Larger vessels containing larger amounts of ingredients can be readily employed but will require longer microwaving times to achieve the same results. During this heating process, esterification was apparent by bubbling and the release of steam from the mixture and the resultant production of a 'muffin' like foam, see Figure 7.1.

Approximately 35–40% of the initial weight of the materials was lost in this esterification process.[46] The final solidified product conformed to the shape of the pan employed and consisted of numerous interlocking bubbles. To complete the esterification process and remove excess moisture, foams were cured in a laboratory oven at 100 °C for 0, 6, 24, 48 or 72 h.[46] Cured citric acid glycerol (CAG) polyesters were ground in a Wiley mill employing a 1 mm screen and compression molded in accordance with American Society

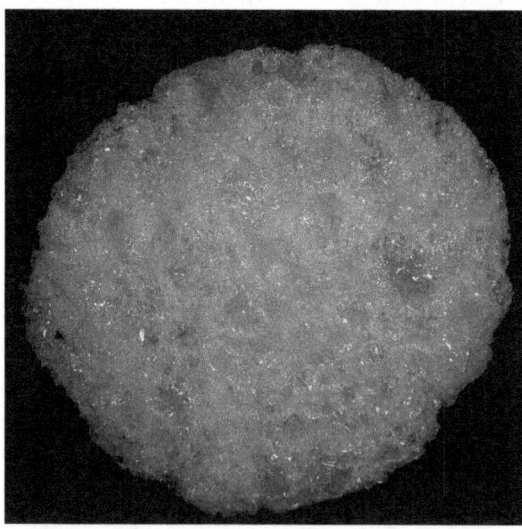

Figure 7.1   Example of a microwaved citrate glycerol polyester.

for Testing and Materials (ASTM) D-638-V standards. Sample powders of 1.5 g were applied to tensile bar wells in a stainless steel mold and heated in a hydraulic press (Model C, Carver Inc., Wabash, IN) at 125–150 °C under a pressure of 89.6 MPa per mold for 20 min. Tensile bars were evaluated using an Instron Universal Testing Machine (Model 4201, Instron Corporation, Norwood, MA) using a cross-head speed of 10 mm min^{-1}, a gauge length of 7.62 mm and a 1 kg load cell.

The mechanical properties [tensile strength (σ_u), Young's modulus (E), and elongation-at-break values (%El)] of the CAGs were affected by the curing process, see Figure 7.2. Additional thermal curing generally resulted in CAGs with significantly higher σ_u and %El values compared to the un-cured CAG samples. However the E values were unaffected by curing, which reflects the high degree of brittleness of the polyesters. Evidently, the uses of additional curing caused additional polymerization through cross-linking and esterification, as evidenced by the improved mechanical properties. It is difficult to compare these mechanical results with the mechanical results of other investigators working with carboxylic acid glycol polyesters because of the differences in ingredients, preparation and heating procedures em-ployed. However, the tensile strength and stiffness properties of the CAG bars are several times greater than those reported in the literature for other carboxylic acid glycol polyesters. However, the elongation-at-break values were much inferior to other carboxylic acid glycol polyesters studies. This is to be expected since high stiffness is at odds with high elongation. Furthermore, the CAG tensile bars exhibited mechanical properties that are comparable with thermoplastic resins, see Table 7.4.

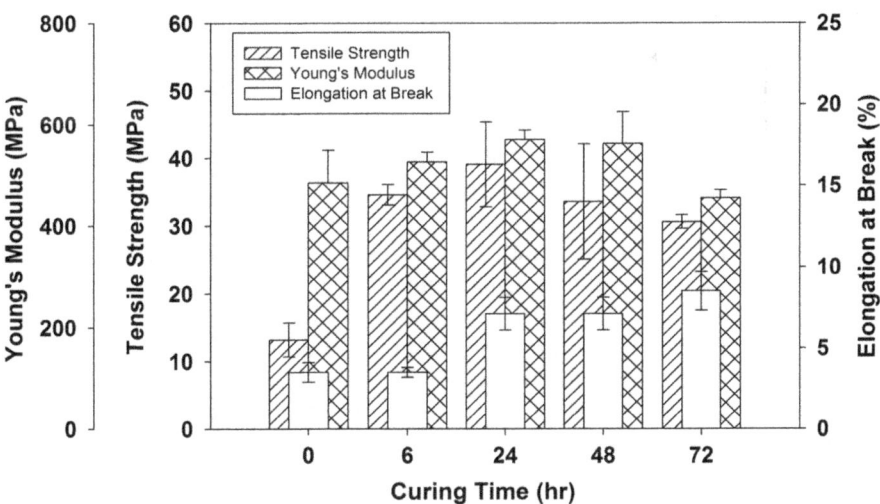

Figure 7.2   Effect of curing duration on the mechanical properties of citrate glycer-ides. Means and standard errors are presented.[46]

Table 7.4 Mechanical properties of various plastics.

Plastic type	σ_u/MPa	E/MPa	%El	Reference
Thermoplastic:				
PP	27	1300	200–700	74
PS	48	3400	3	28
HDPE-extrusion	16.5	155	220	29
HDPE-injection molding	21	340	100	74
PE	20–30	700	20–100	74
PVC	48	3400	200	74
Thermoset:				
Urea Formaldehyde, cellulose filled	38–90	7000–10000	1	74
Epoxy resin, glass filler	68–2000	20 000	4	74
Phenol formaldehyde, mica filled	38–50	17 000–35 000	0.5	74
Biodegradable:				
PLA	60–70	500–570	9–15	29

7.6 Poly(lactic acid)/Citrate Glyceride Blends

PLA can be mixed with various fillers and ingredients to obtain composite blends with novel mechanical or bio-degradable properties, or to reduce the cost.[27,46,62] Blending PLA with thermoplastic starch has become widespread in the bio-plastic community during the last 10 years.[13,42,62,65,66] Citric acid esters in PLA blends serve as plasticizers.[67] Since the cost of the CAG is considerably less than PLA, we conducted a study to evaluate the mechanical properties of PLA-CAG composites. A 1 : 1 molar ratio of citric acid to glycerol polyester was prepared by microwaving heating as previous described.[27,28,46] The CAG polyester was then further cured for 48 h at 100 °C and ground into a powder with a Wiley mill employing a 1 mm screen. Ground CAG powder was mixed with PLA at various concentrations (10–70%) and extruded through with a single-screw extruder (C. W. Brabender, South Hackensack, NJ) at four temperatures (150, 170, 170 and 150 °C) from the feed throat to the die. A 3 : 1 high shear mixing zone screw was employed with a hangar-type die (254 mm×1.5 mm opening) at 150 °C. Ribbons were ground in a Wiley mill through a 3 mm screen and reintroduced into the extruder to facilitate homogenous blending of the CAG polyester in the PLA matrix. PLA-CAG mixtures containing 60 and 70% CAG could not be extruded as ribbons with this die, and had to be removed in order to obtain these blended concentrations which were subsequently ground. Ribbons of PLA-CAG blends containing 10, 20, 30, 40 and 50% CAG (*i.e.*, PLA-10CAG, PLA-20CAG, PLA-30-CAG, PLA-40CAG, and PLA-50CAG) were stamped into ASTM D638-99 Type V tensile bars (60 mm×10 mm). Blends of PLA-CAG containing 60 and 70% CAG (*i.e.*, PLA-60CAG and PLA-70CAG) were compression molded in a Model C Carver hydraulic press at 180 °C under a pressure of 90 MPa per mold for 20 min. The mechanical values of all tensile bars were evaluated using the Instron Model 1122 universal testing machine at testing speed of 5 mm min^{-1}.

Table 7.5   Effect of citrate glyceride (CAG) concentration on the mechanical properties of PLA-CAG composites compared to neat PLA.

Treatment[a]	σ_u/MPa	E/MPa	%El
PLA	60 ± 1.4	607 ± 20	13 ± 0.4
PLA-10CAG	39 ± 0.5	523 ± 35	12 ± 0.5
PLA-20CAG	34 ± 1.3	514 ± 15	9 ± 0.4
PLA-30CAG	22 ± 0.8	445 ± 11	7 ± 0.2
PLA-40CAG	13 ± 0.4	326 ± 6.5	5 ± 0.1
PLA-50CAG	9 ± 0.3	251 ± 3.8	4 ± 0.2
PLA-60CAG[b]	6 ± 0.6	194 ± 12	4 ± 0.2
PLA-70CAG[b]	3 ± 0.3	110 ± 8	4 ± 0.3

[a]Tensile bars were obtained from single-screw extrusions and stamped.
[b]Blends that were extruded and then compression molded.

The mechanical properties (*i.e.*, σ_u, E and %El values) of PLA-CAG composites were significantly different to those of neat PLA, see Table 7.5. PLA is a linear aliphatic polyester thermoplastic derived from lactic acid monomers. CAG is a complex and highly cross-linked polyester thermoset derived from citric acid and glycerol. The presence of the ester linkages in these two plastics allows for hydrolytic degradation. Both CAG and PLA are hydrophilic in nature, however PLA is considerably more hydrophobic compared to CAG.[27] Therefore the PLA-CAG composite has relatively weak interfacial bonds between these two dissimilar ingredients. This in turn is reflected in the mechanical properties of the composite blends, see Table 7.5. Tensile bars of PLA-CAG showed significantly lower σ_u, E and %El values compared to tensile bars composed of neat PLA. Increasing the concentration of CAG in the PLA-CAG composites resulted in progressively lower mechanical properties compared to PLA-CAG composites containing lower or no CAG. For example, PLA-10CAG and PLA-50CAG blends exhibited reductions of σ_u, E and %El of -35, -14, and -8% and -85, -60, -70%, compared to neat PLA, respectively. These results differ from other PLA composite studies where employment of a lignocellulosic fillers such as wood flour typically increases the stiffness (E) while σ_u and %El values only slightly decline.[28] It is obvious that PLA and CAG are incompatible as evidenced by the mechanical tests, see Table 7.5. One positive interpretation on the reduction of mechanical properties obtained by the blending of CAG with PLA is that co-polymerized PLA-CAG materials will degrade much more rapidly than neat PLA. PLA is noted for its slow degradation rate.[68,69]

In a subsequent study, PLA-CAG blends containing 10, 25 and 35% CAG were prepared by extrusion as previous described, stamped into tensile bars and then exposed to 30% (dry/laboratory room conditions) or 95% humidity (wet conditions) for 528 h prior to the mechanical testing. This test was conducted to determine how exposure to moisture influences the mechanical properties of the blends. Weights were periodically measured over the incubation period but no weight equilibration time was achieved indicating that the samples would continue to take up more water if allowed to incubate further. At the end of the incubation period, the average percentage weight gain for PLA, PLA-10CAG, PLA-25CAG and PLA-35CAG was 2.2, 2.6, 6.3 and 6.8%, respectively. Higher average percentage weight increases occurred for PLA-CAG composites containing higher concentrations of CAG. Clearly, CAG is more hydrophilic than the PLA. Apparently, more moisture enters the matrix created by those composites containing higher CAG concentrations. The mechanical properties of PLA, see Figure 7.3, were altered by the moisture treatment.

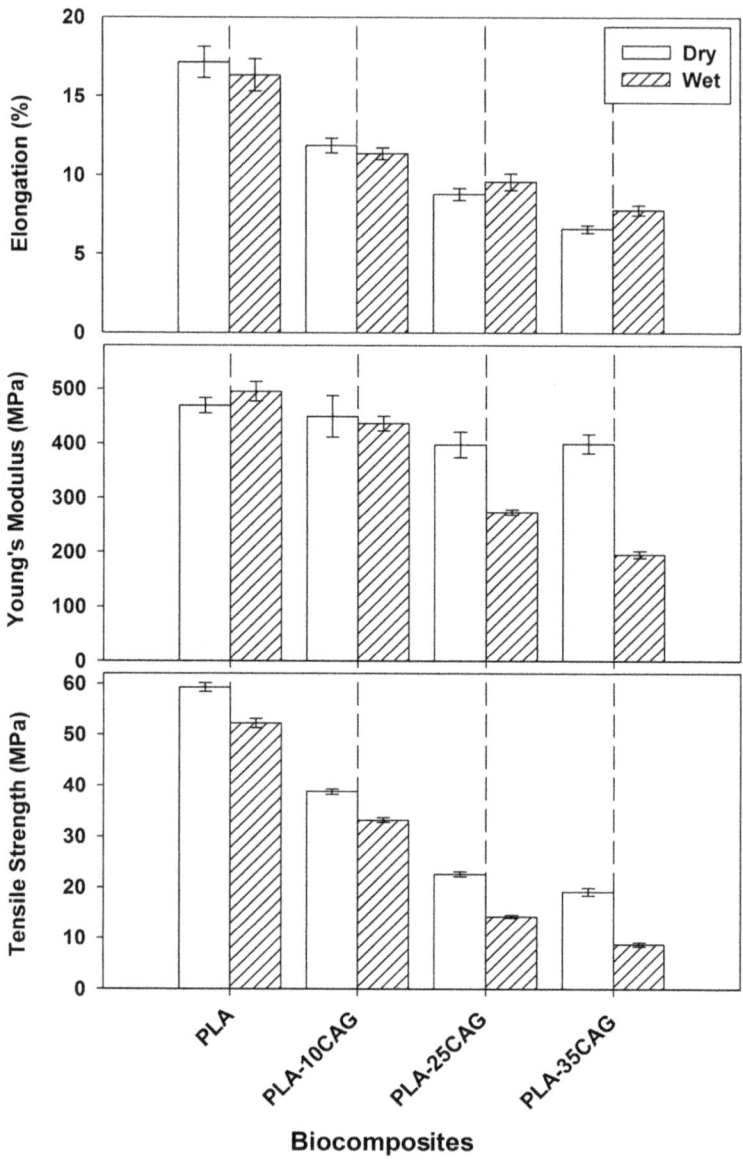

Figure 7.3 Mechanical properties of PLA-CAG composites subjected to dry and wet conditions. Means and standard errors are presented.

For example, the σ_u, E and %El values of wet- or high-humidity-tested PLA were -12, $+6$ and -5%, respectively, compared to dry- or low-humidity-tested PLA. Similarly, the mechanical properties of PLA-CAG composites exposed to wet conditions may differ from composites exposed to dry conditions, see Figure 7.3. This was especially notable for the σ_u and E values of PLA-25CAG and PLA-35CAG blends. For example, the change in percentage in σ_u, E and %El values for PLA-10CAG dry and wet polymers was -14, -3, and -4%, respectively. While the change in percentage in σ_u, E and %El values of PLA-35CAG dry and wet polymers was -54, -50, and $+18\%$, respectively. Significant correlations were found when comparing the percentage CAG composition in the blend with the percentage weight gain and change in mechanical properties. We can attribute the alteration in mechanical properties to moisture uptake due to the concentration of the CAG employed in the blend. These results suggest that increased bio-degradability is conferred to PLA by blending with CAG.

7.7 Conclusions

Citrate glycerol polyesters may have relatively high mechanical properties but these properties are temporary and can be altered by environmental treatment. These polyesters are non-toxic and could be employed in short-term applications where rapid bio-degradability is necessary. One apparent application is bio-medical devices where other polyester materials such as PLA and poly(glycolic acid) are currently employed.[70,71]

References

1. http://dwb4.unl.edu/Chem/CHEM869E/CHEM869ELinks/www. plasticsresource.com/plastics_101/uses/uses.html.
2. R. C. Thompson, S. H. Swan, C. J. Moore and vom F. S. Saal, *Philos. Trans. R. Soc., B*, 2009, **364**, 1973.
3. http://www.plasticseurope.org/information-centre/press-releases/press-releases-2012/first-estimates-suggest-around-4-increase-in-plastics-global-production-from-2010.aspx.
4. D. K. A. Barnes, F. Galgani, R. C. Thompson and M. Barlaz, *Philos. Trans. R. Soc., B*, 2009, **364**, 1985.
5. *Marine Debris in the North Pacific: A Summary of Existing Information and Identification of Data Gaps*, US Environmental Protection Agency, 2011.
6. http://en.wikipedia.org/wiki/Great_Pacific_Garbage_Patch.
7. http://www.plasticsindustry. org/files/about/BPC/Industry%20Overview% 20Guide%20Executive%20Summary%20-%200912%20-%20Final.pdf.
8. http://science.howstuffworks.com/environmental/green-science/future-of-bioplastics. htm.
9. http://www.bccresearch.com/market-research/plastics/bioplastics-technologies-markets-pls050a.html.

10. http://www.ptonline.com/articles/biopolymers-strive-to-meet-price-performance-challenge.
11. http://www.hkc22.com/bioplastics.html.
12. https://www.chem.umn.edu/csp/pdfs/CWJune2012%20article.pdf.
13. Z. Ma, P. R. Chang, J. Yu and M. Stumborg, *Carbohydr. Polym.*, 2009, **75**, 1.
14. T. Saurabh, M. Patnaik, S. L. Bhagt and V. C. Renge, *Int. J. Adv. Eng. Technol.*, 2011, **2**, 491.
15. S. Sullalti, New Eco-friendly Polyesters from Renewable Resources, PhD thesis, Università di Bologna, 2012.
16. P. Saithai, J. Lecomte, E. Dubreucq and V. Tanrattanakul, *eXPRESS Polym. Lett.*, 2013, **7**, 910.
17. S. G. Tan and W. S. Chow, *Polym.-Plast. Technol. Eng.*, 2010, **49**, 1581.
18. H. Deka and N. Karak, *J. Appl. Polym. Sci.*, 2010, **116**, 106.
19. Z. S. Liu, S. Z. Erhan and P. D. Calvert, *J. Appl. Polym. Sci.*, 2004, **93**, 356.
20. Z. Liu and S. Z. Erhan, *Mater. Sci. Eng., A.*, 2008, **483–484**, 708.
21. Y. Quin, J. R. Jia, L. Zhao, Z. X. Huang, S. W. Shao, G. W. Zhang and B. F. Dai, *Adv. Mater. Res.*, 2011, **393–395**, 349.
22. V. Bade, *US Pat.*, 4 071 544, 1978.
23. A. G. Pereira, K. F. Gallagher, P. G. Abend and J. C. Carson Jr., *US Pat.*, 5 597 555, 1997.
24. N. Budhavaram and J. Barone, presented at the 2008 ASABE Annual International Meeting, Rhode Island, 2008.
25. K. M. Doll, R. L. Shogren, J. L. Willett and G. Swift, *J. Polym. Sci., Part A: Polym. Chem.*, 2006, **44**, 4259.
26. D. Gyawali, R. T. Tran, K. J. Guleserian, L. Tang and J. Yang, *J. Biomater. Sci., Polym. Ed.*, 2010, **21**, 1761.
27. B. Tisserat, R. H. O'kuru, H.-S. Hwang, A. A. Mohamed and R. Holser, *J. Appl. Polym. Sci.*, 2012, **125**, 3429.
28. B. Tisserat, R. A. Holser and R. H. O'kuru, *US Pat.*, 8 524 855, 2013.
29. B. Tisserat, N. Joshee, A. K. Mahapatra, G. W. Selling and V. L. Finkenstadt, *Ind. Crops Prod.*, 2013, **44**, 88.
30. R. A. Holser, *J. Appl. Polym. Sci.*, 2008, **110**, 1498.
31. R. Holser, *J. Nat. Fibers*, 2009, **6**, 272.
32. R. A. Holser, *Mater. Chem. Phys.*, 2011, **128**, 10.
33. R. A. Holser, J. L. Willett and S. F. Vaughn, *J. Biobased Mater. Bioenergy*, 2008, **2**, 1.
34. A. S. Kulshrestha, W. Gao and R. A. Gross, *Macromolecules*, 2005, **38**, 3193.
35. I. Djordjevic, N. R. Choudhury, N. K. Dutta and S. Kumar, *Polymer*, 2009, **50**, 1682.
36. N. R. Luman, T. Kim and M. W. Grinstaff, *Pure Appl. Chem.*, 2004, **76**, 1375.
37. D. Himmelsbach and R. A. Holser, *Vib. Spectrosc.*, 2009, **51**, 142.
38. N. Pachauri and B. He, presented at the 2006 ASABE Annual International Meeting, 2006.

39. D. Pramanick, R. Pramanick and G. Betal, *J. Appl. Polym. Sci.*, 2004, **91**, 343.
40. D. Pramanick and T. T. Ray, *Polym. Bull.*, 1987, **18**, 311.
41. D. Pramanick and T. T. Ray, *Polym. Bull.*, 1988, **19**, 365.
42. R. Shi, Z. Zhang, Q. Liu, Y. Han, L. Zhang, D. Chen and W. Tian, *Carbohydr. Polym.*, 2007, **69**, 748.
43. R. Shogren, S. Gonzalez, J. L. Willett, D. Gravier and G. Swift, *J. Biobased Mater. Bioenergy*, 2007, **1**, 229.
44. R. T. Tran, Y. Zhang, D. Gyawali and J. Yang, *Recent Pat. Biomed. Eng.*, 2009, **2**, 216.
45. F. Yao, Y. Bai, W. Chen, X. An, K. Yao, P. Sun and H. Lin, *Eur. Polym. J.*, 2004, **40**, 1895.
46. B. Tisserat, G. W. Selling, J. A. Byars and A. Stuff, *J. Polym. Environ.*, 2012, **20**, 291.
47. I. Djordjevic, N. R. Choudhury, N. K. Dutta and S. Kumar, *Polym. Int.*, 2011, **60**, 333.
48. D. E. Henton, P. Gruber, J. Lunt and J. Randall, *Natural Fibers, Biopolymers and Biocomposites*, ed. A. K. Mohanty, M. Misra and L. T. Drzal, CRC Press, Boca Raton, 2005, ch. 16, pp. 543–577.
49. http://www.icis.com/resources/news/2012/07/02/9573828/bioplastics-surge-towards-commercialization/.
50. http://polymerinnovationblog.com/bio-polyethylene-drop-in-replacement/.
51. http://www.sidel.com/about-sidel/inline-magazine/a-green-future-for-plastic.
52. R. P. Babu, K. O'Connor and R. Seeram, *Prog. Biomater.*, 2013, **2**, 8.
53. http://www.icis.com/resources/news/2011/02/28/9438198/dow-studies-bio-based-propylene-routes/.
54. https://www.plasticsnews.com/resin.
55. http://www.drugs.com/inactive/glycerin-448.html.
56. http://www.oleoline.com/wp-content/uploads/products/reports/Dec2012_462181.pdf.
57. F. Yang, M. A. Hanna and R. Sun, *Biotechnol. Biofuels*, 2012, **5**, 13.
58. http://www.biodieselmagazine.com/articles/9004/report-glycerol-market-expected-to-reach-2-1-billion-in-2018.
59. http://www.chemtotal.com/citric-acid-a-profile.html.
60. A. R. Angumeenal and D. Venkappayya, *LWT–Food Sci. Technol.*, 2013, **50**, 367.
61. K. Sukesh, J. S. Jayasuni, C. N. Gokul and V. Anu, *J. Chem., Biol. Phys. Sci.*, 2013, **3**, 1572.
62. E. Chabrat, H. Abdillahi, A. Rouilly and L. Rigal, *Ind. Crops Prod.*, 2012, **37**, 238.
63. http://renewablechemicals.agra-net.com/2011/10/chinese-citric-acid-industry-begins-to-consolidate/.
64. P. N. Coneski, K. S. Rao and M. H. Schoenfisch, *Biomacromolecules*, 2010, **11**, 3208.
65. Y. Jiugao, W. Ning and M. Xiaofei, *Starch/Staerke*, 2005, **57**, 494.

66. D. R. Lu, C. M. Xiao and S. J. Xu, *eXPRESS Polym. Lett.*, 2009, **3**, 366.
67. L. V. Labrecque, R. A. Kumar, V. Dave, R. A. Gross and S. P. McCarthy, *J. Appl. Polym. Sci.*, 1997, **66**, 1507.
68. *Poly(Lactic Acid): Synthesis, Modification, Processing and Applications*, ed. R. A. Auras, L.-T. Lim, S. E. M. Selke and H. Tsuji, John Wiley & Sons, Hoboken, 2010.
69. J. Ren, *Biodegradable Poly(Lactic Acid): Synthesis, Modification, Processing and Applications*, Springer-Verlag, Heidelberg, 2010.
70. http://www.sigmaaldrich.com/technical-documents/articles/material-matters/resomer-biodegradeable-polymers.html.
71. B. D. Ratner, A. S. Hoffman, F. J. Schoen and J. E. Lemons, *Biomaterials Science: An Introduction to Materials in Medicine*, Elsevier, London, 2nd edn, 2004.
72. http://alibaba.com.
73. http://www.chemceed.com/inquiry.htm.
74. http://www.engineeringtoolbox.com/polymer-properties-d_1222.html.

Development of Bio-based Unsaturated Polyester Resins from Natural Oils or their Derivates

CHENGGUO LIU[a,b] AND YONGHONG ZHOU[*a]

[a] Institute of Chemical Industry of Forest Products, Chinese Academy of Forestry, Nanjing 210042, People's Republic of China; [b] Institute of Forest New Technology, Chinese Academy of Forestry, Beijing, 100091, People's Republic of China
*Email: yhzhou777@sina.com

8.1 Introduction

Due to the uncertainties concerning petroleum supply and prices in the future, as well as environmental pollution problems, there has been a growing interest in the synthesis of chemicals or materials from renewable resources. Natural oils, especially plant oils, have been regarded as one of the most valuable bio-resources because of their low cost, the rich chemistry that their triglyceride structures provide, and their potential bio-degradability.[1–4] Thus far, they have been employed to prepare oil-based unsaturated polyester resins (UPRs) which can be broadly used in coatings, structural plastics, and fiber-reinforced composites.[1–4] Another reason for using plant oils to fabricate UPRs is that plant oils contain the characteristic structures of unsaturated polyester (UPE): unsaturated carbon double (C=C)

RSC Green Chemistry No. 29
Green Materials from Plant Oils
Edited by Zengshe Liu and George Kraus

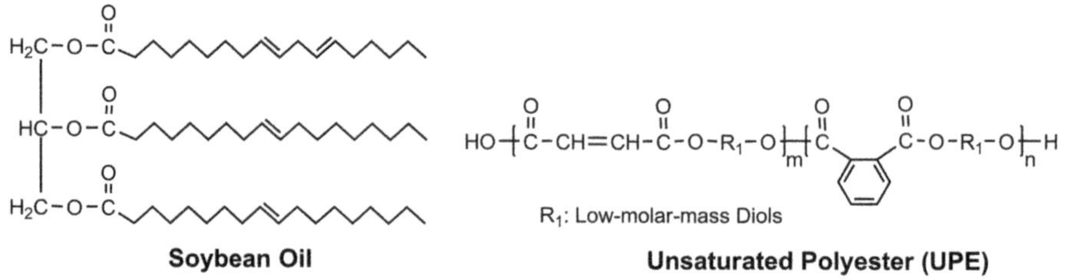

Scheme 8.1 Typical structures of a plant oil triglyceride and a general-purpose UPE resin.

bonds and ester groups (Scheme 8.1), which can be blended with a reactive diluent and cured *via* free-radical co-polymerization.

The conversion of plant oil into industrially useful UPE-like plastic usually proceeds through three pathways. One is through direct co-polymerization of the fatty acid C=C bonds with a variety of alkene co-monomers. For instances, Larock and co-authors[5,6] prepared a range of thermoset plastics by the cationic co-polymerization or the thermal co-polymerization of tung, olive, peanut, sesame, canola, corn, soybean, grape seed, sunflower, safflower, walnut, and linseed oils with divinylbenzene or a combination of styrene and divinylbenzene. The second method is to introduce polymerizable moieties such as maleic anhydride (MA) and acrylic acid (AA) onto the triglyceride or its derivates by using the reactive sites available. Wool and co-authors[1,7,8] developed a broad range of chemical routes to utilize plant oils to make polymers and composite materials that can be used in structural applications. The chemical routes from the initial plant oils to the final UPE-like monomers mainly involved multi-step reactions. Recently, Zhang and co-authors[9] introduced a one-step method for the preparation of acrylated soybean oil by reacting soybean oil and AA directly under the catalysis of boron trifluoride diethyl etherate ($BF_3 \cdot OEt_2$). It was noted that the UPE-like monomer obtained by this method contained a structure that was similar to vinyl ester, but not exactly the same, and was named an "unsaturated ester" (UE). The last, but frequently used method is to blend plant oil or its derivates with general-purpose UPR.[10–16] Due to the poor performance of pure-oil-constructed polymer materials compared to petroleum-based polymer materials, it seems more realistic to develop partially substituted petroleum-based materials. However, the addition of oil or its derivate at a not-so-high content (about 15–20 wt%) caused a great loss of stiffness in the obtained UPR blend materials, although the toughness of the prepared materials was usually improved.[11,12,15]

In our group, we have developed a series of novel oil-based UPE resins, which are promising candidates for use in engineering plastics and composites. For example, using the special Chinese wood oil called as tung oil (TO), tung oil pentaerythritol glyceride maleates (TOPERMA) were obtained by functionalizing TO triglyceride in two basic steps: (1) alcoholysis to produce TO pentaerythritol alcoholysis products; and (2) reaction with MA.[17,18]

We also tried to seek new methods to prepare oil-based UPR. For instance, unsaturated polyester resins terminated with dicyclopentadiene (DCPD-UPR) were modified by TO *via* an intermolecular Diels–Alder reaction in the later stage of melt polycondensation.[19] In order to investigate how the incorporation of a highly-functionalized UE macromonomer would affect the resultant properties of bio-based UPRs, castor oil pentacrythritol glyceridemaleate (COPERMA) model products were employed to fabricate a partially bio-based UPR through a blending method.[20]

This chapter reports the synthesis and characterization of the three obtained UPE resins: TOPERMA, TO-modified DCPD-UPR (DCPD-UPR-TO), and a COPERMA/UPR blend. Moreover, the ultimate properties of these oil-based UPR materials were evaluated. The structure–property relationship, especially the effects of phase separation and cross-link density on mechanical properties, was studied carefully.

8.2   Experimental

8.2.1   Synthesis of TOPERMA

The synthesis of TOPERMA was carried out in two steps: (1) alcoholysis to produce TO pentaerythritol (TOPER) alcoholysis products; and (2) reaction of TOPER with MA, as shown in Scheme 8.2. Diels–Alder products formed between MA and the TO conjugated triene in the maleinization process of TOPER. The Diels–Alder addition not only consumed the MA molecules, but

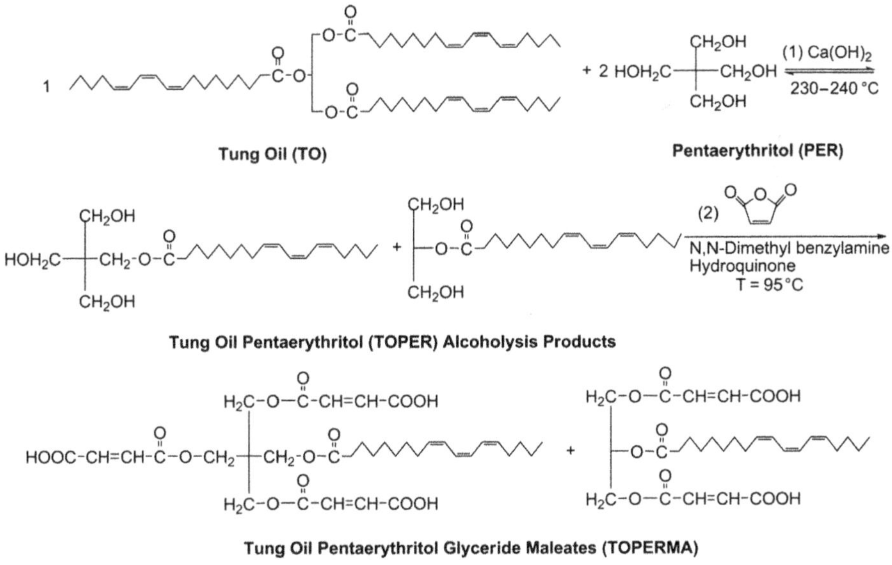

Scheme 8.2   Ideal reaction scheme for the synthesis of TOPERMA from TO triglyceride.

also resulted in the transformation of the TO conjugated trienes into non-conjugated double bonds, which might affect the co-polymerization with styrene. After careful experiment optimization, the best procedure for the maleinization reaction was at a reaction temperature of 95 °C, with a reaction time of 5 h, and a feed ratio of 1 : 2 : 8 (TO : PER : MA).[17] The acid value of the optimized TOPERMA product was 211 mg KOH g^{-1}.

8.2.2   Synthesis of DCPD-UPR-TO

The synthesis of DCPD-UPR-TO polymers was carried out in three basic steps, as shown in Scheme 8.3. First, dicyclopentadiene maleate (DCPD-MA) was synthesized *via* hydrolysis. The best molar feed ratio of DCPD to MA was 1 : 0.8. Second, with the optimized DCPD-MA product, DCPD-modified UPR (DCPD-UPR) was prepared *via* melt polycondensation. Third, when the acid value of the reacting mixture decreased to about 70 mg KOH g^{-1} at the latter stage of melt polycondensation, the reactor was cooled to 160 °C and a certain amount (for example, 16.7 g) of tung oil was added in drops. After that the reactor was heated to 200 °C again and maintained at this temperature until the acid value decreased to the set value. In our experiments, the contents of TO in the obtained DCPD-UPR-TO polymers were 0, 5, 10, 15 and 20% by weight, thus the corresponding DCPD-UPR-TO polymers were labeled as DCPD-UPR, DCPD-UPR-TO5, DCPD-UPR-TO10, DCPD-UPR-TO15,

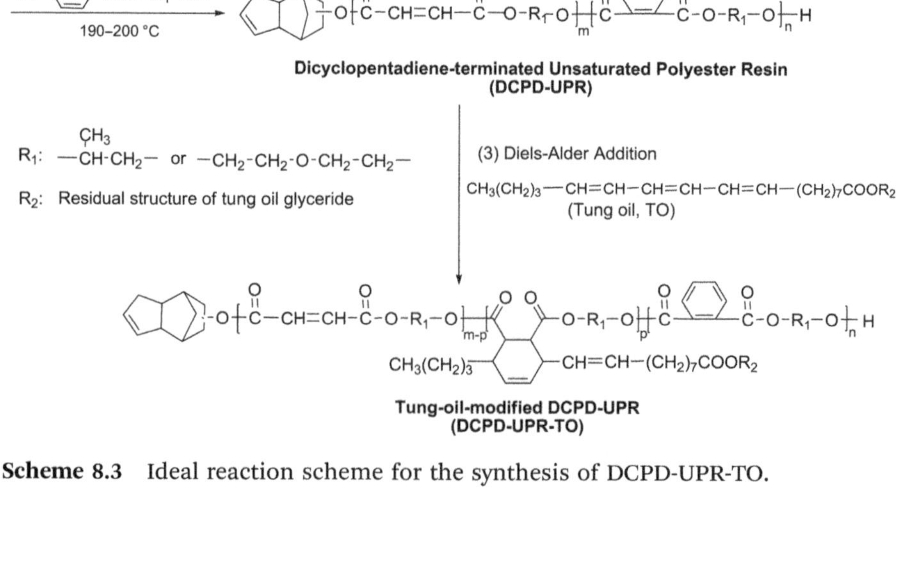

Scheme 8.3   Ideal reaction scheme for the synthesis of DCPD-UPR-TO.

and DCPD-UPR-TO20, respectively. The obtained DCPD-UPR and DCPD-UPR-TO polymer products were blended with styrene and hydroquinone at 90 °C for 1 h to give liquid resins that were light yellow or yellow.

8.2.3 Synthesis of UPR/COPERMA Blends

The synthesis of COPERMA was similar to that of TOPERMA, which followed the procedures described in ref. 8. The COPERMA product (Scheme 8.4) was a light-yellow soft solid at room temperature. The acid value of COPERMA was 255 mg KOH g^{-1}. In order to make the COPERMA solid product easily add into the UPR, the product was first mixed with 35 wt% styrene. Thereafter, the obtained COPERMA/styrene resin was added to a commercial UPR (containing 35 wt% styrene) readily to produce UPR/COPERMA resins at room temperature. The contents of the COPERMA resin in the UPR/COPERMA blends were 0, 5, 10, 15 and 20% by weight, thus the corresponding samples were labeled as UPR, UPR/COPERMA5, UPR/COPERMA10, UPR/COPERMA15, and UPR/COPERMA20, respectively.

8.2.4 Curing of the Oil-based UE Monomers or UPEs with Styrene

The TOPERMA product was cured with styrene as follows: first, TOPERMA was blended with a specified amount of styrene by stirring under N$_2$ atmosphere for 1 h at 70–80 °C. The viscosity of the TOPERMA resin containing 33 wt% styrene was 1050 mPa · s. Then the resin was mixed with the initiator *tert*-butyl peroxy benzoate at 2% of the resin weight and degassed for 10 min. After that, the resin was poured into molds and placed in an oven. The curing temperature was increased from room temperature to 120 °C at a rate of 5 °C min^{-1}. The resin was cured at this temperature for 3 h and post-cured at 150 °C for 1 h. Four polymer samples containing different ratios of TOPERMA to styrene were prepared. The weight ratios of TOPERMA monomer to styrene were 80 : 20, 70 : 30, 67 : 33 and 60 : 40, therefore the TOPERMA/styrene polymers were designated TOPERMA80-ST20, TOPERMA70-ST30, TOPERMA67-ST33, and TOPERMA60-ST40, respectively.

The DCPD-UPR-TO liquid resins were cured according to the following procedure: they were blended with the initiator benzoyl peroxide (2 wt% of

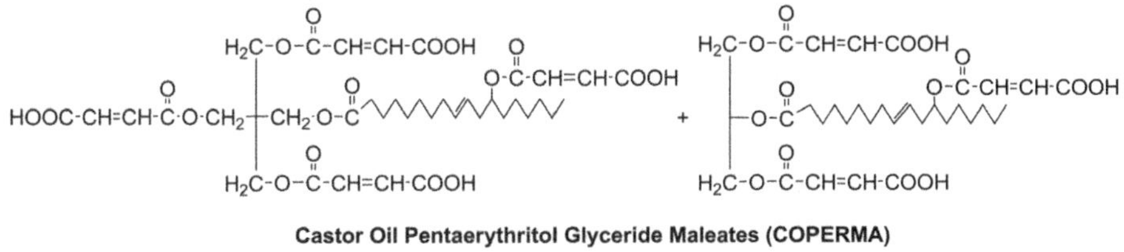

Castor Oil Pentaerythritol Glyceride Maleates (COPERMA)

Scheme 8.4 Ideal chemical structure of the COPERMA macromonomer.

the resin) for 20 min and then with the promoter *N,N*-dimethylaniline (0.2 wt% of the resin) for 2 min, poured into molds, cured at room temperature for 24 h, and post-cured at 60 °C for 3 h.

The UPR/COPERMA resins were cured following a procedure similar to that of the DCPD-UPR-TO resins: they were blended with the initiator benzoyl peroxide (2 wt% of the resin) for 20 min and then with the promoter *N,N*-dimethylaniline (0.2 wt% of the resin) for 2 min, poured into molds, cured at room temperature for 3 h and at 60 °C for 3 h, at finally post-cured at 80 °C for 1 h. The pure COPERMA resin with 35 wt% styrene was also cured by this procedure. However, only semi-rigid COPERMA resin materials were obtained under these curing conditions.

8.2.5   Characterization

8.2.5.1   Acid Value

The A_v values of the oil-based UE macromonomers or UPEs were determined according to GB/T 2895-1982. Typically, about 0.5 g of the product was introduced into 30 mL of a toluene–ethanol (50:50 v/v) solution. Using phenolphthalein as the indicator, the solution was titrated with a KOH/ethanol solution, with an accurate concentration determined using a potassium acid phthalate standard substance. The acid value was calculated as:[15]

$$A_v = \frac{56 \cdot \Delta V \cdot C_{KOH}}{1000 \cdot \Delta m} \tag{1}$$

where ΔV, C_{KOH}, and Δm denote the consumed volume and initial concentration of the KOH solution, and the product weight, respectively.

8.2.5.2   Viscosity

The V_s measurements of the oil-based resins diluted with styrene were performed on a NDJ-8S rotational viscometer (Shanghai Changji Dizhi Instrument Corporation, China).

8.2.5.3   Fourier-transform Infrared Spectroscopy

The FT-IR spectra were recorded on a Nicolet iS10 IR spectrometer (Thermo-Fisher Corporation, USA).

8.2.5.4   Nuclear Magnetic Resonance

The ^{1}H-NMR spectra were recorded on a Bruker DRX-300 Advance NMR spectrometer (Bruker Corporation, Germany).

8.2.5.5   Mass Spectrometry

The mass spectra were recorded on a WATERS Q-TOF Premier mass spectrometer with electrospray ionization (ESI-MS).

8.2.5.6   Gel Permeation Chromatography

Samples were analyzed at 25 °C using a GPC with a 515 pump (Waters Corporation, USA) equipped with a 2414 refractive index detector (Waters). The columns were Styragel HR1 and HR2 (300×7.8 mm, Waters). The flow rate of the eluent tetrahydrofuran (THF) was 1.0 mL min^{-1}. Samples in THF were passed through a 0.45 μm filter (Millipore Corporation, USA) into a 200 μL loop. A series of narrow distributed polystyrene standards (Polymer Laboratories, USA) with molar masses of 580–10^5 g mol^{-1} were used for the calibration of molar mass.

8.2.5.7   Gel Time

The t_{gel} measurements of the cured oil-based UPRs were tested following the procedure specified in GB/T 7193-2008. Typically, 100 parts of the UPR/COPERMA resin and 2.0 parts benzoyl peroxide as initiator were placed together in a beaker and then the beaker was placed in a water bath at 25 ± 0.5 °C. After mixing for 1 h, 0.02 parts *N,N*-dimethylaniline as a promoter was added by a pipette and the time was recorded immediately by a stopwatch. When the resin liquid was able to be drawn out like silk, the t_{gel} of the resin was recorded.

8.2.5.8   Linear Shrinkage

The l_s tests of the cured oil-based UPRs were performed by measuring the lengths of the impact-test samples with a Vernier caliper.

8.2.5.9   Mechanical Properties

Tensile and flexural tests of the polymer matrices were evaluated using a SANS7 CMT-4304 universal tester (Shenzhen Xinsansi Jiliang Instrument Corporation, China). Impact tests of the samples were performed in a XJJY-5 impact tester (Chengde Xinguo Instrument Corporation, China). All the above tests followed the procedure specified in GB/T 2567-2008. Dumbbell specimens with a size of 50×10×4 mm^3 at the narrow middle part were produced for tensile tests at a constant draw speed of 5.0 mm min^{-1}. Cuboid specimens with a size of 100×15×4 mm^3 were produced for flexural tests at a constant cross-head speed of 10 mm min^{-1}. Cuboid specimens with a size of 80×10×4 mm^3 were produced for impact tests. All the specimens were polished with abrasive papers to remove surface defects before testing.

At least five specimens were tested for each polymer sample and three to five close results were statistically averaged.

8.2.5.10   Thermogravimetric Analysis

Thermogravimetric analysis (TGA) was performed on a STA 409PC thermogravimetry instrument (Netzsch Corporation, Germany) at a heating rate of 15 °C min^{-1}.

8.2.5.11   Dynamic Mechanical Analysis

For the TOPERMA/styrene resins, the DMA measurements were taken on a DMA + 450 instrument (01dB-Metravib Corporation, France) under tension mode from −50 to 200 °C at a heating rate of 2 °C min^{-1} using a frequency of 1 Hz. For DCPD-UPR-TO and UPR/COPERMA resins, the DMA tests were performed on a DMA Q800 (TA Instruments, USA) instrument in single-cantilever mode with an oscillating frequency of 1 Hz.

8.2.5.12   Scanning Electron Microscopy

SEM examinations of the fractured samples were performed on S4800 or S3400 scanning electron microscopes (HITACHI Corporation, Japan). The fractured surfaces were coated with gold prior to SEM observation.

8.3   Results and Discussion

8.3.1   Structures of the Oil-based UE Monomers and UPEs

8.3.1.1   TOPERMA Macromonomers

The products of the TO alcoholysis and maleinization reactions were characterized by FT-IR, ^{1}H-NMR, and ESI-MS techniques. Figure 8.1 shows the FT-IR spectra of TO, TOPER, and TOPERMA samples. Tung oil triglyceride had the following characteristic bands: conjugated triene (3012, 991, and 964 cm^{-1}), the strong ester carbonyl peak of the glycerides (1744 and 1050–1290 cm^{-1}), and methylene and methyl groups (2925 and 2854 cm^{-1}, 1463 and 1376 cm^{-1}). The TOPER product carried the characteristic features of both PER and the glyceride structure *i.e.*, the broad hydroxyl band (around 3352 cm^{-1}), the strong ester carbonyl peak (1739 cm^{-1}), and the characteristic features of TO conjugated trienes (3012, 991 and 964 cm^{-1}). The ester carbonyl peak at 1739 cm^{-1} was shifted from the peak of TO at 1744 cm^{-1}, indicating the occurrence of the alcoholysis reaction. Meanwhile the ester carbonyl peak of TOPER had a shoulder compared with the single peak of TO at 1744 cm^{-1}, which indicates that TOPER contains two kinds of ester carbonyl groups. The primary alcohols of TOPER and the residual PER showed bands at around 1045 cm^{-1}. The hydroxyl band of the alcoholysis product at

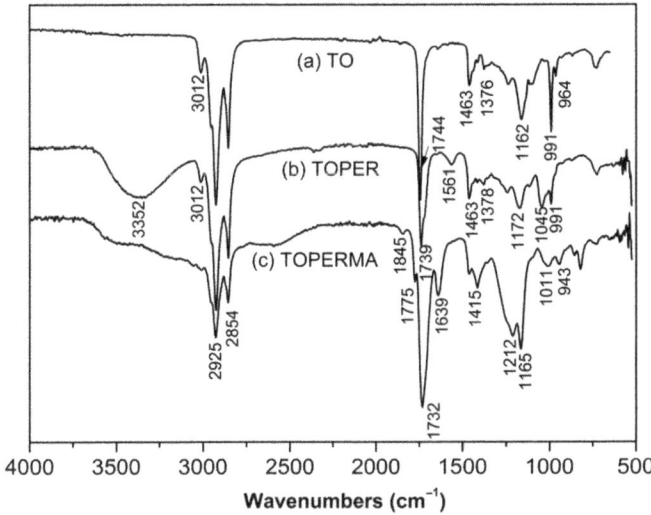

Figure 8.1   FT-IR spectra of (a) TO, (b) TOPER, and (c) TOPERMA.

3352 cm^{-1} was replaced by a broad acid band of maleate half-esters at 2500–3400 cm^{-1} as the maleinization reaction proceeded. The strong ester carbonyl peak of TOPERMA at 1732 cm^{-1} was shifted from the peak of TOPER at 1739 cm^{-1}. This carbonyl peak of TOPERMA had a higher intensity than those of TO and TOPER when the peak intensity of methylene and methyl groups at around 2925 and 2854 cm^{-1} was set as reference. The reason for this is that more ester carbonyl groups were produced in TOPERMA. The peaks at 1639 cm^{-1} mainly denoted the C=C bond on the MA structure. Some unreacted MA molecules were evidenced by the carbonyl (C=O) asymmetric and symmetric stretching vibrations at 1775 and 1845 cm^{-1}. The characteristic features of TO conjugated triene at 3012, 991, and 964 cm^{-1} were undistinguishable in TOPERMA, which may be caused by the Diels–Alder reaction between MA and the conjugated triene.

The ^{1}H-NMR spectra of TO, TOPER, and TOPERMA are shown in Figure 8.2. In the TO spectrum [Figure 8.2(a)], the peaks at 4.0–4.4 ppm and 5.2–5.3 ppm correspond to the protons of the glycerol backbone in the TO triglyceride unit. The multiple peaks at 5.3–6.5 ppm show the protons of conjugated trienes in the triglyceride. For all the ^{1}H-NMR spectra of TO, TOPER, and TOPERMA, the peaks at around 0.9 ppm show that the terminal methyl protons of fatty acids can be taken as a reference, because the intensity of these peaks should not be altered throughout the alcoholysis and maleinization processes. The integral of this peak should show nine protons per TO triglyceride molecule in the original formulation. Referenced against the terminal methyl protons at 0.9 ppm, we calculated that a single TO triglyceride had 15.6 protons in the TO conjugated triene structure, which means a conjugated triene in TO only has 2.6 carbon double bonds. This is because TO also contains some fatty acid chains with non-conjugated double

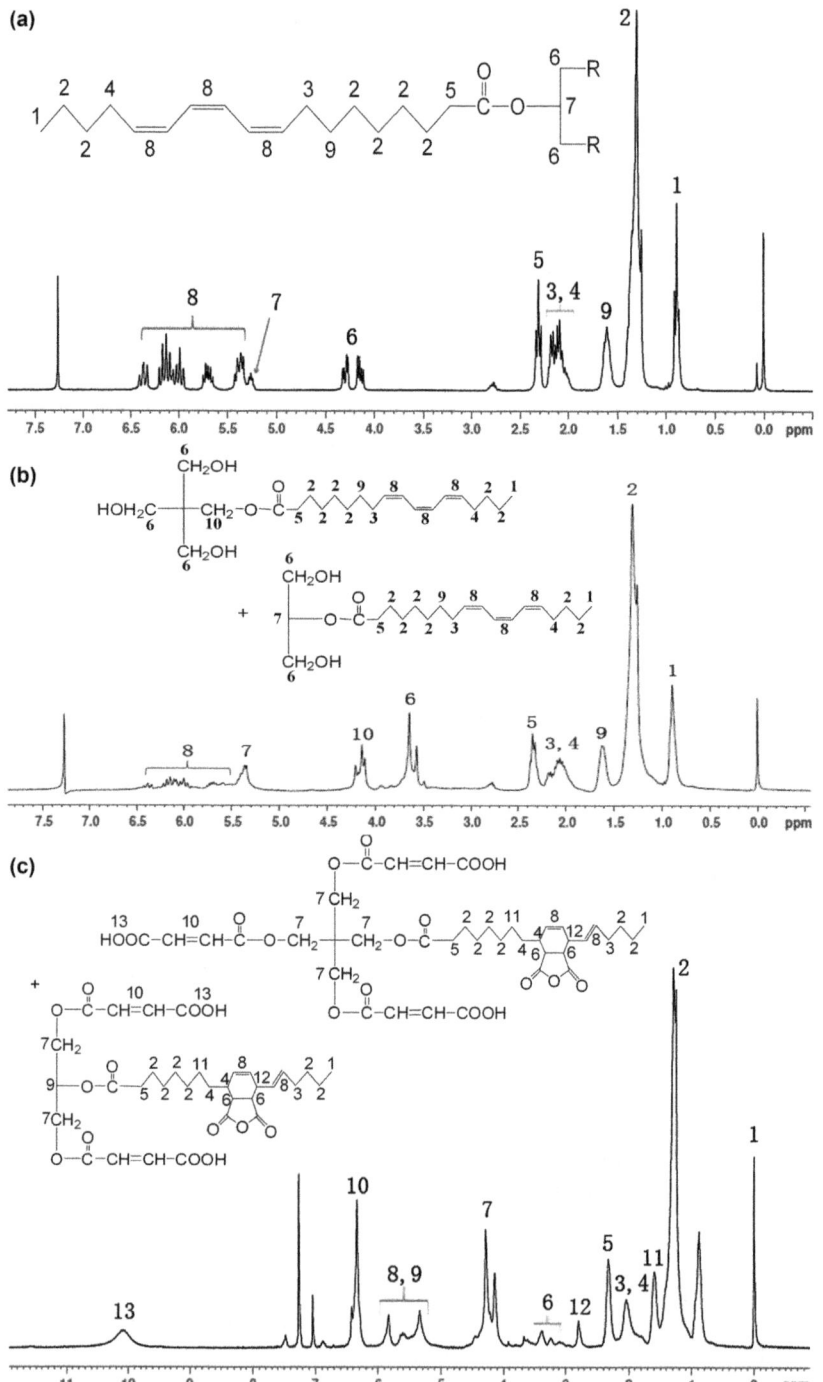

Figure 8.2 ¹H-NMR spectra of (a) TO, (b) TOPER, and (c) TOPERMA.

bonds, which are evidenced by the presence of a minor peak at 2.8 ppm corresponding to the protons in the methylene groups between two non-conjugated double bonds (–CH=CH–CH₂–CH=CH–). In the TOPER spectrum [Figure 8.2(b)], the multiplet peaks in the 3.4–3.8 ppm region belong to the hydroxyl functional methylene protons of the polylol. The peaks at 4.0–4.4 ppm show the methylene protons of the polyol backbones connecting to the ester structure (–CH₂–O–(C=O)–CH₂–). The TOPERMA product was characterized by the maleate vinyl protons at 6.3 ppm and the peak of acid protons at 10.1 ppm [Figure 8.2(c)]. The fumarate (*trans*-isomer of the maleate) peak appeared at 6.9 ppm with a very low intensity, since the maleate–fumarate isomerization favors a higher temperature than the reaction temperature.[21] The peak at 7.1 ppm represents the unreacted MA vinyl protons. As the maleinization reaction of TOPER proceeded, the protons at 3.4–3.8 ppm were converted to those at 4.0–4.5 ppm, the methylene protons of the polyol backbones connecting to the ester structure. The peaks at 2.8 ppm belonging to the protons in the –CH=CH–CH₂–CH=CH– structure became sharp, indicating the occurrence of a new similar structure. This may result from the protons on the conjugated triene connecting to MA *via* a Diels–Alder reaction [Figure 8.2(c)]. The peaks at 3.1–3.4 ppm can be designated as the protons on the MA connecting to the conjugated triene. Hence, the maleinated products shown in Scheme 8.2 were ideal structures of TOPERMA, and the real products should contain the byproducts resulting from the Diels–Alder reaction. The peak that appears at 7.28 ppm in each spectrum represents the residual protons of CDCl₃. The peak at 7.45 ppm shows the aromatic protons of the *N,N*-dimethylbenzyl amine used as a catalyst for the maleinization reaction.

Soft ionization combined with the inherent multiple charging mechanism of ESI has made MS an ideal tool to determine accurate molar masses. Figure 8.3 shows the ESI-MS spectra of TOPER and TOPERMA. The structures assigned to various mass numbers are given in Scheme 8.5. The characteristic peaks and the possible molecular structures of the TOPER products were correlated as $353.2[M_2 + H]^+$, $379.2[M_2 + Na]^+$, $397.2[M_1 + H]^+$, and $435.2[M_1 + K]^+$, indicating the presence of monopentaerythritide and monoglyceride. The peaks and the structures of TOPERMA were correlated as $449.2[M_4-H]^-$, $493.2[M_3-H]^-$, $547.1[M_6-H]^-$, $591.2[M_5-H]^-$, $645.1[M_8-H]^-$, $689.2[M_7-H]^-$, and $787.1[M_9-H]^-$, suggesting the yield of several maleate half-esters. The peaks of $645.1[M_8-H]^-$ obviously revealed the Diels–Alder structure in TOPERMA, because the monoglyceride could not react with three MA molecules without the Diels–Alder reaction. Multiplicity seen in the mass spectra near the intense peaks is due to the natural abundance of various fatty acids in TO.

8.3.1.2 DCPD-UPR-TO Polymers

Data of A_v, V_s, weight-average and number-average molar masses (M_w and M_n), and polydispersity (*PDI*) for DCPD-UPR-TO polymers with different TO

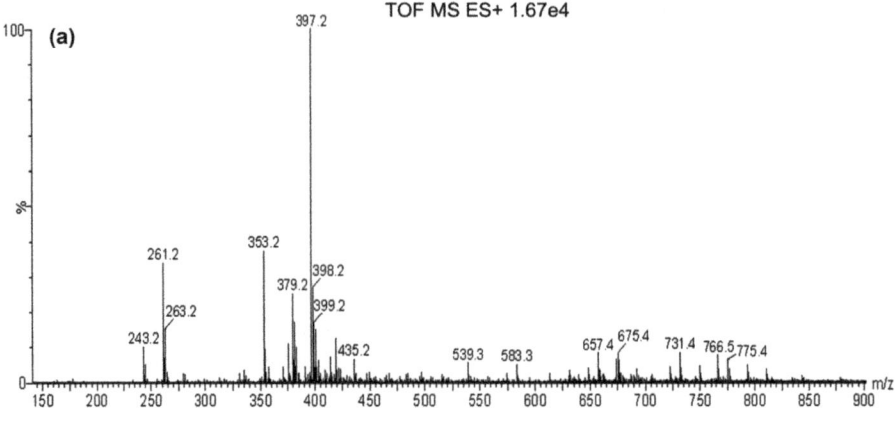

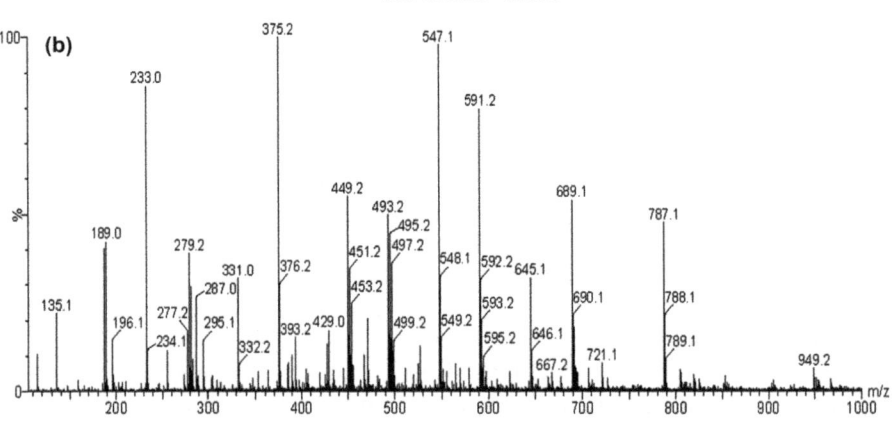

Figure 8.3 ESI-MS spectra of (a) TOPER and (b) TOPERMA.

contents are shown in Table 8.1. The A_v values were very low and almost the same (about 20 mg KOH g^{-1}), indicating the completion of melt poly-condensation for all the DCPD-UPR-TO polymers. With the increase of TO content, viscosities for the DCPD-UPR-TO/styrene resins gradually increased and reached a maximum value of 1370 mPa·s. The A_v and V_s values suggest that these oil-based UPRs are suitable for applications in fiber-reinforced materials.

The FT-IR spectra of TO, DCPD-UPR, and DCPD-UPR-TO5 are depicted in Figure 8.4. The peaks in the TO spectrum were designated in Figure 8.1. Some typical peaks can be seen in the spectrum of DCPD-UPR: methyl and methylene groups (2963 cm^{-1} and 2887 cm^{-1}), the C=O on an ester group (1730 cm^{-1}, strong), and C–O (1000–1300 cm^{-1}). The hydroxyl peak at 3000–3500 cm^{-1}, usually strong for general UPRs, was very weak for DCPD-UPR, which can be attributed to the termination reaction between DCPD and the hydroxyl groups on UPR. After the incorporation of TO into DCPD-UPR, two

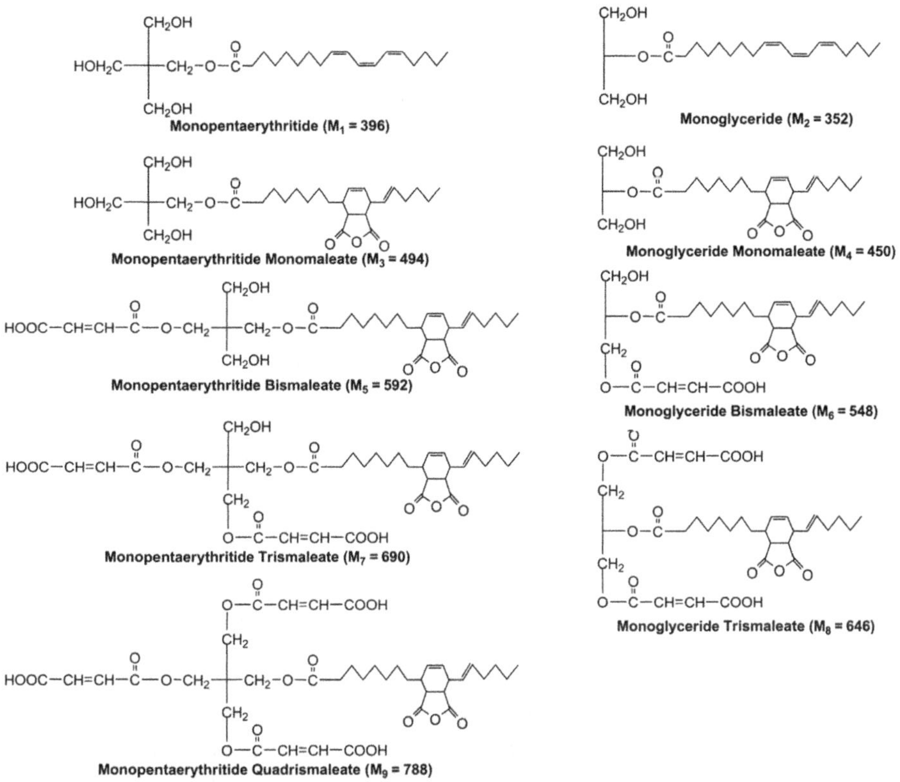

Scheme 8.5 Mass numbers of probable structures expected in TOPER and TOPERMA.

Table 8.1 Acid value (A_v), viscosity (V_s), weight-average and number-average molar masses (M_w and M_n) and polydispersity (*PDI*) of the DCPD-UPR-TO polymers with different TO contents.

Sample	A_v/mg KOH g^{-1}	V_s/mPa·s	M_w/g mol^{-1}	M_n/g mol^{-1}	*PDI*
DCPD-UPR	19.9	840	1275	618	2.06
DCPD-UPR-TO5	16.3	967	2103	737	2.85
DCPD-UPR-TO10	20.3	1370	3145	858	3.67
DCPD-UPR-TO15	22.1	1185	2993	807	3.71
DCPD-UPR-TO20	23.6	1348	3170	818	3.88

obvious variations were found in the spectrum of DCPD-UPR-TO5. First, the characteristic features of the TO conjugated trienes at 3012, 991, and 964 cm^{-1} were not observed. Second, the intensity of the methylene peak at 2963 cm^{-1} greatly increased relative to that of the C=O peak at around 1730 cm^{-1}. These changes indicate that the conjugated trienes on TO had reacted with the C=C bonds on DCPD-UPR.

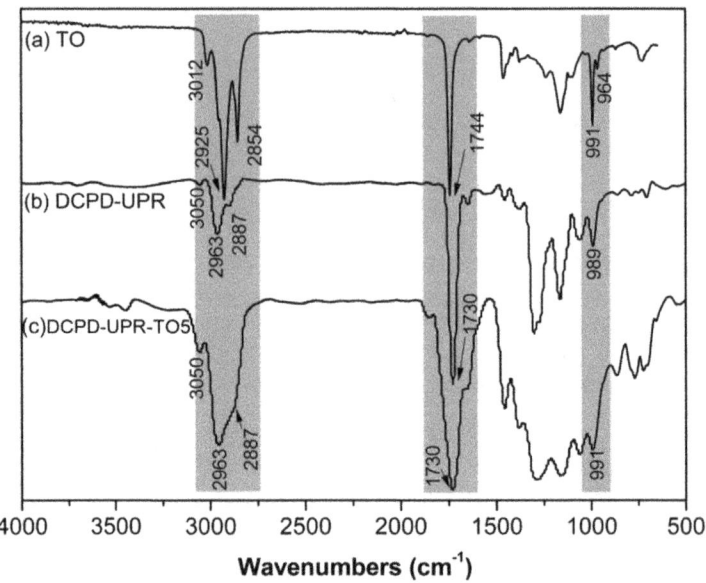

Figure 8.4   FT-IR spectra of (a) TO, (b) DCPD-UPR, and (c) DCPD-UPR-TO5.

The ^{1}H-NMR spectra of DCPD-UPR and DCPD-UPR-TO5 are depicted in Figure 8.5. The double peak at 7.5–8.0 ppm corresponds to the protons on the benzene structure from phthalic anhydride. It can be taken as a reference, because the intensity of this peak should not be altered by the polycondensation process. The peaks at 6.3 ppm and 6.9 ppm represent the maleate and fumarate vinyl protons, respectively. It was found that the ratios of fumarate to maleate vinyl groups were 4.13 and 1.67 for DCPD-UPR and DCPD-UPR-TO5 polymers, respectively, indicating the occurrence of maleate-to-fumarate (*cis–trans*) isomerization. Compared to that of DCPD-UPR [Figure 8.5(a)], the intensity of the peak representing the maleate vinyl protons on DCPD-UPR-TO5 [Figure 8.5(b)] remained almost constant, while the fumarate vinyl proton peak decreased from 0.62 to 0.30. Two important facts are indicated here: first, the TO conjugated triene only reacts with the fumarate structure on DCPD-UPRs in the Diels–Alder reaction; second, about half of the fumarate vinyl groups were consumed by the Diels–Alder reaction for DCPD-UPR-TO5, which means the fumarate vinyl groups can consume 10 wt% tung oil at most. In addition, the peaks at 3.0–3.4 ppm represent the protons at the place where the C=C bond on polyester links to the TO conjugated triene. The peak at 0.9 ppm belongs to the terminal methyl protons of TO fatty acids.

The GPC technique was employed to analyze the molar masses of the obtained DCPD-UPR-TO polymers. The elution chromatographs of GPC are shown in Figure 8.6 and the corresponding results of molar masses are listed in Table 8.1. As shown in Figure 8.6, the area of elution time at 11–13 min, which corresponds to the high-molar-mass section, increased greatly after

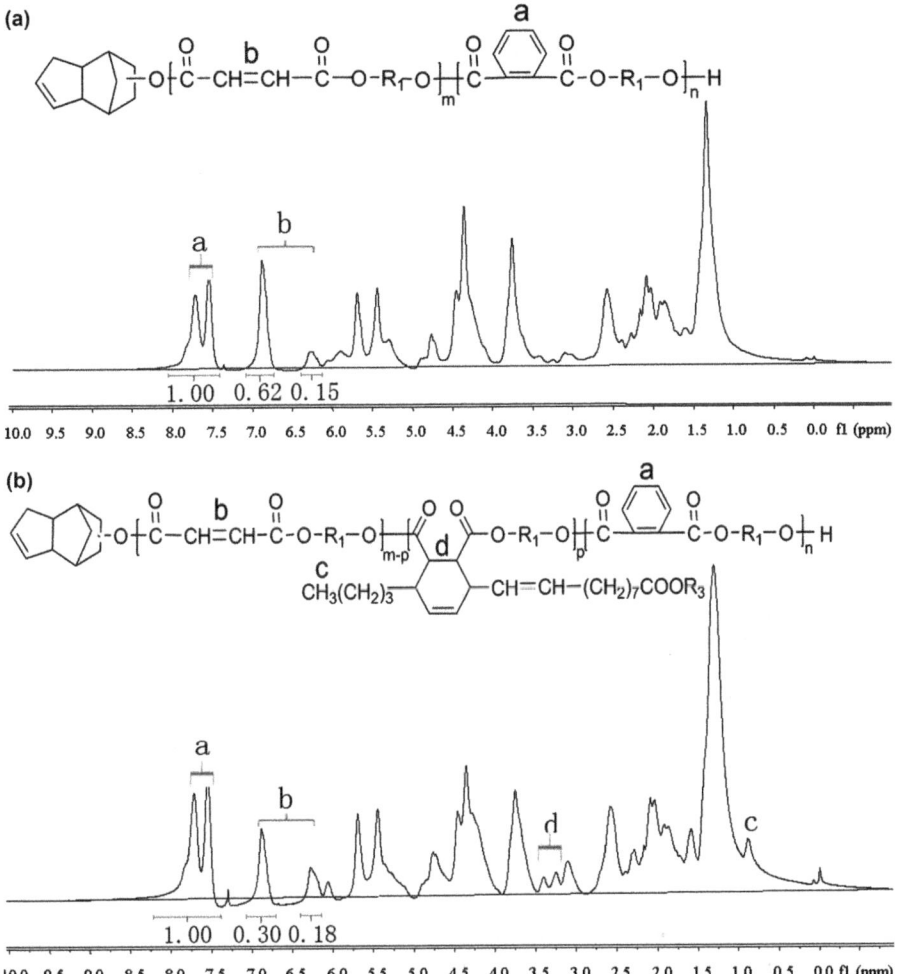

Figure 8.5   ¹H-NMR spectra of (a) DCPD-UPR and (b) DCPD-UPR-TO5.

the incorporation of TO. This fact was further evidenced by the change of molar mass for the obtained DCPD-UPR and DCPD-UPR-TO polymers (Table 8.1). Both the M_w and M_n of DCPD-UPR-TO polymers were larger than those of neat DCPD-UPR. This indicates the occurrence of the Diels–Alder reaction between DCPD-UPR and TO. Moreover, it was found that almost maximum values of M_w and M_n were reached for a TO content of 10 wt%. Thereafter the molar masses varied only slightly. This is because the stoichiometric equilibrium of DCPD-UPR to TO in the Diels–Alder reaction is about 9 : 1 by weight, which is in good agreement with the ¹H-NMR results. When adding more TO (above 10 wt%) to DCPD-UPR, redundant tung oil molecules remain in the DCPD-UPR-TO polymers, thus leading to smaller molar masses (especially M_n) and larger *PDI* values.

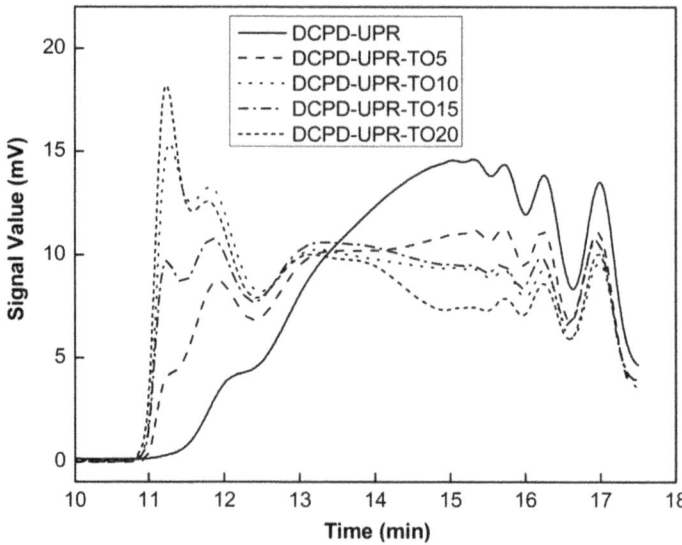

Figure 8.6   GPC elution chromatograms of the DCPD-UPR-TO polymers with different TO contents.

8.3.1.3   COPERMA Macromonomers

Structures of the products from the synthesis of COPERMA were monitored by FT-IR, ^{1}H-NMR, and ESI-MS. The FT-IR spectra of castor oil, COPER, and COPERMA are depicted in Figure 8.7. Several characteristic peaks were indicated in the spectrum of castor oil: hydroxyl groups on the fatty acid chains (around 3368 cm^{-1}), methyl and methylene groups (2925 and 2854 cm^{-1}), and ester carbonyl groups (1744 cm^{-1}).[8] There were also some typical peaks in the spectrum of COPER: hydroxyl groups (around 3368 cm^{-1}), methyl and methylene groups (2923 and 2854 cm^{-1}), and ester carbonyl groups (1734 cm^{-1}). Compared to that of castor oil, the peak intensity of COPER at 3368 cm^{-1} greatly increased, indicating more hydroxyl groups on the COPER structure. The peak representing ester carbonyl groups shifted from 1744 to 1734 cm^{-1}, suggesting the occurrence of the alcoholysis reaction. In the spectrum of COPERMA, several characteristic peaks can be seen: carboxyl groups on the maleate half-esters (2500–3400 cm^{-1}), carboxyl groups on the unreacted MA molecules (1850 and 1776 cm^{-1}), ester carbonyl groups (1729 cm^{-1}), and C=C groups on the MA structure (1633 cm^{-1}). The appearance of the broad peak at 2500–3400 cm^{-1} and the shift of the ester carbonyl groups from 1734 to 1729 cm^{-1} clearly reflect the occurrence of the maleinization reaction.

The ^{1}H-NMR spectra of the COPER and COPERMA products are shown in Figure 8.8. As indicated in the spectrum of COPER [Figure 8.8(a)], the multipeaks at 3.4–3.8 ppm were hydroxyl functional methylene protons of the polylol and the characteristic methine protons on the castor oil fatty acid

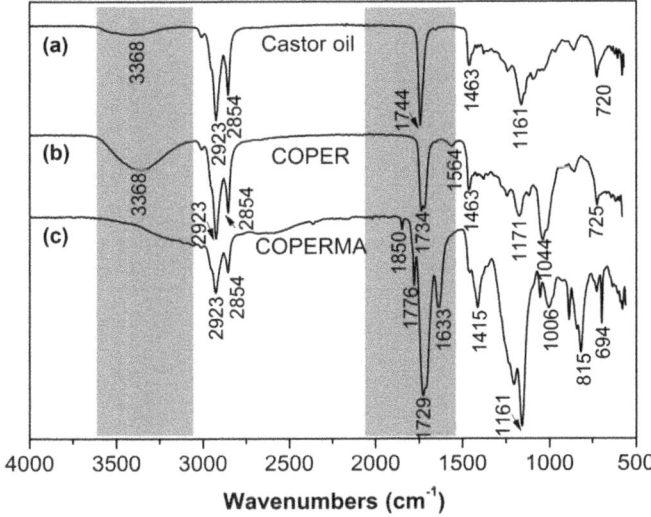

Figure 8.7 FT-IR spectra of (a) castor oil, (b) COPER, and (c) COPERMA.

chains connecting to the hydroxyl groups.[8] The peak at around 4.1 ppm corresponded to the methylene protons of the polyol backbones connecting to the ester structure ($-CH_2-O-(C=O)-CH_2-$). The peak at around 0.9 ppm shows that the terminal methyl protons of fatty acids could also be taken as a reference. The integral of this peak should show nine protons per castor oil triglyceride. In the spectrum of COPERMA [Figure 8.8(b)], the typical peaks at 6.3, 6.9, and 7.1 ppm correspond to the maleate, fumarate, and MA vinyl protons, respectively. These peaks can be used to determine the reaction extent of MA.[17] Using the intensity of the peak at 0.9 ppm as a reference, the maleate, fumarate, and MA vinyl protons were calculated readily. The corresponding proton amounts per castor oil triglyceride were 14.81, 0.87, and 1.16, which corresponds to the maleate C=C bonds of 7.41, 0.44, and 0.58, respectively. Hence, the real maleate C=C functionality of COPERMA was $(7.41 + 0.44)/3 = 2.62$, which is very high compared to other reported oil-based macromonomers.[8,11,12,15,17,22] For example, the maleate C=C functionalities were 1.7, 1.3, 1.59, 0.83, and 1.61 for maleate half-esters of hydroxymethylated soybean oil and sunflower oil, soybean oil pentaerythritol glyceride maleates (feed ratio 1 : 2 : 8), COMA (feed ratio 1 : 3), and tung oil pentaerythritol glyceride maleates (feed ratio 1 : 2 : 8), respectively. As the maleinization reaction of COPER proceeded, the proton bands at 3.4–3.8 ppm were converted to the bands at 4.1–4.5 ppm, the methylene protons of the polyol backbones connecting to the ester structure. In addition, a new peak at around 5.05 ppm appeared showing the change of connecting method of the characteristic methine on the castor oil fatty acid chain. The peak at around 10.5 ppm is the acid proton provided by the ring opening of the MA structure during the maleinization.

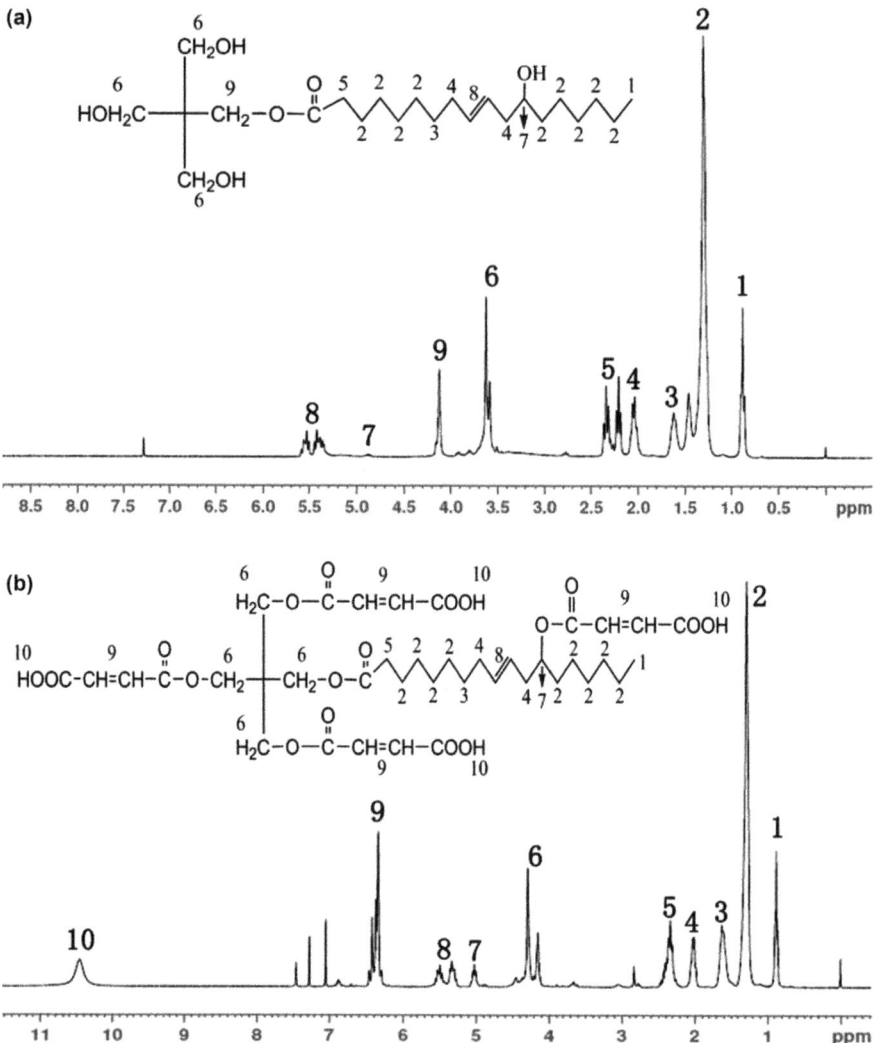

Figure 8.8 ¹H NMR spectra of (a) COPER and (b) COPERMA.

The ESI-MS spectra of the COPER and COPERMA products are shown in Figure 8.9. The main possible structures in the COPER and COPERMA products assigned to various mass numbers are provided in Scheme 8.6. The characteristic peaks and the possible molecular structures of the COPER product were correlated as $415.2[M_1–H]^-$ and $371.2[M_2–H]^-$, suggesting the presence of monopentaerythritide and monoglyceride. The characteristic peaks and the possible molecular structures of the COPERMA product were correlated as $513.1[M_3–H]^-$, $469.1[M_4–H]^-$, $611.1[M_5–H]^-$, $567.1[M_6–H]^-$, $709.1[M_7–H]^-$, $665.1[M_8–H]^{-1}$, and $807.1[M_9–H]^-$. These peaks indicate the yield of many types of maleate half-esters after completion of the

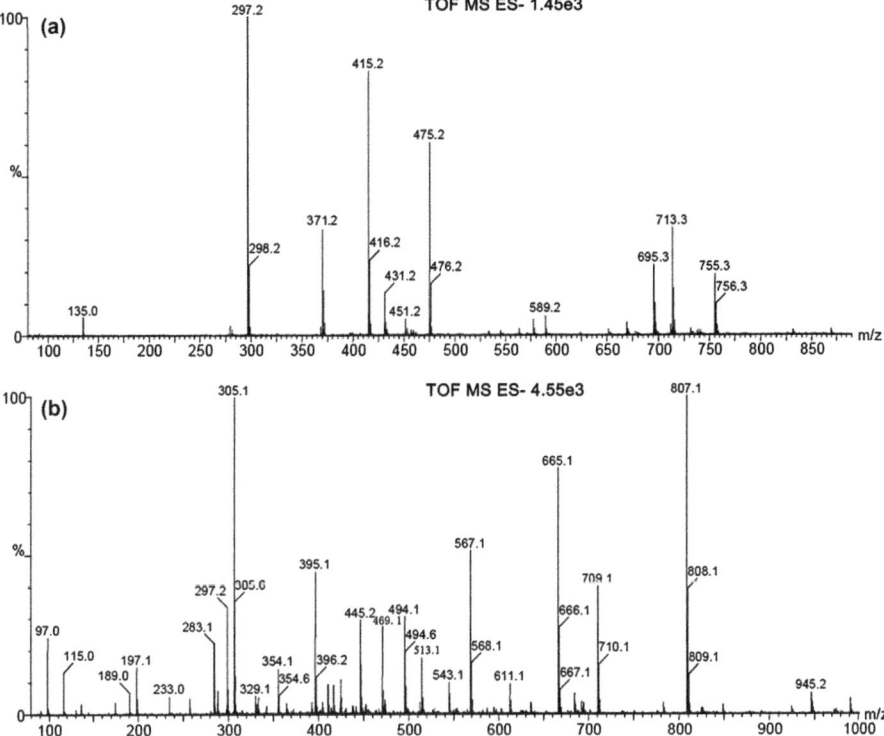

Figure 8.9 ESI-MS spectra of (a) COPER and (b) COPERMA.

maleinization reaction. The intensities of the peaks at 665.1 and 807.1 were relatively higher than those of the other half-ester peaks, indicating the sufficient maleinization of the COPER product into high-functionality castor-oil-based maleate esters. Multiplicity seen in the mass spectra near the intense peaks was caused by the natural abundance of various fatty acids in castor oil.

8.3.2 Structure and Properties of the Cured TOPERMA/Styrene Thermosets

8.3.2.1 Transparency and SEM Analysis

The transparency is directly related to the phase separation of polymer samples.[11,12,14,23] In other words, a lack of transparency is the result of phase separation. Figure 8.10 shows the transparency of the four polymer matrices after curing. Four types of transparency were demonstrated: (A) *opaque* for TOPERMA80-ST20, (B) *quasi-opaque* for TOPERMA70-ST30, (C) *transparent* for TOPERMA67-ST33, and (D) *half-transparent* for TOPERMA60-ST40, respectively. Hence, the order of the extent of phase-separation can be simply listed as (A)>(B)>(D)>(C).

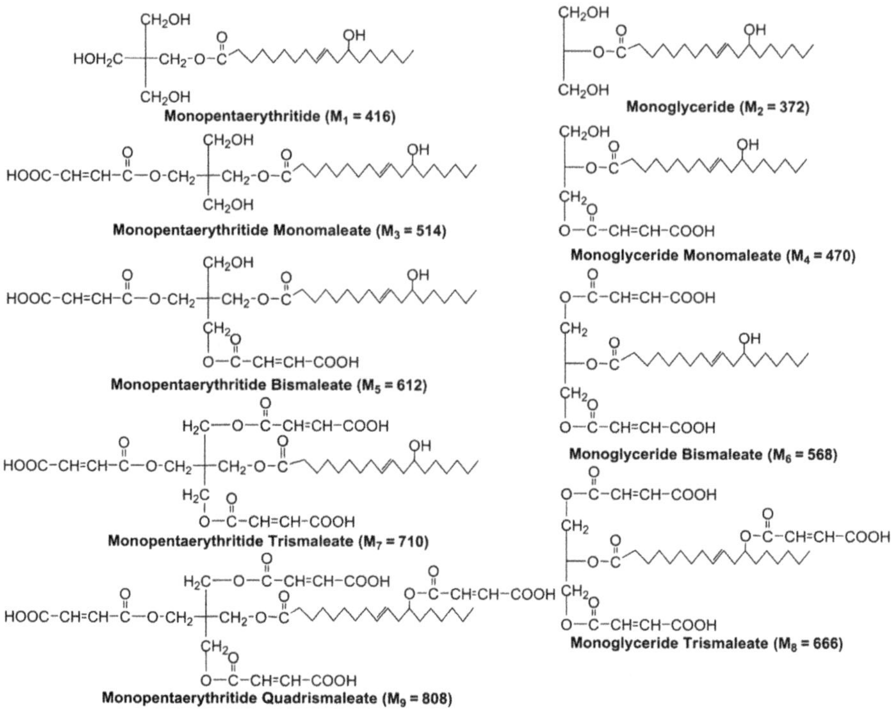
Monopentaerythritide (M1 = 416)
Monoglyceride (M2 = 372)
Monopentaerythritide Monomaleate (M3 = 514)
Monoglyceride Monomaleate (M4 = 470)
Monopentaerythritide Bismaleate (M5 = 612)
Monoglyceride Bismaleate (M6 = 568)
Monopentaerythritide Trismaleate (M7 = 710)
Monoglyceride Trismaleate (M8 = 666)
Monopentaerythritide Quadrismaleate (M9 = 808)

Scheme 8.6 Mass numbers of probable structures expected in COPER and COPERMA.

To understand the phase-separation mechanism on the microscale, the failure surface morphologies of the bio-based polymer samples were observed by SEM. Figure 8.11 demonstrates the failure surface morphologies after completion of the tensile tests. It was seen that the surface became less rough as the styrene concentration increased from 20 to 40 wt%. This may result from an increase in solubilization of the maleinated glycerides in styrene as the styrene concentration increases.[24] The TOPERMA80-ST20 polymer matrix showed the roughest surface with a large number of small resin pieces compared to the other three matrices. The surfaces of TOPERMA67-ST33 and TOPERMA60-ST40 samples were generally flat and featureless, suggesting that the behavior of the bio-based materials is elastic and that cracks propagate in a planar manner under tensile loading. On the other hand, phase separation was observed in the polymer matrices. The second phase presented as craters and holes, which have also been reported by other researchers.[10–12,16] They appeared in different sizes, but had the same composition,[10] which may contain TOPERMA monomer (because maleates do not homopolymerize significantly[8,25,26]), polystyrene, and unreacted oil molecules. From the SEM images (a), (b) and (d), it can be seen that the craters and holes in the second phase became less dense and smaller in size as the styrene concentration increases, while (c) shows a

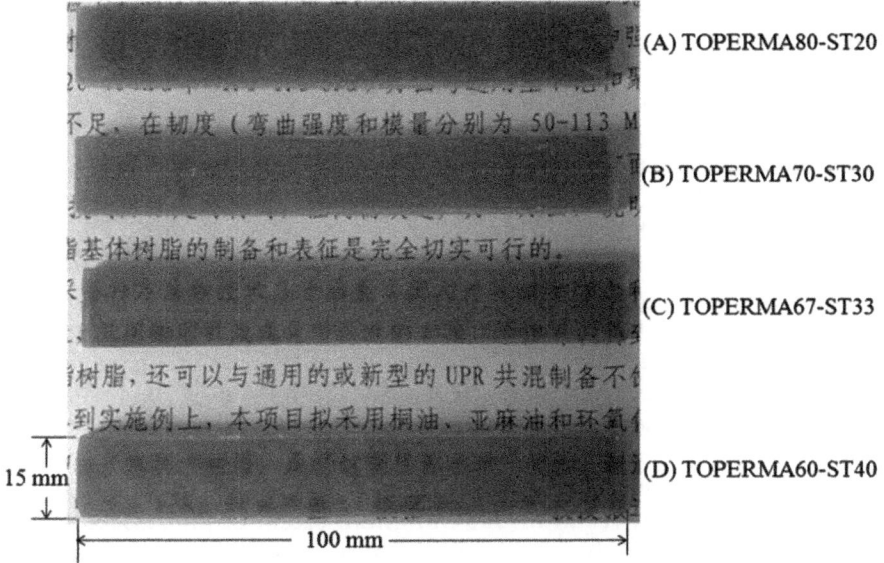

Figure 8.10   Polymer matrices of the TOPERMA resin with (A) 20% (B) 30% (C) 33%, and (D) 40% styrene concentration by weight for flexural testing.

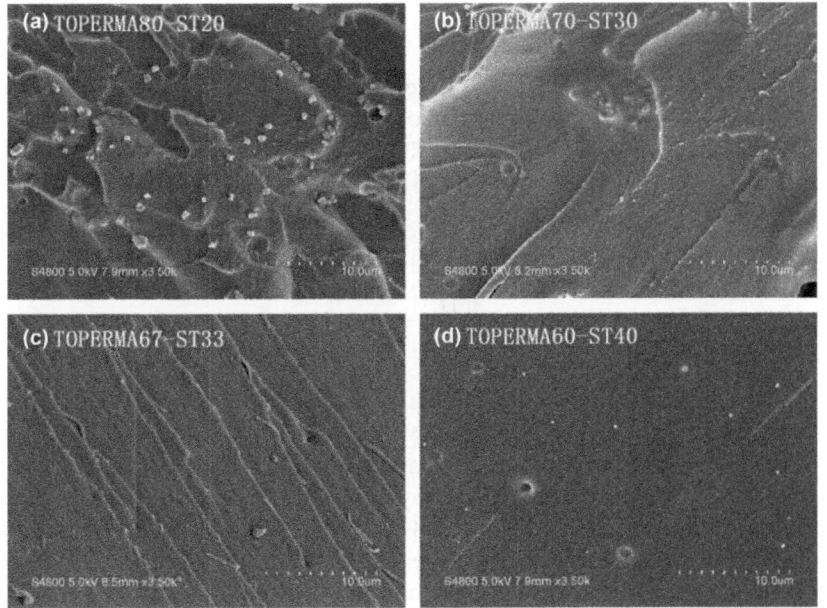

Figure 8.11   SEM micrographs of the tensile fracture surfaces of the bio-based polymers (a) TOPERMA-ST20, (b) TOPERMA-ST30, (c) TOPERMA-ST33, and (d) TOPERMA-ST40.

uniform feature and a planar direction without a second phase. It indicates that the four samples have an order of phase separation extent of (a)>(b)>(d)>(c), which agrees well with the transparency analysis.

The phase separation in solution or matrix can be attributed to the incompatibility of the maleinated oil-based resins and styrene. The acid number of TOPERMA was above 200 mg KOH g^{-1},[17] which is much higher than general-purpose UPEs (about 30 mg KOH g^{-1}). As a consequence, the maleate half-esters have a much higher polarity compared to styrene, leading to the incompatibility between them. Similar results for the mixtures of soybean oil pentaerythritol glyceride maleates (SOPERMA) and styrene were reported.[8] As the styrene content increases, the phase separation extent should decrease gradually due to better solubilization of the maleinated glycerides in styrene. Therefore, why did the TOPERMA67-ST33 matrix show the minimum phase-separation extent after curing? One possible effect is that the high curing temperature improves the collision chance between TOPERMA and styrene, thus accelerating their diffusion to each other. Another effect that may result in this minimum extent of phase separation is the reactivity ratios of the reactive monomers. It is known that each propagation reaction has a characteristic rate constant, K_{mn}, where the first subscript refers to the active center and the second refers to the monomer. Defining the propagation rate constant ratio k_{11}/k_{12} and k_{22}/k_{21} as r_1 and r_2, respectively, one finally obtains:[26]

$$\frac{d[M_1]}{d[M_2]} = \frac{[M_1]}{[M_2]} \cdot \frac{r_1[M_1] + [M_2]}{r_2[M_2] + [M_1]} \tag{2}$$

where $[M_1]/[M_2]$ and $d[M_1]/d[M_2]$ denote the initial feed ratio of monomers and the mole fraction ratio of monomer units in the co-polymer, respectively. Here styrene and TOPERMA were designated the 1st and 2nd monomers, respectively. When $d[M_1]/d[M_2] = [M_1]/[M_2]$, an azeotropic co-polymerization between them is established. As a result eqn (2) can be written as

$$\frac{[M_1]}{[M_2]} = \frac{r_2 - 1}{r_1 - 1} \tag{3}$$

The reactivity ratios of styrene and monoethyl maleate were reported as $r_1 = 0.13$ and $r_2 = 0.035$, respectively.[25] If these values are utilized for the TOPERMA/styrene system, the molar ratio $[M_1]/[M_2]$ is about 1.11, *i.e.*, the weight fraction of styrene is about 32 wt%. This value is almost equal to the styrene content in the TOPERMA67-ST33 feed. Normally, when the feed ratio of two monomers reaches the ratio for an azeotropic co-polymerization, the resulting polymer has a tendency to form a homogeneous system.[6,27] Otherwise matrices appear as complex heterogeneous systems composed of styrene-rich phases (hard domains) and oil-rich phases (soft domains).[5]

8.3.2.2   Dynamic Mechanical Analysis

The temperature dependence of the storage modulus (E') for the TOPERMA polymers is demonstrated in Figure 8.12(a). A distinct feature for the four triglyceride-based resins was that these polymers showed very broad transitions from the glassy to rubbery state. Another feature was that they showed a skew glassy region, unlike most common thermosetting polymers, which have a distinct glassy region in which the modulus is independent of temperature. The broad glass transition mainly results from three effects. First, TOPERMA is a complicated mixture containing many kinds of maleates, as shown in Scheme 8.5. All these maleates can co-polymerize with styrene, thus forming a network with complicated components. Second, the phase separation results in higher glass-transition-temperature (T_g) styrene-rich regions, and lower T_g TOPERMA-rich regions in the polymer matrix. Third, the plasticizing effect of soft fatty acid chains in the TOPERMA mixture may also result in the broad glass transition.[1,24]

At room temperature, all four polymers were already in the transition from the glassy region to the rubbery plateau. The E' values at room temperature are listed in Table 8.2. Based on the kinetic theory of rubber elasticity, the

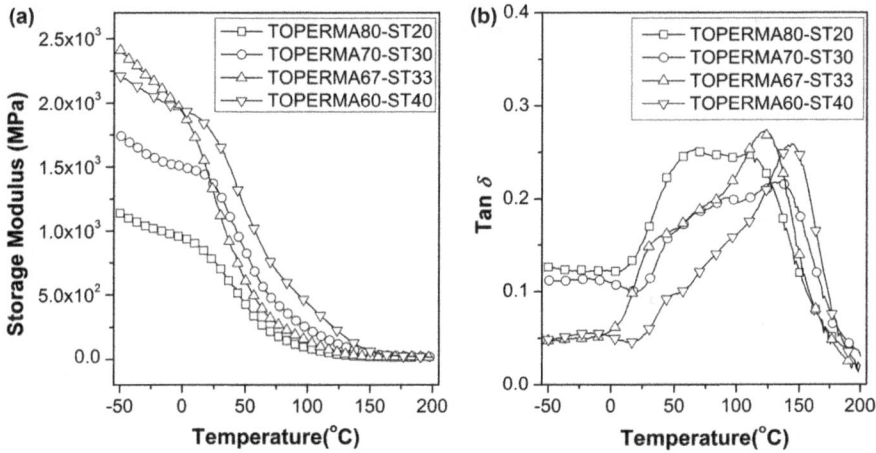

Figure 8.12   Dynamic mechanical analysis results. (a) Storage modulus and (b) loss factor for the four TO-based resins.

Table 8.2   Storage modulus (E'), glass-transition temperature (T_g), cross-link density (ν_e), and effective molar mass between cross-links (M_c) of the TO-based thermosetting polymers.

Sample	E' at 25 °C/MPa	T_g/°C	ν_e/mol m^{-3}	M_c/g mol^{-1}
TOPERMA80-ST20	765	65, 107	1483	748
TOPERMA70-ST30	1360	133	2139	518
TOPERMA67-ST33	1290	121	1775	631
TOPERMA60-ST40	1740	144	2145	527

experimental cross-link density of the co-polymers (ν_e), namely the average number of cross-links per unit volume, can be determined from the rubbery modulus using the following equation:[23,24,28]

$$E' = 3\nu_e RT = \frac{3dRT}{M_c} \tag{4}$$

where E', R, T, d and M_c represent the storage modulus of the cross-linked co-polymer in the rubbery region, the gas constant (8.314 J K^{-1} mol^{-1}), the absolute temperature, the polymer density, and the effective molar mass between cross-links, respectively. The E' values used for the calculations were taken at approximately 50 °C above T_g. The density of the TOPERMA/styrene resins ranged from 1.11 to 1.13 g cm^{-3}. As shown in Table 8.2, the ν_e values of these bio-polymers were 1483–2145 mol m^{-3}, which correspond to M_c values of 522–755 g mol^{-1}. The ν_e values for 20 to 40 wt% styrene concentration were higher than those of SOPERMA/styrene polymers (1050–1550 mol m^{-3}), but lower than those of COPERMA/styrene polymers (1700–4500 mol m^{-3}).[24] A possible reason for this is that the fatty acid chains in SOPERMA barely participate in the free-radical co-polymerization with styrene, while the fatty acid chains in COPERMA are functionalized with MA and can readily co-polymerize with styrene. Also, the fatty acid chains in TOPERMA still have some conjugated trienes remaining (although some of them are consumed by the Diels–Alder reaction), which may participate in the co-polymerization.

The temperature dependence of the loss factor (tan δ) is shown in Figure 8.12(b), in which the tan δ curves also exhibit very broad peaks. The curves for TOPERMA80-ST20 and TOPERMA70-ST30 show two glass transitions due to phase separation. This revealed two T_g values at the tan δ peaks: one was for the TOPERMA-rich region at about 65 °C; the other was for the styrene-rich region in the range of 107–144 °C. This two-glass-transition phenomenon was similar to previous reported results.[1,5,6] As the styrene content increased, the lower transition peak decreased in intensity relative to the higher transition peak, which indicates a content decrease of the TOPERMA-rich phase and a content increase of the styrene-rich phase. The possible effect for this decrease can be attributed to the increase of styrene content producing more effective solubilization for TOPERMA in styrene, which leads to the formation of a less heterogeneous system. tan δ is a sensitive indication of cross-linking. As the cross-link density increases, the tan δ maximum shifts to higher temperatures, the peak broadens and the tan δ values decrease.[24] As a result, the TOPERMA67-ST33 polymer with a lower cross-link density had a lower T_g value than TOPERMA70-ST30 and TOPERMA60-ST40.

8.3.2.3   Effect of Styrene Concentration on Mechanical Properties

To find the appropriate styrene concentration, it is necessary to investigate the effect of styrene concentration on the mechanical properties of the

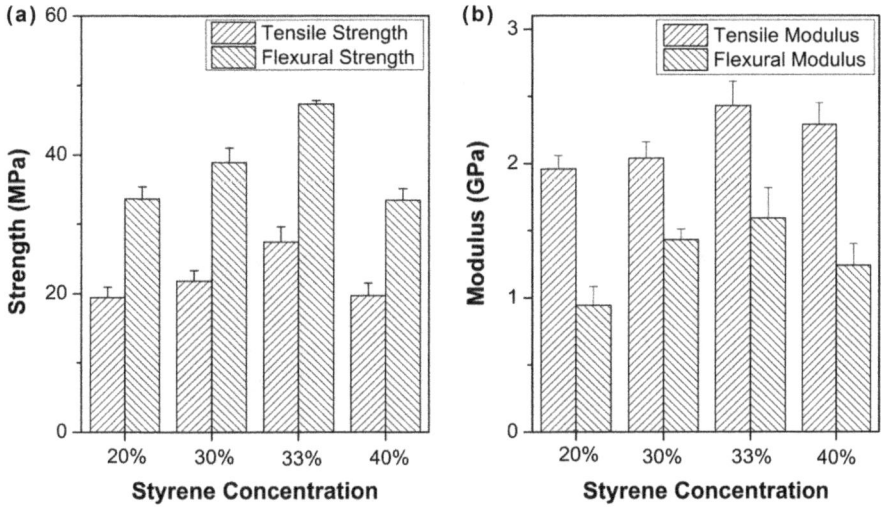

Figure 8.13 Tensile and flexural properties. (a) Strength and (b) modulus of the four TO-based resins with different styrene concentrations.

TO-based polymer. The mechanical properties of TOPERMA polymers (Figure 8.13) varied with styrene concentration in a different manner to the changes seen in the SOPERMA and COPERMA polymers reported.[24] For example, the flexural strength and modulus of the TOPERMA polymers had their maximum values at a styrene content of 33 wt%, while the flexural strength and modulus of both the SOPERMA and COPERMA polymers increased with increasing styrene content. It was explained that these changes were caused by their cross-link densities.[24] However, the variation of cross-link density of those polymers did not match the changes in properties. It seems that some other important factors affecting the mechanical properties were neglected. From our transparency and SEM analysis, we believe that the effect of phase separation is one of these important factors. Hence, the mechanical properties of the four resins were explored, with respect the effects of phase separation and cross-link density.

Figure 8.13 shows the changes in tensile and flexural strength and modulus of the four resins with different styrene contents. These mechanical properties reflect the stiffness of a polymer matrix. It was observed that all these properties had a similar trend: increasing with the increase of styrene content until 33 wt% styrene, then decreasing after this point. This trend was similar to the change of phase separation, but did not agree with the change of cross-link density. In other words, the phase separation seems to have a more pronounced effect on the stiffness of the polymers, than the cross-link density. The introduction of a second, rubber-like, oil-rich phase improves the toughness of a polymer matrix but sacrifices its stiffness, while the increase of cross-link density shows the opposite effect.[8,10,14,23,24,28] Due to the absence of phase separation, the TOPERMA67-ST33 matrix achieved

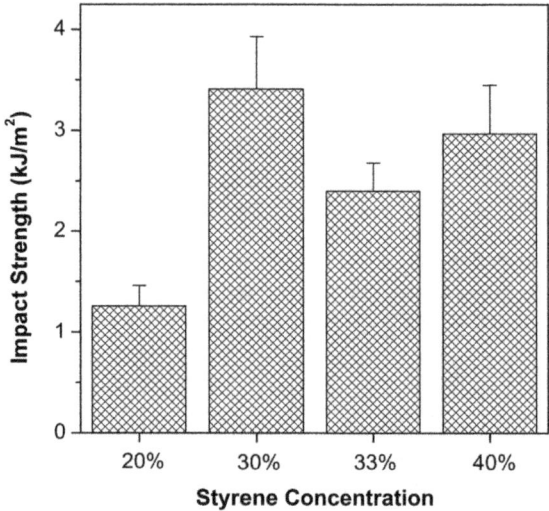

Figure 8.14 Impact strength of the four TO-based resins with different styrene concentrations.

the best stiffness in the four resins, although it had a moderate cross-link density (Table 8.2). Hence, phase separation obviously had the dominant effect on the tensile and flexural properties of the four resins rather than cross-link density.

The change in impact strength of the four resins is depicted in Figure 8.14. The impact strength can reflect the toughness of a polymer matrix. The TOPERMA70-ST30 and TOPERMA60-ST40 polymers had higher impact strengths than the TOPERMA67-ST33 polymer, although they also had higher cross-link densities. This is because the rubber TOPERMA-rich phase absorbs the impact energy and improves the impact strength of the two polymers.[10,23] The TOPERMA80-ST20 matrix had the lowest cross-link density and the highest extent of phase separation, but showed the worst toughness, which may be caused by having the lowest content of styrene and the highest plasticizing effect of the pendant fatty acid chains.[8,24] Overall, the effect of phase separation plays an important role in the mechanical properties of TO-based thermosets.

8.3.3 Structure and Properties of the Cured DCPD-UPR-TO Thermosets

8.3.3.1 SEM Analysis

The surface morphologies of the matrices obtained from DCPD-UPR-TO polymers with different TO contents were investigated by SEM. Figure 8.15 shows the failure surface morphologies after the completion of impact tests. In Figure 8.15(a), the surface of the DCPD-UPR matrix was generally flat and

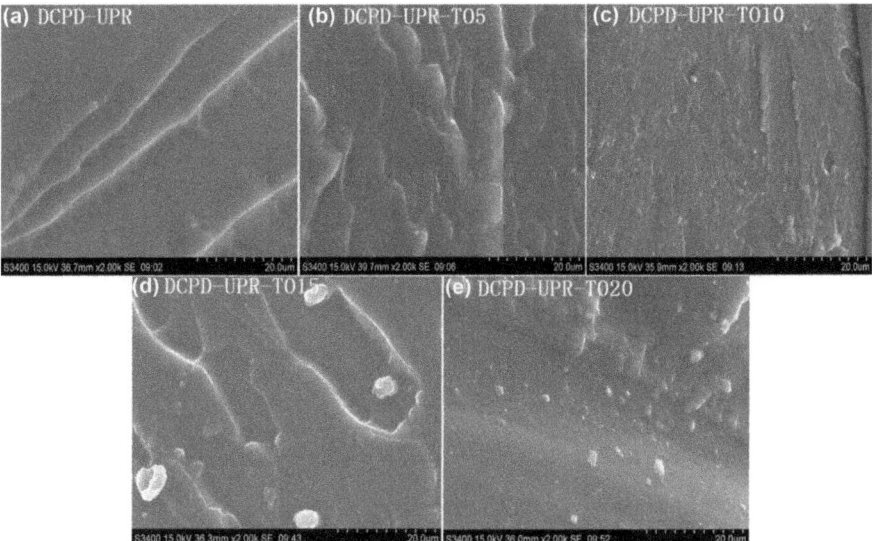

Figure 8.15 SEM micrographs of the impact fracture surfaces of the polymer matrices obtained from DCPD-UPR-TO polymers with different TO contents.

featureless. This suggests that the behavior of the matrix is linear elastic, and the crack propagates in a planar manner under impact loading. However, the surface of the TO-modified DCPD-UPR matrices became rougher due to the soft structure provided by the pendant fatty acid chains and unreacted TO triglycerides. It has been reported that the rougher surface is identical to enable it to dissipate more energy due to shear deformation during crack propagation.[23] Thus it is expected that the increased roughness will have a pronounced toughening effect on the brittle DCPD-UPR matrix. In addition, the DCPD-UPR-TO matrices with TO content from 0 to 10 wt% were generally homogeneous, while the matrices with a TO content higher than 10 wt% clearly demonstrated phase separation such as small resin pieces. This phenomenon also indicates that TO can completely react with DCPD-UPR chains up to a TO content of about 10 wt%, which is in good agreement with the ^{1}H-NMR and GPC results. This phenomenon is also expected to affect the properties of the TO-modified DCPD-UPR bio-materials. It is noteworthy that the exact TO weight fraction in the cured DCPD-UPR-TO10 matrix was about 7.4%. This value was much higher than that reported by Das *et al.*,[16] in whose work, through a simple physical-blending approach, the prepared UPR/TO matrix showed clear phase separation at a TO content of about 2 wt%.

8.3.3.2 Dynamic Mechanical Analysis

DMA was conducted to study the thermomechanical properties of the obtained bio-plastics. The results are illustrated in Figure 8.16 and Table 8.3.

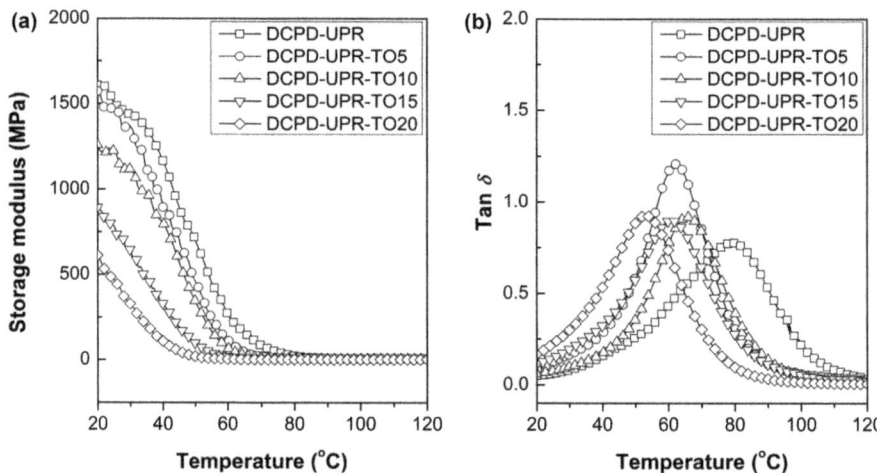

Figure 8.16 Dynamic mechanical analysis results. (a) Storage modulus and (b) damping parameter for the polymer matrices obtained from DCPD-UPR-TO polymers with different TO contents.

Table 8.3 Storage modulus (E'), glass-transition temperature (T_g), cross-link density (ν_e), and effective molar mass between cross-links (M_c) and 10 and 50% weight-loss temperatures (T_{10} and T_{50}) of the neat DCPD-UPR and DCPD-UPR-TO polymer matrices.

Sample	E' at 25 °C/GPa	T_g/°C	$\nu_e/10^3$ mol m^{-3}	$M_c/$ g mol^{-1}	T_{10}/°C	T_{50}/°C
DCPD-UPR	1.53	79.3	2.81	391	294.5	382.0
DCPD-UPR-TO5	1.26	61.8	2.73	403	294.5	384.5
DCPD-UPR-TO10	1.24	65.6	2.59	424	302.0	389.5
DCPD-UPR-TO15	0.802	60.7	2.40	458	294.5	387.0
DCPD-UPR-TO20	0.490	53.1	2.29	480	292.0	389.5

The storage moduli of all the DCPD-UPR-TO bio-plastics at 25 °C were lower than that of the neat DCPD-UPR, which means the dynamic mechanical properties of the DCPD-UPR-TO matrices were decreased by the addition of TO. Based on eqn (4), the ν_e and M_c values were determined, as shown in Table 8.3. The E' values used for the calculations were taken at approximately 50 °C above the T_g. The density of the DCPD-UPR-TO resins ranged from 1.098 to 1.10 g cm^{-3}. The ν_e values were in the range of 2290–2810 mol m^{-3}, which corresponds to M_c values of 391–480 g mol^{-1}. It was found that E' at 25 °C and ν_e decreased as the TO content increased. This is mainly because the intermolecular Diels–Alder reaction between polyester and TO reduces the number of reactive C=C bonds on the polyester chains. T_g is determined from the peak of tan δ. It can be seen from Figure 8.16(b) and Table 8.3 that the T_g value of the DCPD-UPR-TO bio-plastic was reduced by the incorporation of TO and had a minimum decrease of 13.7 °C for DCPD-UPR-TO10 compared to neat DCPD-UPR. This may be caused by the combined effect of phase separation and cross-link density: the decrease in cross-link density results in the decrease of T_g, but the phase separation phenomenon at a TO content of 10 wt% shields the DCPD-UPR-TO10 matrix from this effect.

8.3.3.3 Thermogravimetric Analysis

Figure 8.17 shows TGA thermograms of the cured DCPD-UPR-TO polymer matrices. All the bio-plastics were thermally stable in nitrogen gas below 180 °C and exhibited a three-stage thermal degradation above this temperature. The first stage degradation (about 180–320 °C) was attributed to the evaporation and decomposition of unreacted oil or other soluble components in the bulk material.[5] The second stage (about 320–460 °C) was fastest and corresponded to degradation and charformation of the cross-linked polymer structure, while the last stage (>460 °C) corresponded to gradual degradation of the char residue.[5] Table 8.3 summarizes the thermal data for these bio-based materials, including 10 and 50% mass-loss temperatures (T_{10} and T_{50}). Both of the temperatures increased to maximum values at a TO content of 10 wt%, which is mainly caused by the phase separation effect as mentioned above. Compared with the neat DCPD-UPR matrix, the DCPD-UPR-TO10 matrix showed a maximum increase of 7.5 °C in both T_{10} and T_{50}. Hence, the DCPD-UPR-TO10 matrix shows slightly better thermal stability than the pure DCPD-UPR matrix.

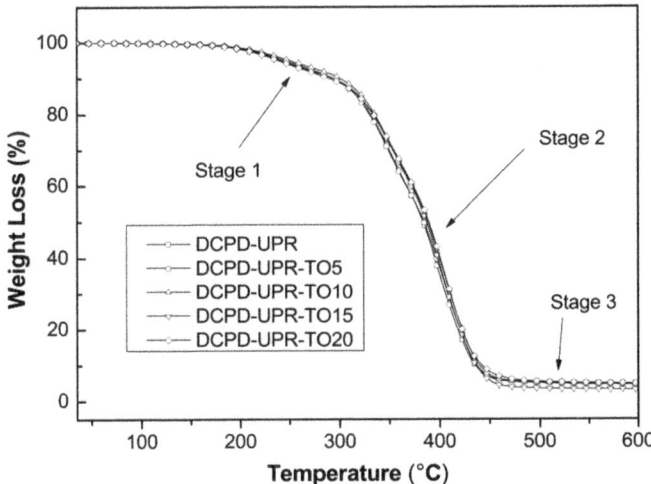

Figure 8.17   TGA curves of the polymer matrices obtained from DCPD-UPR-TO polymers with different TO contents.

8.3.3.4   Effect of TO Content on Mechanical Properties

The changes in tensile and flexural strength and modulus of the matrices obtained from DCPD-UPR-TO polymers with different TO contents are illustrated in Figure 8.18. All these properties decreased as the TO content increased, indicating a decrease in stiffness for all DCPD-UPR-TO polymer matrices. This is mainly due to the decrease in cross-link density. Interestingly, the decrease in stiffness showed two distinct regions: at TO contents below 10 wt%, the stiffness of DCPD-UPR-TO bio-materials gradually decreased; at TO contents above 10 wt%, the stiffness decreased rapidly. This phenomenon can be attributed to the phase separation, as indicated in the SEM images (Figure 8.15). As mentioned above, the introduction of a second rubber-like, oil-rich phase would sacrifice the stiffness of a polymer matrix. Compared to the neat DCPD-UPR matrix, the DCPD-UPR-TO10 matrix demonstrated decreases of 5.3% and 18.8% in tensile and flexural strength, and 19.3% and 16% in tensile and flexural modulus, respectively.

The changes in impact strength and tensile failure strain of the DCPD-UPR-TO polymer matrices are shown in Figure 8.19. All these properties increased as the TO content increased suggesting an increase of toughness for all the DCPD-UPR-TO polymer matrices. Similar to the stiffness shown above, the increase of toughness could be divided into two distinct regions, which was also caused by the effect of phase separation. Compared to the neat DCPD-UPR matrix, the DCPD-UPR-TO10 matrix showed increases of 72.5 and 91.3% in impact strength and tensile failure strain, while the increases for the DCPD-UPR-TO20 matrix were 373 and 875%, respectively. Hence, the toughness of the DCPD-UPR matrix was greatly improved by the chemical approach we employed. Such a large improvement in toughness is

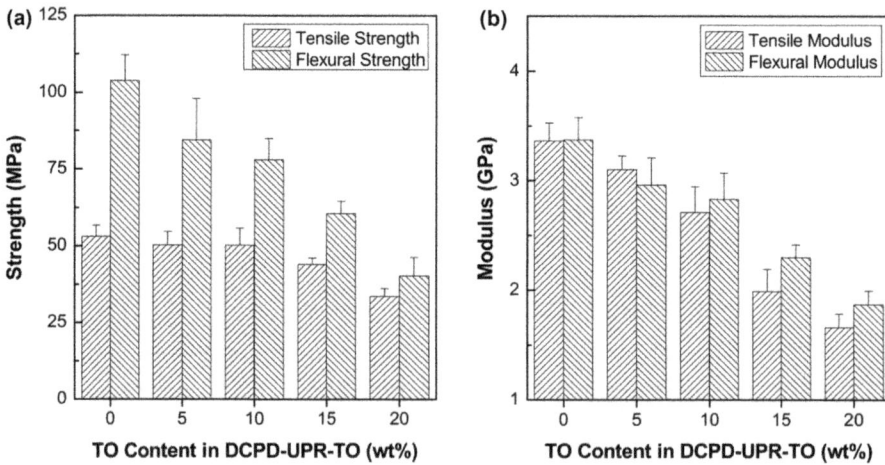

Figure 8.18   Tensile and flexural properties. (a) Strength and (b) modulus of the polymer matrices obtained from DCPD-UPR-TO polymers with different TO contents.

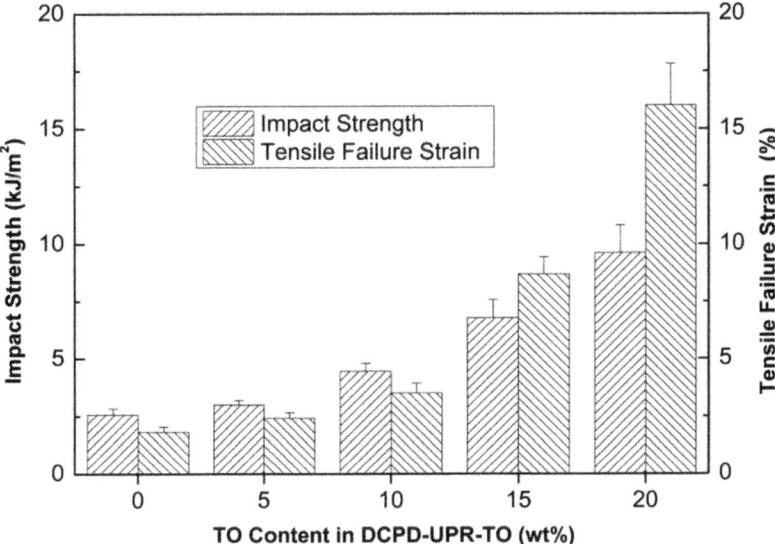

Figure 8.19 Impact strength and tensile failure strain of the polymer matrices obtained from DCPD-UPR-TO polymers with different TO contents.

due to the occurrence of phase separation and the decrease of cross-link density, as analyzed above.

8.3.4 Structure and Properties of the Cured UPR/COPERMA Thermosets

The physical properties of the obtained UPR/COPERMA resins are shown in Table 8.4. It can be seen that the viscosities and gel times of the UPR/COPERMA resins are in a range suitable for liquid molding processing. Moreover, the addition of COPERMA into UPR reduced the linear shrinkage of the resulting polymer matrix. This could be attributed to the abundance of acid groups on COPERMA, which leads to better wetting and spreading in the resin transfer molding process and good adhesion between the resins and steel molds. The synthesized COPERMA or TOPERMA resins should be thickened easily with divalent cations (*e.g.*, MgO) because of the complexation of MgO with their carboxylic acid groups when they are used as sheet molding compound resins.[28] The high A_v values of COPERMA and TOPERMA may also increase their compatibility with some fillers, like wood flour and some untreated reinforced fibers, due to interactions between the acid groups and the hydroxyl groups on the surfaces of fillers or fibers.[29] The flexural strength of the COPERMA polymer matrix was 104.6 MPa. Its tensile strength, which was tested using a matrix cured by the same procedure as Can *et al.*,[24] was about 38.5 MPa. These values are very high compared to those of other reported oil-based monomers/styrene co-polymers.[18,22,24]

Table 8.4   Viscosity (V_s), gel time (t_{gel}), and linear shrinkage (l_s) of the neat UPR, COPERMA, and UPR/COPERMA resins.

Sample	V_s/mPa·s	t_{gel}/min	l_s/%
UPR	440	7–8	0.52 ± 0.09
COPERMA	2680	>30	—
UPR/COPERMA5	570	8–9	0.41 ± 0.07
UPR/COPERMA10	680	9–10	0.31 ± 0.06
UPR/COPERMA15	810	10–11	0.12 ± 0.04
UPR/COPERMA20	950	11–12	0

On the other hand, the COPERMA resin did not generate rigid bio-plastics using the curing procedure for general-purpose UPRs, as shown in the experimental section. Therefore, it will be valuable to partially substitute petroleum-based UPRs to create blends with this relatively new bio-resin.

8.3.4.1   Dynamic Mechanical Analysis

The DMA technique was employed to investigate the thermomechanical properties of the obtained bio-materials. The results are shown in Figure 8.20 and Table 8.5. The storage moduli at 35 °C of UPR/COPERMA5 and UPR/COPERMA10 were higher than that of neat UPR, while it was the opposite for UPR/COPERMA15 and UPR/COPERMA20. A maximum increase of 17.6% was observed for UPR/COPERMA5. The ν_e and M_c values were calculated according to eqn (4), and are presented in Table 8.5. The values of E' used for the calculations were taken at approximately 40 °C above the T_g in these materials. The density of the UPR/COPERMA resins ranged from 1.099 to 1.103 g cm^{-3}. The ν_e values were in the range of 2270–2570 mol m^{-3}, which corresponds to molar masses of 429–484 g mol^{-1}. With the addition of COPERMA into UPR, the ν_e increased slightly, which was quite different from the reduction caused by the incorporation of other reported bio-additives.[11,12] This ν_e increase may lead to a pronounced effect on the mechanical properties of the obtained polymers. The temperature dependence of tan δ is shown in Figure 8.20(b). The tan δ curves showed only a single T_g, suggesting that phase separation in the obtained UPR/COPERMA resins is not obvious. T_g was determined from the peak of tan δ, as illustrated in Table 8.5. A slight increase was observed for the bio-based resins. The maximum increase of about 3.4 °C was seen for UPR/COPERMA15, which can be attributed to it having the maximum ν_e increase.

8.3.4.2   Thermogravimetric Analysis

Figure 8.21 shows TGA thermograms and their derivative curves for the cured neat UPR and UPR/COPERMA polymer matrices. It can be seen that

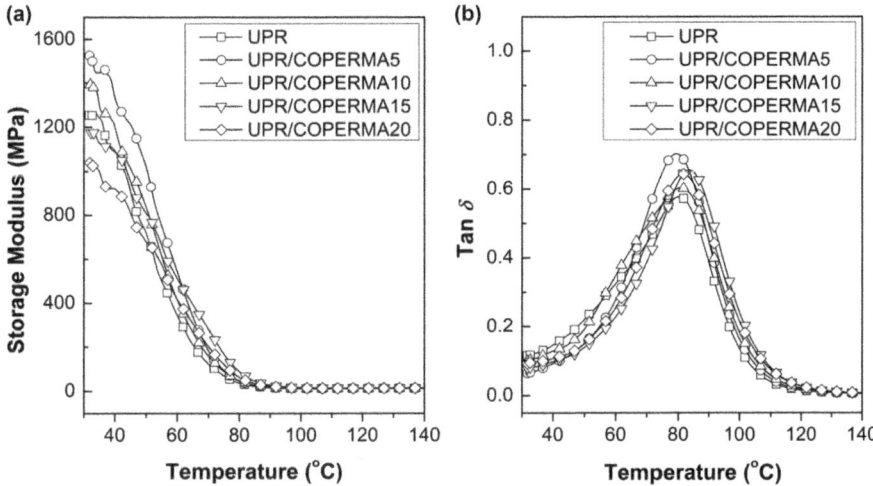

Figure 8.20 Dynamic mechanical analysis results. (a) Storage modulus and (b) damping parameter for the cured neat UPR and UPR/COPERMA resins.

Table 8.5 Storage modulus (E'), glass-transition temperature (T_g), cross-link density (ν_e), and effective molar mass between cross-links (M_c), 10 and 50% weight-loss temperatures (T_{10} and T_{50}), and temperature at the maximum weight-loss rate (T_p) of neat UPR and UPR/COPERMA resins.

Sample	E' at 35 °C/GPa	T_g/°C	ν_e/ 10^3 mol m^{-3}	M_c/ g mol^{-1}	T_{10}/°C	T_{50}/°C	T_p/°C
UPR	1.26	80.3	2.27	484	348.5	401.2	403.7
UPR/COPERMA5	1.48	79.4	2.36	466	344.4	399.4	399.2
UPR/COPERMA10	1.32	80.9	2.48	444	343.4	400.6	401.6
UPR/COPERMA15	1.14	83.7	2.57	429	340.4	400.7	401.0
UPR/COPERMA20	1.01	81.6	2.42	454	328.2	401.4	400.2

all the bio-materials were thermally stable in an N_2 atmosphere below 150 °C and exhibited a three-stage thermal-degradation process above this temperature similar to the above DCPD-UPR-TO resins. Table 8.5 summarizes the thermal property data of these bio-based plastics, including T_{10}, T_{50}, and the temperature at maximum weight-loss rate (T_p). As the content of COPERMA resin increased, T_{10} decreased while T_{50} and T_p almost remained unchanged. The decrease of T_{10} can be mainly attributed to the decomposition of the COPERMA, the co-polymer of which was cured using the method in Can *et al.*[24] and showed a T_{10} value of 279.8 °C (its TGA curve is not included in this chapter). However, the T_{10} decrease was slight when 0–15 wt% of the UPR was replaced with the COPERMA resin. Hence, the prepared bio-based materials demonstrated only slightly inferior thermal stability compared to the pure UPR matrix.

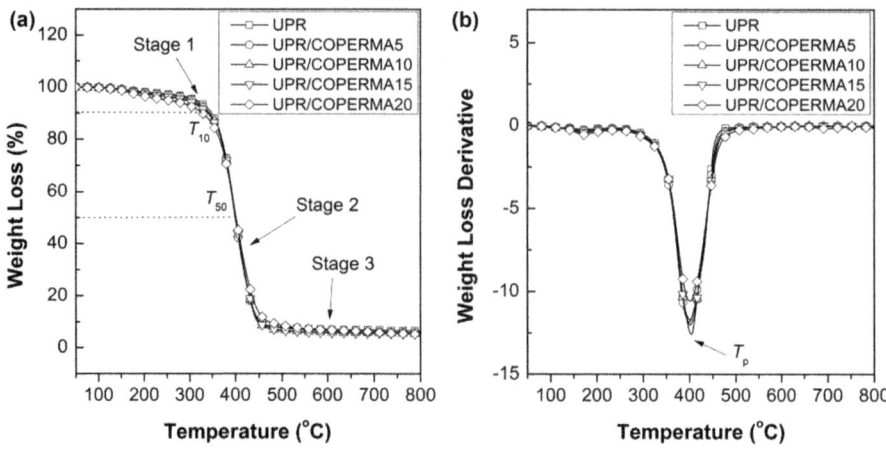

Figure 8.21 TGA curves (a) and their derivatives (b) of the cured neat UPR and UPR/ COPERMA resins.

8.3.4.3 *Effect of COPERMA Content on Mechanical Properties*

The changes in tensile and flexural properties of the cured neat UPR and UPR/COPERMA resins are shown in Figure 8.22. The influence of the COPERMA content on the tensile strength of the UPR/COPERMA resins is presented in Figure 8.22(a). The tensile strength remained constant for up to 10 wt% COPERMA, thereafter it began to decrease as the concentration increased. Compared with neat UPR, there were decreases of 18.6 and 23.0% in tensile strength for UPR/COPERMA15 and UPR/COPERMA20, respectively. However, no obvious changes in flexural strength were observed in resins when the COPERMA content varied from 0 to 20 wt% [Figure 8.22(a)]. The tensile and flexural moduli for the UPR/COPERMA resins were larger than those of neat UPR [Figure 8.22(b)]. For example, the tensile modulus for the neat UPR was 3.02 GPa, while the modulus increased by 14.6 and 16.6% for UPR/COPERMA5 and UPR/COPERMA15, respectively. The flexural modulus increased by 52.6 and 47.4% in UPR/COPERMA5 and UPR/COPERMA15, respectively, compared with that of neat UPR. In summary, after the incorporation of the COPERMA resin into UPR, the stiffness of the obtained bio-based resins was comparable to that of neat UPR, meeting the requirements for engineering plastics or fiber-reinforced materials. The reason for this can be attributed to the addition of the high-functionality COPERMA, which improves the cross-link density of the resultant materials. Other functionalized oil products do not have this effect.[11,12,15]

The changes in tensile failure strain and impact strength of the cured neat UPR and UPR/COPERMA resins are depicted in Figure 8.23. The change in tensile failure strain [Figure 8.23(a)] showed a similar change to tensile strength: no obvious difference was observed when 10 wt% of UPR was replaced by COPERMA resin. Composites containing more than 10 wt%

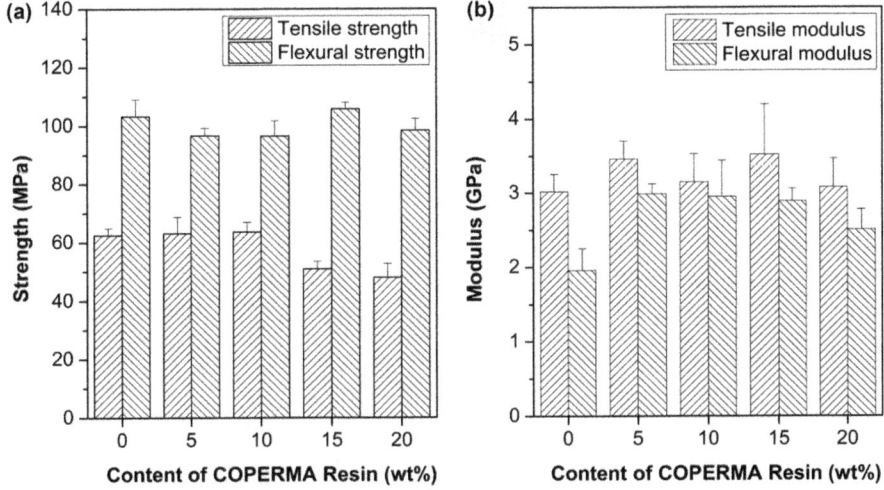

Figure 8.22 Tensile and flexural properties. (a) Strength and (b) modulus of the cured neat UPR and UPR/COPERMA resins.

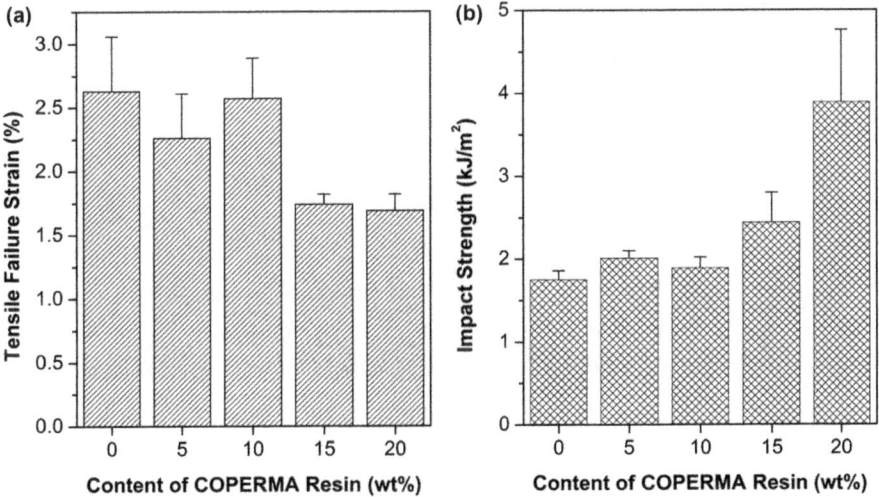

Figure 8.23 (a) Tensile failure strain and (b) impact strength of the cured neat UPR and UPR/COPERMA resins.

COPERMA resin showed a decrease in breaking strain from 2.63 to 1.69%. The impact strength of the neat UPR was 1.75 kJ m^{-2}. No obvious difference in impact strength was observed when 0–10 wt% of UPR was replaced by the COPERMA resin. After more than 10 wt% of the UPR was replaced by the COPERMA resin, the impact strength began to increase. A maximum increase of 122% was observed in UPR/COPERMA20.

8.3.4.4   Morphology of Impact Failure

To understand the mechanism behind the variation of impact strength, surface morphologies at the impact failure of the cured samples were investigated by SEM. Figure 8.24(a)–(e) demonstrate the SEM micrographs of the impact failure surfaces of neat UPR and UPR/COPERMA samples. In Figure 8.24(a) and (b), the failure surfaces of the neat UPR and UPR/COPERMA5 samples were generally flat and featureless. This indicates that the behavior of the polymer matrices is linear elastic, and the crack propagates in a planar manner under the impact loading. In fact, the impact strengths of these bio-based UPRs containing 0–5 wt% of the COPERMA resin were almost constant. In contrast, the failure surface of the

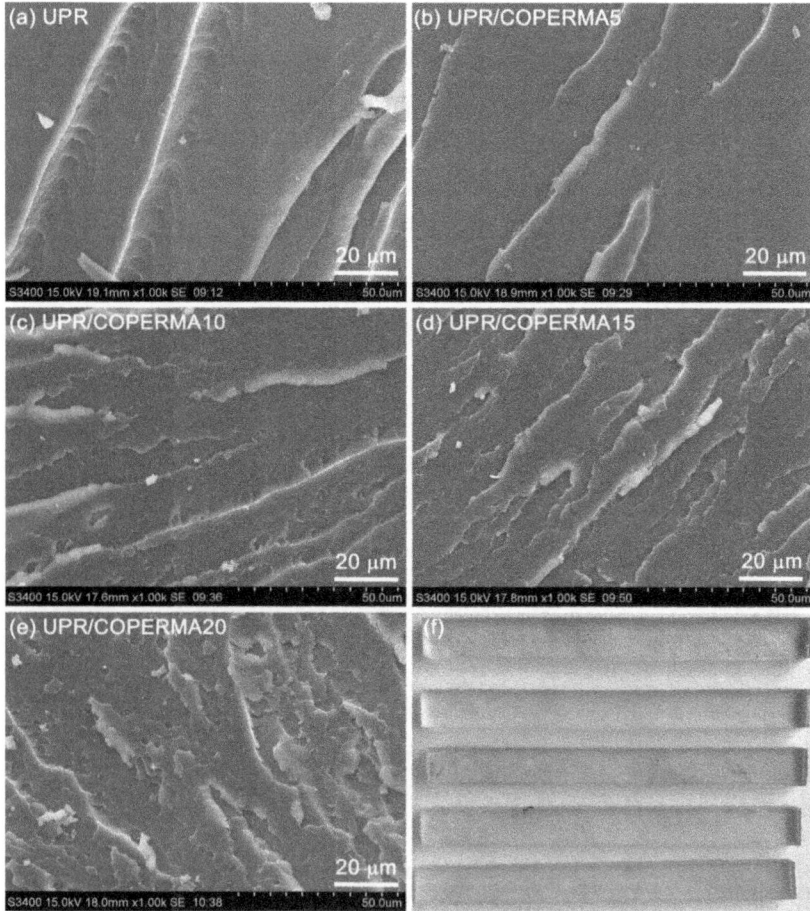

Figure 8.24   SEM micrographs of the impact fracture surfaces of the cured (a) neat UPR, (b) UPR/COPERMA5, (c) UPR/COPERMA10, (d) UPR/COPERMA15, and (e) UPR/COPERMA20 resins. (f) Photographs of the impact test samples of (a) to (e) from top to bottom.

UPR/COPERMA resins became rougher as the COPERMA-resin content increased from 10 to 20 wt%.

The increased roughness suggests that all constituents of UPR and COPERMA are not homogeneously mixed at the molecular level, which may result from the incompatibility of the petroleum-based UPR and COPERMA, probably generated by the large difference in acid group numbers. As mentioned above, the increase of cross-link density would lead to a decreasing toughness of the resulting thermosets, while a rougher surface corresponds to the dissipation of more impact energy, due to shear deformation during crack propagation.[10,11,14,15,18,23] Hence, although the slight increase in cross-link density (Table 8.5) decreases the toughness of the UPR/COPERMA polymer matrix a little bit, the increase in roughness leads to a pronounced toughening effect and compensates for the loss of toughness caused by the increase of ν_e. When the roughness effect overwhelms the ν_e effect on toughness, the impact strength of UPR/COPERMA resin begins to increase. This is the reason why an increase in toughness was observed from UPR/COPERMA10 to UPR/COPERMA20.

All of the samples tested were transparent or quasi-transparent from the visual inspection of the impact test samples [Figure 8.24(f)], indicating no obvious phase separation of ingredients in the material. This agreed well with the DMA results. When the content of COPERMA resin was increased from 0 to 5 wt%, the samples were transparent and almost colorless. However, as the content of COPERMA resin was increased from 10 to 20 wt%, the samples were not as transparent as the pure UPR and UPR/COPERMA5 and the color of them changed to light yellow. These results agree with the SEM observations of surface roughness changes of the UPR/COPERMA samples.

8.4 Conclusions

Currently the development of oil-based UPRs from natural oils or their derivates is widely advocated due to the requirements of "green" chemistry all around the world. In this chapter, novel oil-based UPRs including TOPERMA resin, DCPD-UPR-TO resin, and UPR/COPERMA blends have been successfully fabricated by functionalizing plant oil triglycerides or blending with petroleum-based UPRs. These materials show promise for applications as engineering plastics and in composite fields. In the development of DCPD-UPR-TO polymers, the method employed was not only a simple blending method, but also involved the chemical modification of UPE *via* a Diels–Alder reaction between C=C bonds on polyester and TO conjugated trienes. This method avoids the occurrence of phase separation in the resultant polymer matrix, up to a TO content of 7.4 wt%. Moreover, by the incorporation of highly functionalized COPERMA into UPR, we obtained a bio-based UPR blend with physical, mechanical, and thermal properties that were comparable to, or even slightly better than, those of neat UPR. Therefore, blending a high-functionality UE monomer with petroleum-based UPR is an effective approach to fabricate a high-performance bio-based UPR for structural applications. These examples give us new inspiration in the development of oil-based UPRs.

On the other hand, the properties of the prepared oil-based UPR materials were found to be affected by phase separation and cross-link density. In the study of the above three oil-based UPR resins, we found that the occurrence of phase separation (or the increased roughness) improved the materials' toughness whilst sacrificing their stiffness, while the increase in cross-link density showed the opposite effect. This provides a new explanation for the structure–property relationships of these oil-based amorphous polymer materials.

Acknowledgements

The authors are grateful for financial support from the National Natural Science Foundation of China (No. 31300489) and the International S&T Cooperation Program of China (No. 2011DFA32440).

References

1. R. P. Wool and X. S. Sun, *Bio-Based Polymers and Composites*, Elsevier, Amesterdam, 2005.
2. F. S. Güner, Y. Yağcı and A. T. Erciyes, *Prog. Polym. Sci.*, 2006, **31**, 633.
3. L. M. de Espinosa and M. A. R. Meier, *Eur. Polym. J.*, 2011, **47**, 837.
4. M. A. R. Meier, J. O. Metzger and U. S. Schubert, *Chem. Soc. Rev.*, 2007, **36**, 1788.
5. D. D. Andjelkovic, M. Valverde, P. Henna, F. Li and R. C. Larock, *Polymer*, 2005, **46**, 9674.
6. F. Li and R. C. Larock, *Biomacromolecules*, 2003, **4**, 1018.
7. J. La Scala and R. P. Wool, *Polymer*, 2005, **46**, 61.
8. E. Can, R. P. Wool and S. Küsefoğlu, *J. Appl. Polym. Sci.*, 2006, **102**, 2433.
9. P. Zhang and J. W. Zhang, *Green. Chem.*, 2013, **15**, 641.
10. G. Mehta, A. K. Mohanty, M. Misra and L. T. Drzal, *Green Chem.*, 2004, **6**, 254.
11. H. Miyagawa, A. K. Mohanty, R. Burgueno, L. T. Drzal and M. Misra, *Ind. Eng. Chem. Res.*, 2006, **45**, 1014.
12. H. Miyagawa, A. K. Mohanty, R. Burgueno, L. T. Drzal and M. Misra, *J. Polym. Sci., Polym. Phys. Ed.*, 2007, **45**, 698.
13. M. Haq, R. Burgueño, A. K. Mohanty and M. Misra, *Composites, Part A*, 2009, **40**, 540.
14. M. Haq, R. Burgueño, A. K. Mohanty and M. Misra, *Composites, Part A*, 2011, **42**, 41.
15. S. Ghorui, N. R. Bandyopadhyay, D. Ray, S. Sengupta and T. Kar, *Ind. Crop Prod.*, 2011, **34**, 893.
16. K. Das, D. Ray, C. Banerjee, N. R. Bandyopadhyay, A. K. Mohanty and M. Misra, *J. Appl. Polym. Sci.*, 2011, **119**, 2174.

17. C. G. Liu, X. H. Yang, J. F. Cui, Y. H. Zhou, L. H. Hu, M. Zhang and H. J. Liu, *Bioresources*, 2012, **7**, 447.
18. C. G. Liu, Y. Dai, C. S. Wang, H. F. Xie, Y. H. Zhou, X. Y. Lin and L. Y. Zhang, *Ind. Crop. Prod.*, 2013, **43**, 677.
19. C. G. Liu, W. Lei, Z. C. Cai, J. Q. Chen, L. H. Hu, Y. Dai and Y. H. Zhou, *Ind. Crop Prod.*, 2013, **49**, 412.
20. C. G. Liu, J. Li, W. Lei and Y. H. Zhou, *Ind. Crop Prod.*, 2014, **52**, 329.
21. E. Can, S. Küsefoğlu and R. P. Wool, *J. Appl. Polym. Sci.*, 2001, **81**, 69.
22. T. Eren and S. H. Küsefoğlu, *J. Appl. Polym. Sci.*, 2004, **91**, 4037.
23. H. Miyagawa, M. Misra, L. T. Drzal and A. K. Mohanty, *Polym. Eng. Sci.*, 2005, **45**, 487.
24. E. Can, R. P. Wool and S. Küsefoğlu, *J. Appl. Polym. Sci.*, 2006, **102**, 1497.
25. F. M. Lewis and F. R. Mayo, *J. Am. Chem. Soc.*, 1948, **70**, 1533.
26. K. Ishizu and X. X. Shen, *Polymer*, 1999, **40**, 3251.
27. D. L. Hull and J. P. Kennedy, *J. Polym., Sci. Polym. Chem.*, 2001, **39**, 1515.
28. J. Lu, S. Khot and R. P. Wool, *Polymer*, 2005, **46**, 71.
29. M. Mosiewicki, J. Borrajo and M. I. Aranguren, *Polym. Int.*, 2005, **54**, 829.

CHAPTER 9

Towards Green: A Review of Recent Developments in Bio-renewable Epoxy Resins from Vegetable Oils

RONGPENG WANG AND THOMAS SCHUMAN*

Department of Chemistry, Missouri University of Science and Technology, Rolla, MO 65409, USA
*Email: tschuman@mst.edu

9.1 Introduction

Epoxy resin is a compound or pre-polymer normally containing more than one equivalent of oxirane per mole of compound. Oxiranes, also known as "epoxides", are highly reactive due to their strained ring and polar bond structure and can afford a large variety of chemical reactions. Epoxy resins can react with themselves, through anionic or cationic homo-polymerization, or with a variety of curing agents, often called "hardeners" or "cross-linkers". Common curing agents include polyamines, anhydrides, or phenols. An enormous number of epoxy formulations fitting various applications are possible through a down-selection of the epoxy resin, curing agent, additive(s) and curing conditions. The cured epoxy resins exhibit excellent thermal and mechanical strength, outstanding chemical resistance, high adhesive strength and low shrinkage. Since the first commercial debut in about the 1940s, epoxy resins have become one of the most

RSC Green Chemistry No. 29
Green Materials from Plant Oils
Edited by Zengshe Liu and George Kraus
© The Royal Society of Chemistry 2015
Published by the Royal Society of Chemistry, www.rsc.org

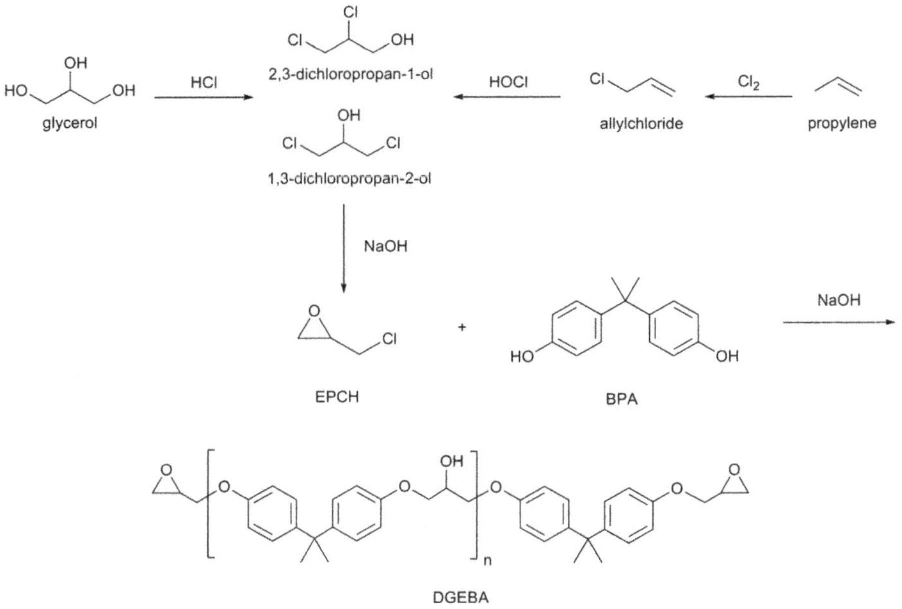

Scheme 9.1 The synthesis of an epichlorohydrin and DGEBA epoxy resin.

important monomers for synthesizing thermoset polymers and are widely used in coatings, adhesives and composites.[1]

Epoxy resins can be roughly divided into three classes: aliphatic, cycloaliphatic and aromatic. By far, the diglycidyl ether of bisphenol A (DGEBA) structure, which is made from the condensation reaction of bisphenol A (BPA) and epichlorohydrin (EPCH), is the most common, commercially available epoxy resin. EPCH is traditionally produced from propylene in a multi-step process. Glycerol, which is an effluent byproduct from the biodiesel industry, can also be used to produce EPCH and has been recently commercialized (Scheme 9.1).[2] Since BPA is classified as an endocrine disruptor, which may have a negative impact on human health, several countries have banned BPA use in infant bottles and considerable research has been focused on using compounds derived from wood/lignin,[3–6] rosin,[7–12] tannins,[13,14] sugar,[15,16] cardanol,[17] or itaconic acid[18,19] to replace BPA and, at the same time, to contribute toward sustainable development in the polymer industry as bio-based thermoset polymers. However, some of these epoxies still have unresolved issues such as limited production, low purity, complex structures and lack of structural control, hydrophilicity, brittleness, and/or unknown toxicity. The efficient and economical synthesis of a bio-based epoxy is a challenge and is still strongly dependent on future developments.[20]

9.2 Epoxidized Vegetable Oils

Apart from renewability, availability and relatively low and stable prices, vegetable oils (VOs) such as soybean or linseed oil, of diverse chemical structure and high synthetic potential, could be attractive and feasible re-sources in the synthesis of bio-based chemicals.[21,22] VOs are major agri-cultural commodities with a total production about 159 million tons in 2012. While their production has continuously increased in recent years,[23] only a small portion of VOs are used as oleochemicals for surfactants, lubricants, coatings, paints and bio-diesels. The industrial exploitation of VOs is mostly based on chemical modification of the carboxyl and C=C groups present in triglycerides, *i.e.*, the glycerol esters of fatty acids.

There are five dominating types of free fatty acids that range in length from 14 to 18 carbon atoms, with 0 to 3 double bonds within the chain. The common unsaturated fatty acids are oleic, linoleic and linolenic acids, containing one (C18:1), two (C18:2) or three (C18:3) double bonds, respect-ively. The most common saturated fatty acids are palmitic acid (C16:0) and stearic acid (C18:0). One must bear in mind that saturated fatty acids show no reactivity except through a telechelic carbonyl group. Highly unsaturated fatty acids are desirable for thermoset polymer applications since double bonds provide opportunities for development through highly cross-linked structures, hence better thermal and mechanical strength.

Epoxidized vegetable oils (EVO) have been a frequently studied polymer precursor in recent years.[22] Vernonia oil is a naturally occurring EVO that is obtained from the seeds of a plant native to Africa, *Vernonia galamensis*. The seeds contain up to 40 wt% oil by weight, with typical fatty acid distributions averaging 6% oleic acid, 12% linoleic acid, and 80% vernolic acid (Scheme 9.2).[24] Possessing low viscosity, vernonia oil has been used as a reactive diluent in epoxy coating formulations[25] or in cationically cured blends with commercial epoxy.[26]

Epoxidized soybean oil (ESO) and epoxidized linseed oil (ELO) are cur-rently the only bio-renewable epoxies that reach industrial-scale production. World annual production of EVOs is greater than 200 000 tons.[27] EVO can be prepared by the epoxidation of the double bond of fatty acids using peracids, and such processes have been utilized since the 1940s.[28,29] Performic acid and peracetic acid are commonly employed by the industry and are formed *in situ* from hydrogen peroxide and the corresponding acid in the presence of a strong acid catalyst such as sulfuric acid (Scheme 9.3).[30] However, strong acids also catalyze the ring-opening reaction of the desired product, oxirane. In order to improve epoxidation selectivity and reduce side-reactions, acidic ion-exchange resins,[31] heterogeneous transition metal catalysts,[32] and en-zymes[33,34] have been used as peracid catalysts for the epoxidation of VOs. The latter have proved to be very effective for the epoxidation of VOs with extremely high yields and fewer side-reactions. The epoxy content of VOs

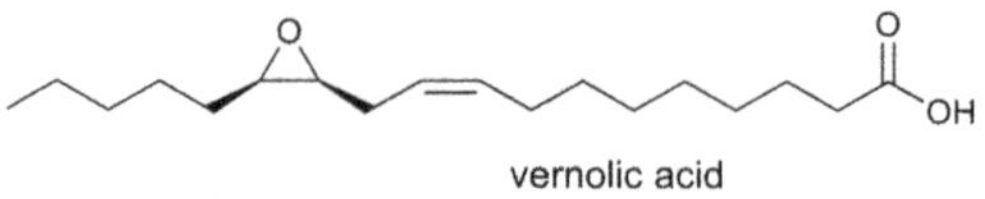

vernolic acid

Scheme 9.2   The structure of venolic acid in vernonia oil.

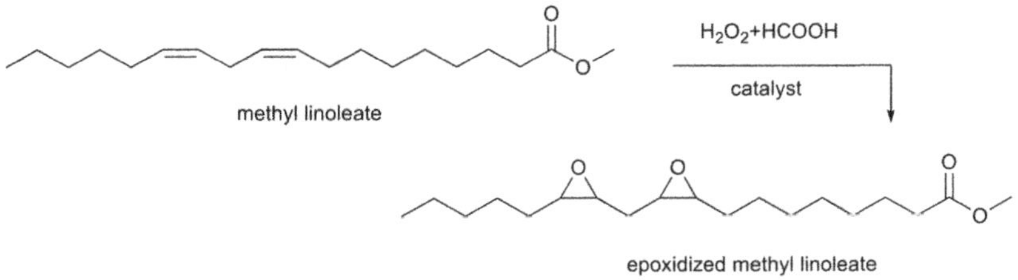

Scheme 9.3   The epoxidation of VOs or their fatty acid derivatives.

depends on the epoxidation methodology and origin of the VO, *i.e.*, the extent of epoxidation and starting iodine value. The oxirane oxygen concentrations of ESO and ELO are approximately 7 and 9%, respectively.

EVOs are industrially applied in polyvinyl chloride (PVC) plastics as a secondary plasticizer and scavenger for hydrochloric acid liberated during the heat treatment of PVC. EVO offers promise as an inexpensive, renewable material (about US$1500 per ton in 2013) for many epoxy applications because EVOs share many of the characteristics of conventional petroleum-based epoxies. The epoxy group of EVO is versatile as a reactive intermediate to provide other functionalities suitable for polymer synthesis, *e.g.*, polyols for polyurethanes,[35] and maleinized and/or acrylated VOs for sheet molding compounds.[36] Such processes have been well established and are also commercialized. These applications, however, fall outside the scope of this chapter where only the direct use of fatty-acid-derived epoxies as monomers for preparing epoxy thermoset polymers is considered.

Despite their promise and versatility, EVOs are not able to compete with analogous petroleum-based epoxy polymers in many structural applications. Direct use of EVO as an epoxy resin[37,38] dates back to the 1950s, but has since shown limited success. EVOs lack a stiff, aromatic or cycloaliphatic structure which confers greater strength, as found in other commercial epoxy thermoset polymers. Their internal, secondary carbon oxiranyl groups possess relatively lower reactivities compared to common polyamine or anhydride curing agents.

In what could have been regarded as a stalled field, a strong revitalization has occurred in recent years. Firstly, a "green chemistry" emphasis, also called "sustainable chemistry", has entered the polymer industry. For instance, the reputed RSC publishing journal *Green Chemistry*, which focuses on research into alternative/sustainable technologies, debuted in 1999. The utilization of natural, renewable products is considered one of the most important approaches to conducting green chemistry. Use of an EVO not only takes advantage of the synthetic potential of nature but can also reduce our environmental footprint through a reduction in consumption of non-renewable resources such as petroleum.

Secondly, broader applications of EVOs are being actively sought. Different thermoset polymers have been synthesized from EVOs through new curing agents and curing conditions. New formulation approaches for composites, coatings and toughening agents are being continuously developed, as will be detailed in later sections. Lastly, but most importantly, new epoxy monomers derived from VOs have been successfully synthesized and show greater promise than common EVOs in terms of polymerization reactivity and thermal and mechanical strength. New monomers are a step towards more advanced applications, such as structural composites, which are made possible through improved chemical structure, reactivity, compatibility with other monomers, and proper choice of formulation conditions. Properties are the avenue by which to provide the opportunity for bio-based epoxy resins to replace or supplement their petroleum-based counterparts. Without appropriate properties, there can be no commercial opportunities.

9.3   Vegetable-oil-derived Epoxy Monomers

Commercial EVOs, as the major epoxy resins derived from VOs, have inherent problems that derive from their chemical structure, flexibility and hindered reactivity. Most thermoset polymers derived from EVO have very low glass-transition temperatures (T_g) and are mainly of a rubbery state, which inevitably limits their applications. As a result, it has been of interest to synthesize VO-derived epoxies of enhanced polymerization rate and with stiffer polymer backbones.

Functionalized oils have been prepared from linseed oils and 1,3-butadiene, cyclopentadiene or dicyclopentadiene through Diels–Alder reactions (Scheme 9.4). Epoxynorbornane linseed oils (ENLOs) were prepared using hydrogen peroxide with a catalyst.[39,40] The produced cycloaliphatic structure was expected to improve polymer tensile strength, toughness, and T_g and be suitable for cationic polymerization. However, the double bond conversion of linseed oil had to be limited, *e.g.*, <30%, otherwise only high-viscosity liquids or soft solids were obtained. Reactive diluents were required to reduce the viscosity of the formulation, accelerate the rate of cationic polymerization, and increase their final conversions.[41]

Epoxidized sucrose esters of fatty acids (ESEFAs), highly functional epoxy compounds with reasonably well-defined structures, have been synthesized by Webster and co-workers (Scheme 9.5).[42,43] Anhydride-cured ESEFAs showed better thermal and mechanical strengths than EVOs. ESEFAs still possess internal epoxy groups, which are less reactive with common anhydride and amine curing agents than the terminal epoxy groups analogous to DGEBA. Internal epoxy groups are more useful for cationic-cured coating applications.[44] A drawback to the ESEFA approach is the relatively large viscosity increase compared to EVOs that hampers some applications.

Polyepoxides were derived from poly(vinyl ether of soybean oil fatty acid esters) (poly-VESFA) through the transesterification of soybean oil with ethylene glycol vinyl ether (Scheme 9.6). Poly-VESFA has an increased number of fatty branches per molecule compared to native soybean oil, thus the epoxidized poly-VESFA showed faster curing kinetics and improved T_g

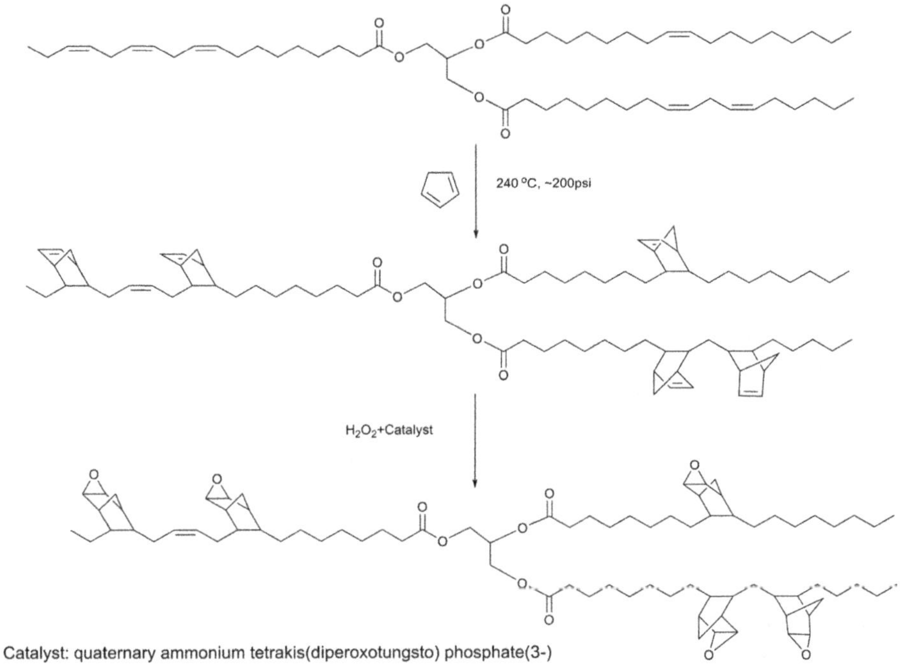
240 °C, ~200psi
H2O2+Catalyst
Catalyst: quaternary ammonium tetrakis(diperoxotungsto) phosphate(3-)

Scheme 9.4 The synthesis of ENLO derived from linseed oil and cyclopentadiene.

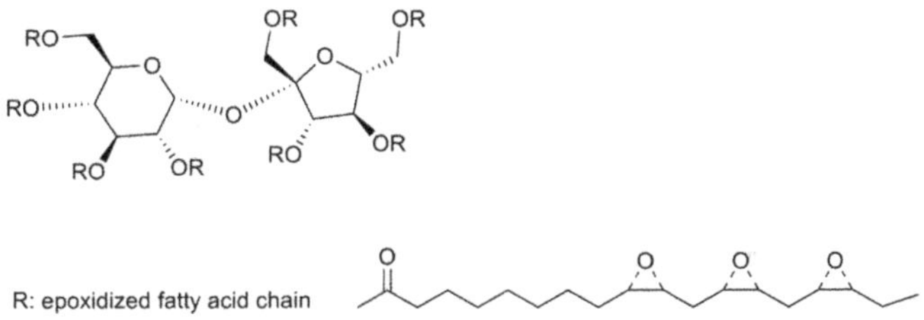
R: epoxidized fatty acid chain

Scheme 9.5 The molecular structure of ESEFA.

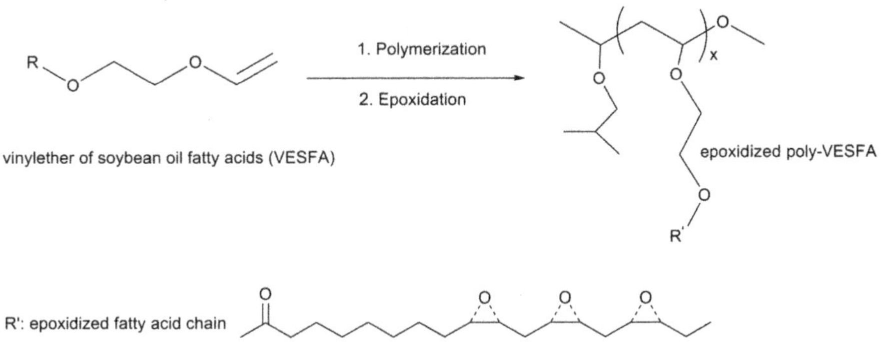
1. Polymerization
2. Epoxidation
vinylether of soybean oil fatty acids (VESFA)
epoxidized poly-VESFA
R': epoxidized fatty acid chain

Scheme 9.6 The synthesis of epoxidized poly-VESFA.

compared to ESO due to the higher number of epoxy groups per molecule.[45] The much higher viscosity, a reduced molecular mobility associated with a polymeric structure, as compared with EVOs, and the presence of only internal epoxy groups limits the use of epoxidized poly-VESFA in coating applications.[46]

Unlike the internal oxiranes of monomers such as EVOs, terminal epoxies such as glycidyl show improved reactivities during nucleophilic curing reactions. For instance, the terminal epoxy groups of epoxidized triglyceride esters of undecylenic acid (Scheme 9.7) have been synthesized and successfully used in epoxy–amine or epoxy–anhydride curing.[47,48] The prepared coating compounds also exhibited UV stability due to the predominance of aliphatic structures.[49] The epoxidation rate of the terminal electron-deficient alkenes in undecylenic acid by peracid is much lower than for the internal double bonds of natural fatty acids.[50–52]

Undecylenic acid is produced by the pyrolytic cracking of castor oil under pressure. The non-natural fatty acids then must be reacted with glycerol to reform triglyceride esters and increase the cross-link density. The maximum 3 oxiranes per epoxidized triglyceride ester of the undecylenic acid molecule is still lower than that of ESO with about 4.5 oxiranes per triglyceride. Cured thermoset polymers offer few advantages over their similar, but more readily available, ESO or ELO counterparts with the exception of minimized pendant alkyl chain content.

Terminal epoxies of glycidyl esters synthesized from dimer or trimer fatty acids have been commercially available for some time.[53–55] Recently, both dicarboxylic acids and a tricarboxylic acids were synthesized by Huang *et al.*[56] using Diels–Alder addition onto tung oil (eleostearic) fatty acid with acrylic acid and fumaric acid, respectively. The corresponding diglycidyl or triglycidyl esters were prepared using base and EPCH (Scheme 9.8). Both epoxies showed higher reactivities and improved performance compared to ESO. The triglycidyl ester version displayed comparable strength, modulus, and T_g to a DGEBA control.

Using readily available soybean oil or its free fatty acids, a fatty glycidyl ester epoxy was synthesized by Wang and Schuman.[57] The versatility of transesterification and epoxidation reactions (Scheme 9.9) provides several routes towards the synthesis of the glycidyl esters of epoxidized fatty acids (EGSs). EGS merits include higher oxirane content and lower viscosity than commercial ESO, ELO or DGEBA. A structure–property relationship study measured the effects of oxirane content and presence of saturated fatty acids

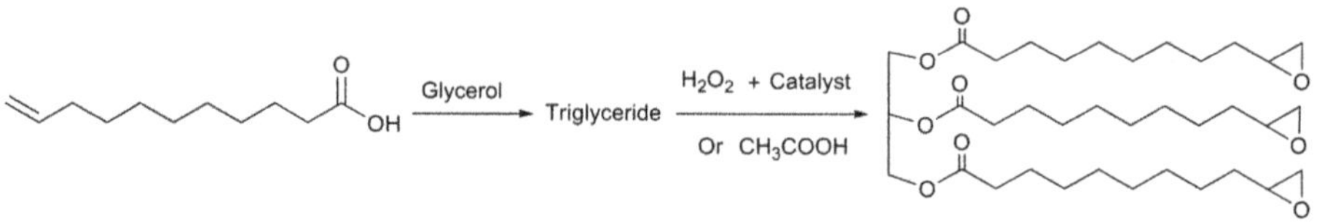

Scheme 9.7 The synthesis of the epoxidized triglyceride ester of undecylenic acid.

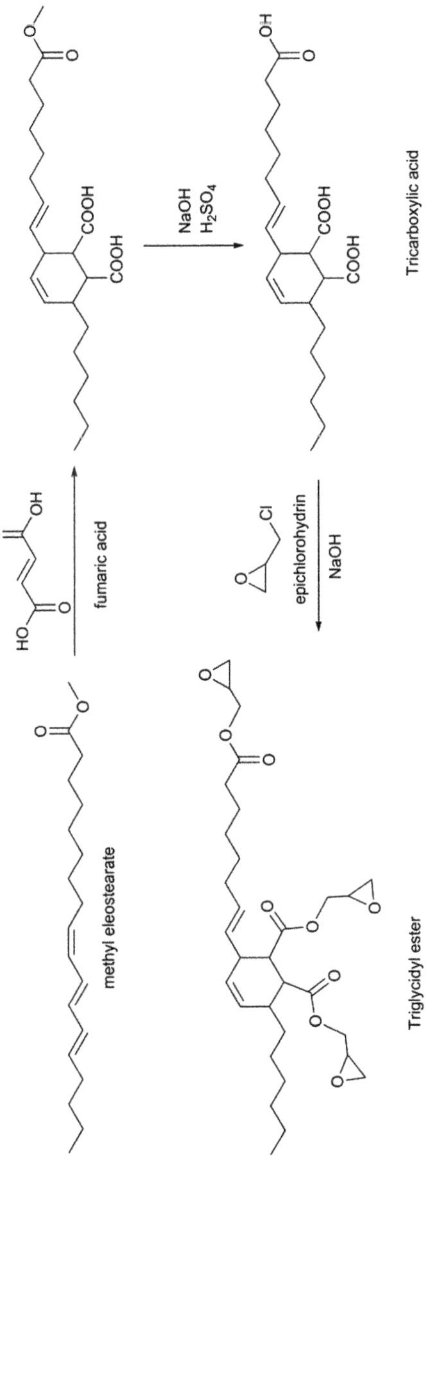

Scheme 9.8 The synthesis of triglycidyl esters derived from tung fatty acid and fumaric acid.

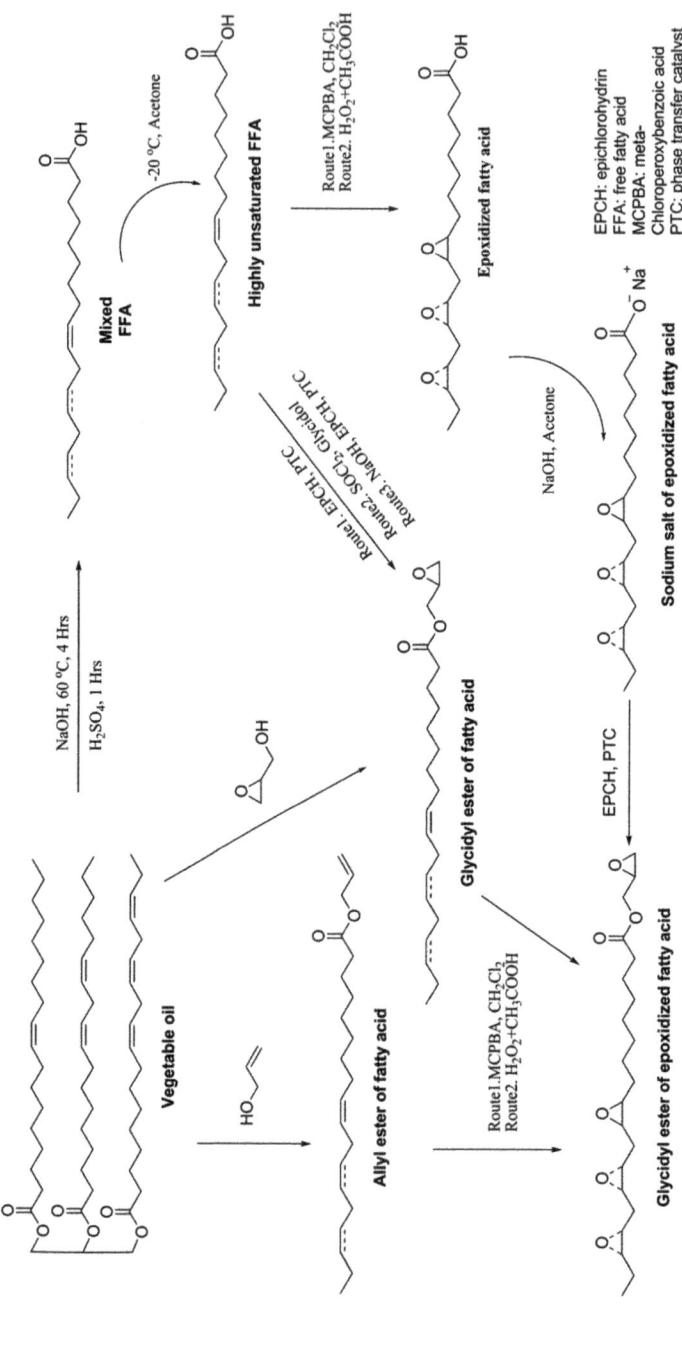

Scheme 9.9 The synthesis of glycidyl esters of epoxidized fatty acid (EGS and EGL).

on polymer properties. EGS had its unreactive saturated fatty acid components removed. Upon curing cationically, neat EGS polymer displayed T_g values well above room temperature. Even higher T_g values *e.g.*, greater than 100 °C, and improved mechanical properties, compared to other bio-based systems, were obtained through careful selection of curing agents and catalysts.

9.4   Curing Reactions of Epoxidized Vegetable Oils

The polymeric materials of EVOs are three-dimensional cross-linked networks prepared through the use of curing agents. There are two types of curing agent, catalytic and co-reactive. The catalytic curing agent initiates polymerization of EVOs themselves, *i.e.*, through homopolymerization, whereas the co-reactive curing agent behaves as a co-monomer for EVO. The curing process is bond formation through a combination of step-growth and/or chain-growth mechanisms. Due to the polarity of C–O bonds, the electron-deficient carbon of oxirane constitutes an active site for nucleophilic reaction, while the electron-rich oxygen atom affords an electrophilic reaction site. The rate of curing is dependent on temperature, curing agent and thus mechanism, as well as the type and number of epoxy groups present in the chemical structure.

Despite a large volume of literature on the reactivity of EVOs, some of which is conflicting, there is consistency in its general conclusions: due to sterically hindered and electron-donating alkyl substituents, the rate of reaction of EVOs with nucleophilic curing agents is lower than glycidyl (terminal) epoxies, while the rate is higher with electrophilic curing agents. For instance, EVOs react especially sluggishly with common polyamine curing agents.[48,58] It is not uncommon for some EVOs to show no, or a reduced degree, of curing due to their low reactivity and/or low oxirane content. In many ways, the curing behaviors of EVOs are analogous to commercial cycloaliphatic epoxies rather than DGEBA. Polyacids and their derivative anhydrides, plus cationic catalysts, are commonly used curing agents for EVOs.

9.4.1   Addition with Polyamines

Polyamines are very frequently used curing agents for epoxy resins. The overall reaction rate of amine with epoxy resin is influenced by the structure and electronic properties of the amine. The nucleophilic reactivity of amines generally follows the order: aliphatic > cycloaliphatic > aromatic. Where the EVO molecules have long, flexible, and aliphatic structures, cycloaliphatic or aromatic amines can compensate for this shortcoming through their rigid structures. Cyclic curing agents favor applications with high thermal and mechanical strength requirements[59–61] but require higher curing temperatures and longer reaction times.[62]

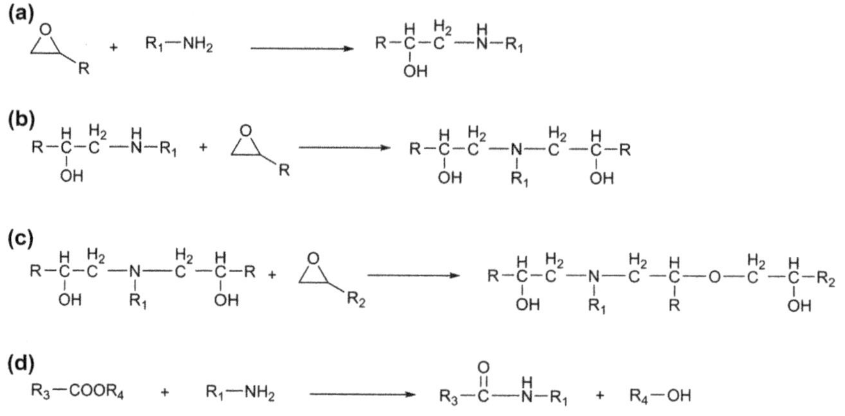

(a)

(b)

(c)

(d)

R$_3$—COOR$_4$ +   R$_1$—NH$_2$ $\longrightarrow$ R$_3$—C(=O)—N(H)—R$_1$ +   R$_4$—OH

Scheme 9.10 The mechanism of primary amine cure of an epoxy resin: (a) through a primary amine; (b) through a secondary amine; (c) through a hydroxyl group generated from reactions (a) and (b); and (d) through an ester-aminolysis reaction.

The main reaction of polyamines with epoxy is through a step-growth polymerization mechanism without formation of byproducts (Scheme 9.10). A primary amine with two active hydrogens can consume two epoxy groups, while a secondary amine will only consume one. A tertiary amine group, which has no active hydrogens, is not bond-forming with epoxy but instead behaves as a catalyst to accelerate epoxy–amine reactions. Thus, curing with polyamines is an autocatalytic process. However, polyamines are less efficient curing agents for EVOs because of the lower reactivity of internal epoxies as mentioned above. Accelerators or high temperatures are required to cure EVOs, even for nucleophilic aliphatic amines, which can cure DGEBA at room or low temperatures. In addition, ester groups will react with primary amines and form alcohols and amides, *i.e.*, *via* an ester-aminolysis reaction.[63] Epoxy monomers may also be attacked by the hydroxyl group of the reaction product, especially under high temperatures, a source of uncertainty during formulation.

Autocatalytic curing behaviors were observed by Manthey *et al.*[64,65] in curing epoxidized hemp oil (EHO) with triethylenetetramine (TETA) and/or isophorone diamine (IPD). The addition of IPD was found to increase the curing rate of the EHO with TETA. A modified Kamal autocatalytic model indicated a decrease of reaction order with increase in temperature, and a negative activation energy (E_a) was also observed. The authors believed this was due to an unidentified competitive reaction at higher temperatures. Two different mechanisms, depending on the temperature for the epoxidized methyl oleate (EMO) and aniline system, were postulated by del Río *et al.*[66] The mechanism was autocatalytic at lower temperatures and non-autocatalytic at higher temperatures, which favored an ester-aminolysis reaction and led to thermoset polymers of poor quality.

The reaction mechanism of EVO with polyamines has been determined by Fourier-transform infrared spectroscopy and nuclear magnetic resonance. Wang *et al.*[67] found only one of two adjacent epoxy groups in the same fatty acid chain takes part in the ring-opening reaction, due to the steric hindrance. While internal epoxies have higher reactivity with primary amines than ester groups, partially cross-linked ESO structures were broken by aminolysis reactions. Secondary amines are unreactive with ester groups. Miao *et al.*[68] also found that some ESO epoxy groups showed low reactivity, especially after partial curing with isopropanol amine. Lu[69] found the secondary amine of bis(4-aminocyclohexyl) methane (PACM) was left unreacted in the cured ESO network, which led to a lower cross-link density and the extent of aminolysis side-reaction was decreased by the lower amine concentration.

The fact that hydroxyl compounds, water, alcohol, phenol, acid, *etc.*, can accelerate the reaction between epoxides and amine compounds is widely recognized in commercial epoxy formulations. Hydroxyl groups catalyze the reaction through the formation of a trimolecular complex, which facilitates nucleophilic attack of the amino group.[70] Interestingly, these strategies are less frequently applied toward EVO-amine curing systems. In contrast, Lewis acids have been used to catalyze EVO-amine reactions.

Harry-O'kuru[71] found the ring-opening of internal epoxies in EVO with dibutylamine under anhydrous $ZnCl_2$ catalysis was facile and the reaction proceeded smoothly at moderate temperatures with only trace amounts of the amide byproduct. Stannous octoate also can significantly reduce the onset and peak exothermic temperature of ELO curing with 4,4′-methylene-dianiline (MDA).[48] BF_3-amine has been used to accelerate the reaction of ESO with cycloaliphatic and aliphatic amines.[69,72] The chemistry of the Lewis-acid-catalyzed cure process is rather complex, both step-growth and chain-growth mechanisms are operative. In addition to the amine curing reaction, homopolymerization of epoxies and ester-aminolysis may also take place, depending on the curing agents and curing conditions.

9.4.2   Addition with Anhydrides

Anhydride reagents are the principal curing agents for EVOs due to their good reactivity with internal epoxies. The reaction of anhydrides with epoxy groups is complex and several competing reactions take place at the same time.[73] However, without an accelerator the reaction is both slow and incomplete. The anhydride initially reacts with a hydroxyl group [Scheme 9.11(a)] and the newly formed carboxyl group reacts with an epoxy group to form a hydroxyl diester [Scheme 9.11(b)]. The hydroxyl diester can react with anhydride to generate another carboxyl group for reaction propagation [Scheme 9.11(c)]. The hydroxyl–epoxy reaction occurs, especially at high temperature [Scheme 9.11(d)]. If using polyacids directly as curing agents, the initial mechanistic steps are not necessary as the reaction can be initiated by the protonation of epoxy groups followed by attack of carboxylic

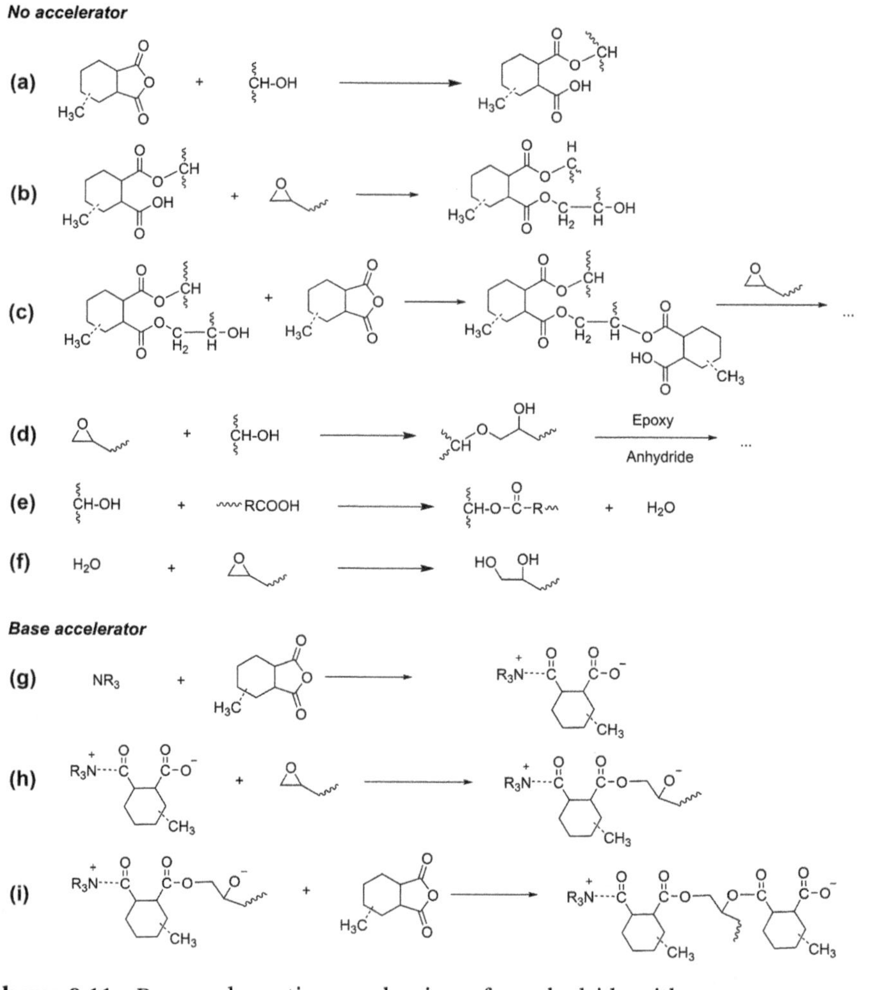
No accelerator
(a)
(b)
(c)
(d)
(e)
(f)
Epoxy
Anhydride
Base accelerator
(g)
(h)
(i)

Scheme 9.11 Proposed reaction mechanisms for anhydride with epoxy.

acid in a step-wise manner. At high temperatures, esterification between carboxylic acids and hydroxyl groups will occur [Scheme 9.11(e)] and the generated water can hydrolyze epoxy groups [Scheme 9.11(f)].[74] Under alkaline catalysis, such as with tertiary amines or imidazole, the carboxylate ion, which is generated by deprotonation of the acid at the beginning of the reaction, will act as a nucleophile in the epoxy ring-opening reaction. While etherification and condensation esterification reactions require the presence of unreacted epoxide or carboxyl groups, the former reaction is faster and the latter generally requires higher temperatures.[75]

Unlike epoxy-acid curing, Lewis base-catalyzed epoxy-anhydride reactions proceed much faster through a chain-growth manner including initiation, propagation, and termination or chain-transfer steps.[76] The initiation mechanism with tertiary amines or imidazoles is not well understood and appears complex. The suggested curing mechanism[1] follows: base accelerators catalyze curing reactions by the generation of carboxyl anions with anhydride [Scheme 9.11(g)]. The carboxylate ion then acts as a nucleophile in the ring-opening of the epoxide, resulting in an alkoxide [Scheme 9.11(h)]. The alkoxide anion in turn ring-opens an anhydride group to generate a carboxylate anion [Scheme 9.11(i)].[77] Continuation of these alternating steps results in a polyester. Etherification between epoxies and alkoxide anions is less likely.[78]

Boquillon and Fringant[79] modeled the cure kinetics of an ELO-tetrahydropthalic anhydride system catalyzed with 2-methylimidazole using differential scanning calorimetry (DSC) and an nth-order rate equation. The curing reaction of their systems followed first-order kinetics at extents of cure above 0.7. Liang and Chandrashekhara[80] studied the catalyzed soya epoxy-anhydride curing system where the curing showed autocatalytic behavior. The overall reaction order was approximately two, based on Kamal's autocatalytic model using DSC and rheology results. Using the same model, Tan *et al.*[81] studied a methylhexahydrophthalic anhydride (MHHPA) cured ESO system in the presence of a 2-ethyl-4-methylimidazole (EMI) catalyst. The EMI content and curing temperature had a significant influence on the reaction rate constant and reaction order. The overall reaction order ranged from 1.5 to 3 and the E_a values decreased inversely with EMI catalyst concentration.

Kinetic analysis of a similar 1-methyl imidazole catalyzed ELO-methyl nadic anhydride system by *iso*-conversion methods found that E_a increased at the beginning of the curing and decreased as cross-linking proceeded.[82] The increased E_a might be due to the slow initiation mechanism by the catalyst and the decrease in E_a by gelation and vitrification or autocatalysis. The curing kinetics of EMO and epoxidized bio-diesel (of sunflower and linseed oil origin) with *cis*-1,2-cyclohexanedicarboxylic anhydride catalyzed by triethylamine were investigated by Nicolau *et al.*[83] Their results indicated E_a was related to the oxirane content and to the locations of the oxiranes in the fatty acid structure. The oxirane at (C9–C10), which is close to the ester group, showed a higher E_a than those at positions C12–C13 or C15–C16. The difference may be due to steric hindrance.

9.4.3 Cationic Polymerization

Catalytic ring-opening of EVOs by Lewis acids is well known[84] and improves reactivity compared to either polyamines or anhydrides alone. Boron trihalides, super acids, have been widely used for cationic cure of EVOs. Due to their high reactivity and concomitant difficulty in handling, these catalysts are generally added as latent complexes, which are inert under normal conditions, such as ambient temperature, but release active species upon external stimulation, such as with heating or photo-irradiation. A boron trifluoride ethylamine complex $(BF_3 \cdot NH_2C_2H_5)$ is used extensively in commercial epoxy formulations. Catalytic polymerization of ESO by boron trifluoride diethyl etherate (BF$_3$·OEt$_2$) and the superacid of fluoroantimonic acid hexahydrate (HSbF$_6$·6H$_2$O) has been well developed by Liu *et al.*[85–88] The bio-degradable polymers prepared by this method find applications in personal care/healthcare upon further chemical functionalization.

Due to the various advantages, such as lack of oxygen inhibition and "dark" reaction post-polymerization (occurring after photo-irradiation has ceased), photo-induced cationic curing of epoxy resins is a rapidly growing method for the application of coatings, inks, and adhesives.[89] The newly developed photo-initiated system by Tehfe *et al.*[90] showed high efficiency even when induced in air *via* solar irradiation. Photo-sensitive onium salts, such as the aryliodonium or triarylsulfonium salts of group VA elements, are promising photo-initiators in curing EVOs.[91] The photolysis of onium salts produces a mixed radical-cation species upon UV irradiation. The superacid species will activate epoxies as oxonium ions, which are attacked by other epoxies and propagates a chain-growth mechanism (Scheme 9.12).

Crivello[26] reported the polymerization rate when catalyzed by strong acids derived from photo-initiators followed the order: HSbF6 > HAsF6 > HPFB, since with the lower nucleophilicity of the counteranion SbF$_6^-$ the tendency for chain termination is minimized.[92] The rate of the cationic photo-polymerization of EVO could be enhanced by the addition of hydroxyl groups or the presence of moisture (humidity), which can reduce the E_a and shift the curing toward an activated monomer mechanism due to the higher nucleophilicity of a hydroxyl group compared to an epoxide (Scheme 9.13).[93] Ortiz[94] reported that alcohol or water promotes a more rapid transfer of protonated oxonium species to monomers to speed up the entire propagation process. Due to the presence of both epoxy and hydroxyl groups,

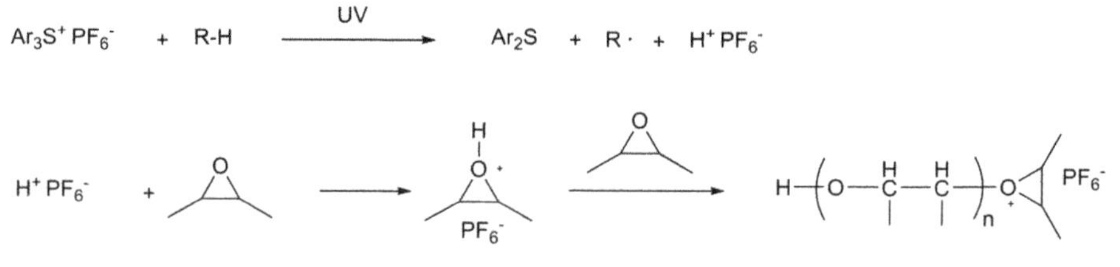

Scheme 9.12 Proposed mechanism for photo-initiated cationic polymerization of epoxies.

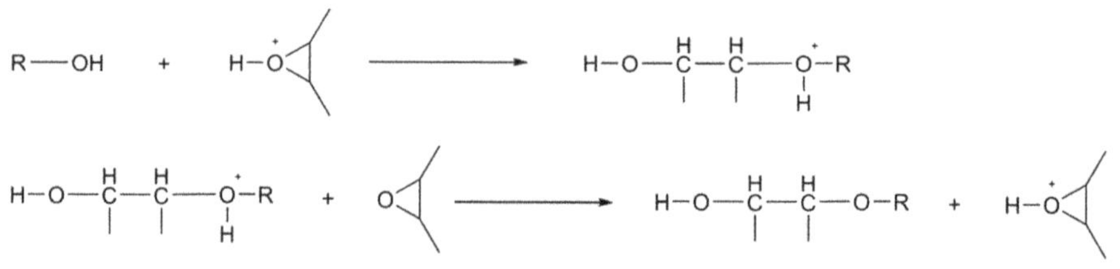

Scheme 9.13 Proposed acceleration mechanism of hydroxyl on the cationic polymerization of epoxies.

epoxidized castor oil (ECO) has been shown to have better reactivity than ELO or ESO when using diaryliodonium salt photo-initiators.[95] However, too many hydroxyl groups or water can also act as chain-transfer agents thus retarding the chain growth process and leading to softer polymer structures.[96,97]

Park *et al.*[98–100] used *N*-benzylpyrazinium hexafluoroantimonate (BPH) and *N*-benzylquinoxalinium hexafluoroantimonate (BQH) as thermally latent catalysts to cure ESO and ECO. BQH showed comparable curing activity for ECO at slightly lower temperatures than that of BPH. Compared to ESO, ECO polymerization initiated at lower temperatures when using the BPH catalyst. The authors also proposed that an observed variation in the thermal and physical properties of the resulting polymers was due to the different activities of catalysts.

ENLO showed higher curing rates than ELO during UV-initiated cationic polymerization, but was still slower than the polymerization of the cyclo-aliphatic epoxide, 3,4-epoxycyclohexylmethyl-3,4-epoxycyclohexane carboxylate. The lower reactivity of ENLO compared to cycloaliphatic epoxide was attributed to a greater steric hindrance present in the epoxybornyl groups and to a higher viscosity.[40] During cationic photo-polymerization, the relative reactivity of the oxiranes was found to be not as important as the viscosity of the reacting system. The polymerization rate was observed to be diffusion controlled, where adding diluents such as divinyl ethers markedly accelerated the curing rate and overall conversion rate of epoxies.[41]

9.4.4   Miscellaneous Curing Agents

Epoxies can also be polymerized in an anionic fashion for precise control of molecular weight and polydispersity as well as chain functionality.[101] Tertiary amines, imidazoles, and ammonium salts are commonly used anionic catalysts for epoxy resin homopolymerization, although their induced curing mechanisms are very complex and not universally accepted.[102] Boonkerd *et al.*[103] successfully synthesized a bio-based elastomer using post-living anionic polymerization of poly(butadienyl) lithium and ESO; however, the strongly nucleophilic anions preferentially cleaved ester groups rather than induced ring-opening of epoxies. Due to a higher oxirane content, ESO is more reactive than EMO for anionic epoxy ring-opening polymerization. While a pyridine-initiated epoxy reaction between ESO and 4-methylpyridine and poly(4-vinylpyridine) has been reported by Öztürk and Küsefoğlu,[104] no homopolymerization of epoxy groups, as initiated by pyridine was observed. Instead, pyridine addition, followed by rearrangement to a pyridone derivative was observed.

Del Rio *et al.*[105] used coordination catalysts to polymerize EMO. Two main polymerization mechanisms, cationic and the ionic-coordinative, were observed with the former being predominant. The polymers produced were a mixture of cyclic and linear structures with different end-groups depending on the initiator used, but a higher molecular weight was obtained than with conventional cationic catalysts. Transesterification side-reactions led to the formation of branched structures containing ester groups in the main chain. The prepared polymers could be used as polyether polyols for polyurethane applications.

Dicyandiamide (DICY) is one of the most popular curing agents for DGEBA. Zhao *et al.*[106] reported rapid cures of either neat ESO or ESO-DGEBA blends using DICY at 190 °C or 160 °C, respectively. Carbonyldiimidazole was used as an accelerator. The optimum stoichiometric molar ratio of epoxy : DICY was found to be 3 : 1. The first two epoxy units reacted with the amine groups of DICY to produce a secondary alcohol and a secondary amine. The produced secondary amine does not attack another internal epoxy and the remaining epoxy unit was linked to the DICY nitrile group.

9.5   Polymer Structure and Properties

A good understanding of structure–property relationships is critical when designing VO-based epoxy thermosets for various applications.[57] However, elucidating a VO-based thermoset polymer structure is quite difficult due to the heterogeneous content of the monomers and of the cured polymers. For instance, the fatty acid composition of VO varies not only from plant to plant but also within oils of the same plant. Unreacted monomers, dangling chains, and intra-cross-linking are common for VO-based thermoset polymers.[107–110] The structure and distance between cross-linked positions, in terms of cross-link density, is an important characteristic when describing the structure of a thermoset polymer. Dynamic mechanical analysis (DMA) has been widely used to calculate cross-link density based on the rubber network elasticity theory. The T_g, which is unique for each epoxy system, also reflects cross-link density where it is generally observed that an increased cross-link density increases T_g. That cured VO resins range from soft rubbers to hard plastics mainly depends on not only the chemistries and structures of the epoxy monomers but also the curing agents. Other factors include the polymerization conditions, monomer ratios, and the catalyst, *e.g.*, as described in previous examples.

9.5.1   Epoxy Resins

For VO-based epoxy monomers, the types of epoxy structure and oxirane content greatly influence the thermoset polymer thermal and physical properties.[99,111] A terminal epoxy and/or high oxirane content can lead to rapid gelation and high cross-link density.[42,48] EVOs of low oxirane values either are not reactive or impart waxy properties of poor strength to the polymer system.[37] The effect of oxirane content of ESO on the mechanical properties of anhydride-cured polymers was investigated by Tanrattanakul and Saithai.[112] The tensile modulus, strength or toughness, and tear strength of ESO thermoset polymers are controlled by the cross-link density and chain flexibility. Fully epoxidized ESO monomers of the highest oxirane content had cured polymers with the lowest elongation-at-break but higher storage modulus, thermal stability, and T_g than their less epoxidized, lower oxirane content counterparts. Due to its rich linolenic content, ELO possesses the highest oxirane content among common EVOs. Thus, cured ELO thermoset polymers generally have higher cross-link densities, T_g values and moduli.[113]

A series of epoxy resins with different structures, oxirane contents and contents of saturated fatty acids was synthesized by Wang and Schuman[57] in order to examine their structure–property relationships. Both anhydride-cured ESO and ELO showed broader T_g regions, indicating a broader distribution of chain environments and more heterogeneous polymer structures due to the lower reactivity of internal oxiranes and the presence of saturated fatty acids. The T_g values of either MHHPA or BF$_3$-amine cationically cured thermoset polymers were observed to increase fairly linearly with oxirane value (Scheme 9.14). Linseed-oil-based epoxies, such as ELO and EGL, had much higher T_g values compared to their respective ESO or EGS counterparts of lower oxirane content. Removal of saturated components greatly increased the T_g. A 30 and 20 °C increase in polymer T_g was observed for MHHPA-cured EGS and EGL compared to EGS-S and EGL-S, EGS and EGL monomer resin without removal of saturated fatty acids, respectively. Such trends were also observed in cross-link density measurements, as a significant increase of cross-link density upon removal of saturated components. Due to the loss of glycerol as a cross-linking site and liberated saturated fatty acid esters, the addition of an unreactive functional group to the ester end, *e.g.*, allyl (EAS) or methyl group (EMS), generated even lower T_g values and polymer cross-link densities, though the oxirane values were similar to those of ESO. The results reiterated the reported, very low T_g value of an epoxidized bio-diesel polymer.[114]

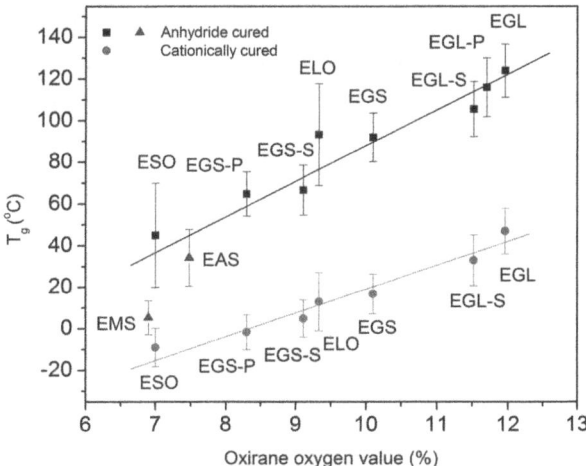

Scheme 9.14   Measured T_g as a function of oxirane content where -P: partially epoxidized and -S: monomer where the saturated fatty ester content was not removed. NOTE: the vertical 'error bars' indicate the breadth of glass transition and polymer heterogeneity.[57]

9.5.2 Curing Agents

The influence of the curing agent is just as critical to the final properties of thermoset polymers as the epoxy resin component. Since the curing agent will become part of the cross-linked network structure, special attention should be paid to its structure and stoichiometry. Lu[69] found that aromatic polyamines were unable to react with ESO. Polymers produced with aliphatic TETA were rubbery with a T_g of 15 °C. Cycloaliphatic polyamines, such as PACM, reacting with the same ESO monomer enhanced the T_g to 58 °C, and the highest flexural strength was achieved at a ESO/PACM molar ratio of 0.53 : 1. Juangvanich[72] also found that the reaction of diaminodiphenyl sulfone or MDA with ESO did not occur even at high temperatures. Aromatic amines, *e.g.*, *p*-phenylenediamine, reacted to a smaller extent compared to curing with more nucleophilic aliphatic amines. An imperfect network was formed when using *p*-phenylenediamine as a curing agent due to intramolecular cross-linking.

Anhydride is one of the most important curing agents for EVOs. Gerbase *et al.*[115] investigated the mechanical and thermal behavior of ESO reacted with various anhydrides in the presence of tertiary amine accelerators. Thermoset polymers showed higher T_g, storage modulus, and cross-link density when the system was cured with the more rigid phthalic, hexahydrophthalic, or maleic anhydride than the more flexible dodecenylsuccinic or succinic anhydride. Similar results were also reported by Rösch.[116] Due to the effects of steric factors and the rigidity of the formed diester segment, ELOs cured by phthalic anhydride and *endo*-3,6-methylene-1,2,3,6-tetrahydrophthalic anhydride in the presence of 2-methylimidazole showed lower cross-linking densities than those cured with *cis*-1,2,3,6-tetrahydrophthalic anhydride.[79]

In addition to the structure of anhydride, a variation in stoichiometric ratios of epoxy to anhydride was also found to have significant effect on the resulting network structure and performance of thermoset polymers. From the epoxy–anhydride polyesterification curing mechanism,[117–119] the maximum cross-linking degree, storage modulus, and T_g may be achieved at a stoichiometric ratio $R = 1.0$. However, in practical formulations, less than stoichiometric ratios are commonly used to achieve balanced properties and also account for competitive reactions such as epoxy homopolymerization. The reaction of ELO with *cis*-1,2,3,6-tetrahydrophthalic anhydride catalyzed by imidazole indicated complete conversion at the stoichiometric ratio of $R = 0.8$. The increase in anhydride to $R = 0.8$ caused an increase in T_g and stiffness but with a sacrifice in chain mobility. The T_g was reduced for $R > 1$ due to the reduced cross-link density.[79]

9.5.3 Catalysts

Due to the low reactivity of EVOs during nucleophilic curing reactions, the choice of catalyst and the amount used are critically important factors that strongly influence the cross-link density, network morphology/structure, and ultimate performance.[77,120] Lewis acid catalysts are commonly used for the EVO-amine curing reaction, *e.g.*, stannous octoate catalyst in an ELO and MDA curing system.[48] The onset and peak temperature of the reaction exotherm were significantly reduced while the polymer T_g was increased by more than 20 °C. Tertiary amines, imidazoles, and quaternary ammonium salts are commonly used catalysts for polyacid- or anhydride-curing of EVOs.

The use of imidazoles has advantages compared with tertiary amines in improving the T_g value,[121] which may due to a reaction of imidazole with the epoxy.[122] Supanchaiyamat *et al.*[75] reported that the mechanical and thermal properties of diacid-cured ELO films were significantly influenced by the type of amine catalyst selected. Both 1-methylimidazole and 4-dimethyl-aminopyridine (DMAP) can significantly enhance the mechanical properties of the resulting films. For DMAP, etherification may occur due to good nucleophilicity. The curing speed is highly sensitive to the catalyst amount, where the optimum DMAP catalyst concentration was 1 wt% of the total ELO and cross-linkers. Further increase of the DMAP concentration decreased the Young's modulus.

In an EMI-catalyzed ESO-MHHPA curing system, Tan and Chow[123] found the rate of polyesterification, the degree of conversion, T_g, storage modulus, and cross-link density were improved at higher EMI concentrations. However, a continued increase in the catalyst concentration led to rapid gelling, but reduced the conversion due to hindered monomer/oligomer diffusion.[79] Tan and Chow[124] also compared the type and concentration of catalysts on the fracture mechanics of MHHPA-cured ESO thermoset polymers. The improvement in fracture toughness with catalyst concentration was due to an increase in the degree of cure, while extreme cross-link densities led to catastrophic brittle fracture and low fracture toughness. For the EMI catalyst, fracture toughness increased with an increase in concentration of EMI, whereas a reduction of fracture toughness was observed when using tetraethylammonium bromide as the catalyst and its concentration exceeded 0.5 wt%.

9.6   Polymer Blends of Epoxidized Vegetable Oils

As mentioned above, an EVO can be polymerized with a variety of curing agents. The cured thermoset polymers, however, generally show low thermal/mechanical performance and cross-linking density due to their flexible structure and lower reactivity compared to DGEBA and cycloaliphatic epoxy. Commercially available epoxies such as DGEBA and cycloaliphatic epoxies possess stiffer structures, thus EVOs can be blended with these petroleum-based epoxy monomers to mutually improve their mechanical and thermal properties. EVOs generally have lower viscosity, so EVOs or their derivatives can be used as reactive diluents for DGEBA resins, which are relatively high-viscosity liquids or solids, to decrease the overall cost and improve the processability. Due to a less homogeneous structure, ESO is less efficient in reducing the viscosity of epoxy resin compared to many petroleum-based reactive diluents.

Strictly speaking, EVOs are not always "reactive". There can be an especially large difference between the reactivities of EVOs and DGEBA, and heterogeneous structures, such as phase inversion, may form and inevitably lead to a significant decrease in performance of the cured polymer. Therefore, few reports of high concentrations (>50 wt%) of EVOs in DGEBA exist, because the low oxirane content and the unreactive saturated components in EVOs both lead to a lower cross-link density upon curing and saturated fatty chains affect the miscibility between EVOs and DGEBA. Compositions with low EVO diluent content mostly preserve the undiluted polymer thermal and mechanical properties, *e.g.*, of the neat petroleum-based epoxy polymer. The blends can, however, improve the impact strength of the pure epoxy polymer, which may be brittle (*e.g.*, see Section 6.3).

9.6.1   Structure and Morphology

Epoxy monomer blends that contain EVOs have produced heterogeneous structures such as phase separation or semi-miscibility[125] due to the different reactivities of the epoxies as a function of the concentration of monomers and curing conditions. In addition, the initial miscibility/compatibility between epoxy monomers also plays an important role in the formation of heterogeneous structures. The Flory–Huggins equation combined with the Hilderbrand solubility parameter was used to assess the compatibility of DGEBA with ESO and EGS.[57] Compared to ESO, the solubility parameter of the EGS monomer was more similar to that of DGEBA compared to ESO. EGS monomers of lower molecular weight and higher reactivity produced EGS-DGEBA blends with improved homogeneity and mechanical strength compared to analogous ESO-DGEBA systems.

ESO less efficiently dissolves into and plasticizes, the rigid DGEBA matrix but still becomes part of the cross-linked structure at low concentrations. At higher ESO concentrations *e.g.*, >70 wt%, a faster gelation of DGEBA occurring at low degrees of conversion of ESO can lead to phase separation or defect structures. Transparent ESO-DGEBA blends can still be produced by catalyst selection or choice of curing conditions (Scheme 9.15). A transparent morphology does not necessarily indicate homogeneity.

Transparent ESO-DGEBA blends cured by methyltetrahydrophthalic anhydride were prepared by Altuna.[126] Despite optical clarity, phase separations at 40 wt% and 60 wt% ESO concentration were observed by SEM or through a change in intensity of the transmitted light. The phenomenon was ascribed to a match of refractive index between the dispersed ESO phase and the continuous phase. In a similar system but with methyl nadic anhydride as the curing agent, Chen *et al.*[127] observed phase-separated structures at only 20 wt% ESO concentration by SEM. The two-phase structure was explained as being a result of the different reaction rates of ESO and DGEBA under the applied curing conditions. Transparent ESO-DGEBA blends of single glass transition were observed by Karger-Kocsis *et al.*[128] Atomic force microscopy inspection of the plasma-etched samples clearly indicated

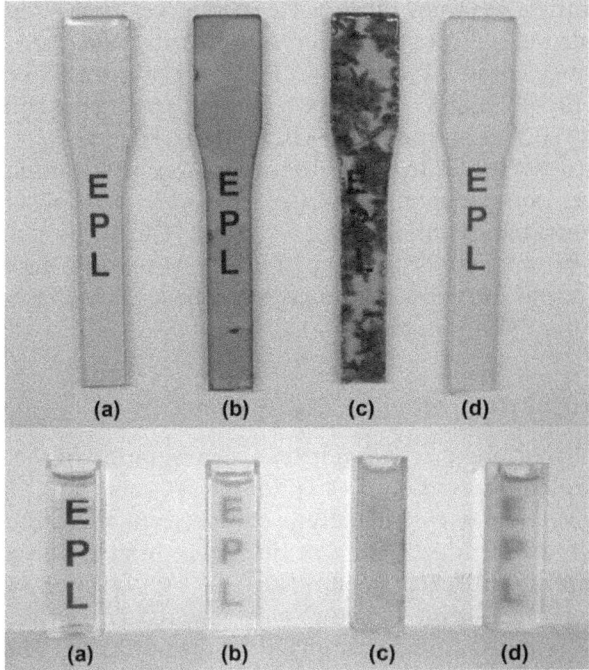

Scheme 9.15   Physical appearance of MHHPA-cured EGS/ESO-DGEBA polymers and uncured monomer blends: (a) EGS-DGEBA (90:10); (b) ESO-DGEBA (90:10) pre-cured at 145 °C for 10 min; (c) ESO-DGEBA (90:10) without pre-curing; and (d) pure ESO inducted for 12 h.

a two-phase structure with a dispersed domain size of about 100 nm, *i.e.*, smaller in size than the wavelength of visible light.

Due to the even lower reactivity of EVOs with amine curing agents than anhydrides, compared to that of DGEBA, an EVO component in EVO-DGEBA blends is either not reactive or proceeds *via* different reaction mechanisms. Therefore, heterogeneous structures such as phase separation are more common than in anhydride polymerization but are still related to the EVO structure, concentration, reaction process and conditions. Using epoxidized crambe oil as an epoxy monomer and an MDA curing agent, Raghavachar *et al.*[129] found the epoxidized crambe oil was only partially compatible with the DGEBA and formed a two-phase structure after direct mixing and curing. ESO of molecular weight lower than epoxidized crambe oil is more compatible with DGEBA.

Using TETA as curing agent, a plasticizing effect was observed when directly mixing ESO into DGEBA followed by polymerization. Phase separation could be induced by two-stage mixing where ESO was first reacted with TETA to form pre-polymers.[130] Similar research was also conducted by Sarwono *et al.*[131] in an epoxidized palm oil (EPO)-DGEBA system cross-linked by xylylenediamine. Mixing 10 wt% EPO into DGEBA or EPO pre-polymers reacted with amine less than 2 h showed opacity, *i.e.*, phase separation. More transparent blends were obtained by synthesizing EPO pre-polymers and reacting for longer than 2 h at 120 °C. Frischinger and Dirlikov[132] prepared liquid rubber pre-polymers of EVO with amines, producing rubbery particles of 15 to 30 wt% rubber, randomly distributed in a rigid DGEBA matrix. Phase inversion was observed at higher, intermediate rubber contents of 30 to 35 wt%. However, homogeneous morphologies were also observed at either lower or higher rubber contents, *i.e.*, >70 wt%. The authors concluded that the particle size and concentration of phase inversion depended on the miscibility between the rubbery and DGEBA phases, which was regulated by the nature of the EVO pre-polymer.

9.6.2   Thermal and Mechanical Properties

Although a polymer blend is a simple idea, combination of the advantages of EVOs and petroleum-based epoxies is not always successful. As with earlier discussions, lower reactivity and oxirane content generally shift the onset and peak reaction temperatures higher. At the same time, a decrease of reaction heat or an increase in E_a has been observed.[128] Unreactive EVO monomers and/or an inherently flexible structure plasticize rigid epoxy matrices such as DGEBA.[133] At high EVO concentrations, the polymerization may occur in two stages to form heterogeneous structures. EVO components not only reduce the cross-link density, but also behave as weak points or flaws where fracture is prematurely initiated by stress concentrations obtained when mechanically straining non-uniform materials' structures. Adding EVOs into petroleum-based epoxies has been frequently observed to decrease mechanical strength, T_g, thermal stability, and chemical resistance.[134]

The concentration of EVO in polymer blends, cost, and acceptable property loss are important considerations during epoxy formulation. To retain optimum properties, the concentration of ESO was limited. An addition of <40 wt% ESO into DGEBA produced storage moduli and T_g values comparable to neat DGEBA polymers, but with a 38% increase in impact strength and without loss of transparency.[126] An abrupt decrease in T_g and flexural strength/modulus at high ESO concentrations, *e.g.*, >50 wt%, was observed by Wang and Schuman.[57] Degradation of physical and thermal properties appear to be predominately effected by the incompatibility of ESO, *i.e.*, to form heterogeneous structure of less synergy between ESO and DGEBA. A non-linear transition of properties and lack of synergy between epoxy monomers was observed in an anhydride-cured ESO-DGEBA system. The Gordon–Taylor equation was applied to account for the T_g–composition relationship of this system. The interaction between ESO and DGEBA was of only medium strength.[135]

The optimum EVO concentration was also related to the structure of the EVO, especially the oxirane content. The heat distortion temperature (HDT) and tensile strength of EVO-DGEBA polymer blends were almost identical for either ESO or epoxidized lard oil at up to the 20 wt% level. Since the oxirane content of ESO is much higher, at higher EVO concentrations the HDTs and strengths of the epoxidized lard oil blends decreased more rapidly than those of the ESO blends.[38]

EPO is richer in saturated fatty acids and thus possesses lower oxirane content and has a greater plasticizing effect than ESO. EPO-DEGBA blends showed significantly reduced T_g with an increase in EPO concentration. Polymer blends also showed higher coefficients of thermal expansion and tan δ due to increased free volume and chain flexibility in the cross-linked networks.[136] To retain thermal and mechanical performance, a low EPO concentration, $\leq$10 wt%, is necessary.[131]

ELO shows the highest oxirane content among common EVOs. Miyagawa *et al.*[137] observed that cross-link densities, T_gs, and storage moduli of MHHPA cured diglycidyl ether of bisphenol F (DGEBF) systems remained relatively constant or were slightly decreased at up to 70 wt% ELO loading and then started increasing again upon further increase in the ELO content. This abnormal phenomenon was ascribed to higher oxirane content of ELO such that more curing agent was required for proper formulation. In addition, ELO is rich in linolenic acid content that facilitates dense cross-linked structures. Thus, it was found possible to replace petroleum-based epoxy with ELO while still maintaining high performance.

However, amine-cured ELO-DGEBF systems showed a completely different trend.[138] The cross-link densities, T_gs, and storage moduli of blends decreased continuously with an increase of ELO concentration. The reduction of storage modulus was especially significant and T_g was close to room temperature for ELO concentrations of greater than 20 wt%. The trend is due to the much lower reactivity of ELO with amine curing agent and the unreacted ELO can plasticize the rigid DGEBF matrix. A decrease of reaction exotherm, thermal stability and mechanical strength of IPD cured DGEBA with increase in epoxidized rapeseed oils or ESO concentration was also reported by Czub[139,140] where polymer blends of high EVO contents were highly flexible and properties were dominated by the EVO content.

Though cationic cross-linking of EVO generally shows higher reactivity than curing with amine or anhydride, it results in rubbery polymers since all the networks are composed of flexible fatty acid components. The addition of stiffer petroleum-based epoxies, cycloaliphatic or DGEBA can increase the EVO hardness and modulus.[141,142] Adding EVO into DGEBA as a diluent not only reduces the viscosity but can shift polymerization temperature lower since the cationic reactivity of EVO is higher than that of DGEBA. Decker *et al.*[91] found that the addition of 20 wt% ESO accelerated the photo-initiated cationic curing process of DGEBA and formed a relatively tight polymer network of better chemical resistance. Park[100] found polymerization of ECO-DGEBA blends initiated using BPH as catalyst had a maximum onset decomposition temperature at 10 wt% ECO content due to an optimum network structure. However, further increases in the EVO content still led to decreased T_g, thermal stability, and cross-link density and increased coefficient of thermal expansion.[143]

9.6.3   Epoxidized Vegetable Oils as Toughening Agents

Epoxy thermoset polymers may suffer low toughness or brittleness due to stiff structures with high cross-link densities. Various methods, include the addition of a either rigid or soft secondary phase, the chemical modification with a flexible backbone, or a lowering of the cross-link density of the polymer, have been attempted to improve epoxy toughness.[144] The addition of rubbery compounds to form phase-separated inclusions has been proved to be one of the most effective methods for toughening epoxy to avoid major deterioration of thermal and mechanical properties. The toughening mechanism is generally thought due to increased shear yielding of the rubber phases at low strain rate and cavitation at high strain rates.[145]

EVOs ability to form heterogeneous phases with petroleum-based epoxies has been found beneficial as reactive toughening agents in epoxy or other engineering plastics.[146,147] As mentioned above, for EVO-DGEBA polymer blends the mechanical and physical properties of an EVO toughened epoxy are closely related to the network structure in terms of the EVO compositions, phase morphology, cross-link density and the chains flexibility. Most polymerized EVO are of rubbery state but also depend on the curing system. A cross-linked EVO structure can efficiently absorb, transform and dissipate fracture energy through deformation of molecular networks analogous to common rubbery compounds. Researchers have shown that EVO polymers possessed better toughness than those of stiff, neat epoxies and that the incorporation of EVOs into epoxy can improve impact strength.[135,139,140,148,149]

Shabeer *et al.*[150] found the fracture toughness of anhydride-cured DGEBA polymer was greatly improved, by more than 200%, with substitution of 75 wt% DGEBA with epoxidized allyl soyate (EAS). Increase in fracture toughness of the blend was attributed to a lesser degree of cross-linking. Ductile fracture behavior at a high concentration of EAS resin was observed. However, this improvement was also associated with a greatly reduced storage modulus, T_g, and cross-link density, *e.g.*, the T_g of 75 wt% EAS blend was only 40.5 °C compared to 90 °C for the neat DGEBA polymer. Anhydride-cured ELO-DGEBF showed a single-phase structure and no apparent improvement in toughness with up to 50 wt% ELO. A further increase in ELO content even resulted in a decrease of fracture toughness and Izod impact strength[137] while 30 wt% ESO showed improvement in toughness due to phase separation of rubbery ESO particles within the rigid DGEBF matrix.[151]

Tan and Chow[152] indicated the plasticizing effect of EPO, which is rich in saturated components, improved the fracture toughness of DGEBA by enhancing flexibility through cavities occupied by unreacted EPO that increase resistance to deformation, crack initiation and propagation. The cross-link density and water absorption capability of the EPO-DGEBA polymer decreased with the increase in loading of EPO, but other thermal and mechanical properties were not disclosed. Under thermally latent catalysis, Jin and Park[153] showed the Izod impact strength of a 60 wt% ESO blend was

58% higher than neat DGEBA, but the flexural strength was also reduced more than 40%.

Amine-polymerized ESO with 4,4′-tetradiglycidy diaminodiphenol methane resulted in a two-phase structure due to the incompatibility between epoxy monomers. The critical stress intensity factor was improved by 54% at 10 wt% ESO content. The flexural strength was also increased but the thermal stability, cross-link density, and T_g of the blends were slightly decreased with the addition of ESO, due to the incomplete curing reaction of ESO in the blend system.[154] An amine-cured ELO-DGEBF polymer showed improved Izod impact compared to neat DGEBF, although no clear phase separation was observed, and although the cross-link densities and storage moduli of the polymer blends were decreased, the T_g dropped almost 50% for only 30 wt% ELO, this is probably due to the plasticizing effects of the less reactive ELO.[138]

When directly mixing EVOs into epoxy resulted in a single-phase structure, the improvement in toughness was less impressive and was commonly associated with decreased cross-link density, T_g, modulus of elasticity, and yield stress.[155] On the other hand, introduction of EVO liquid rubber prepolymers into the epoxy resin has obvious advantages for the preparation of a two-phase thermoset polymer over directly incorporating the ESO monomer.[130] Through proper choice of curing profile, a two-phase structure can be formed. The impact strength of DGEBA can be markedly improved at a relatively low concentration of EVO with marginal sacrifice of thermal and mechanical properties.

A two-phase thermoset polymer that consisted of randomly distributed small ESO rubbery particles (0.1–5 µm) in a rigid epoxy matrix was prepared by Frischinger and Dirlikov.[156] The liquid rubber pre-polymers were prepared from a stoichiometric mixture of ESO and MDA. The diamine molecules at the interface can react with the epoxy groups of both DGEBA and EVO, thus forming a strong interfacial bond between the two phases where no ejection or disbonding of rubbery particles was observed under shear deformation. These EVO rubber-toughened epoxies showed slightly lower T_g and Young's modulus but remarkably improved toughness comparable to commercial, carboxyl-terminated butadiene-acrylonitrile rubber toughening agents.

Ratna[130] compared the effects of direct mixing (single-stage) *versus* prepolymer mixing (two-stage) on an amine-cured DGEBA-ESO thermoset polymer morphology and studied the resulting thermal, flexural and impact properties. A two-stage mixing showed a milder decrease in T_g and flexural and tensile strength compared to single-stage mixing, in which high ESO concentrations usually led to drastic reductions. Network polymers made by a single-stage process showed only modestly increased impact energies, a benefit of nearer single-phase morphology. A more significant increase in impact energy was obtained for modified networks made by the two-stage process at 20 wt% ESO which resulted in a phase-separated morphology.[157]

9.7   Epoxidized Vegetable Oil Paints and Coatings

Due to their versatility, excellent adhesion to a wide range of substrates, and corrosion and chemical resistance, epoxy resins are widely used in coating applications.[78] EVOs have promise as alternatives or supplements to petroleum-based epoxies because of a combination of attributes: low viscosity, low cost, and epoxy functionality. Use of EVOs in coatings not only provides sustainable chemical content but also offers a way to reduce volatile organic compounds and, as just discussed, improve the flexibility or toughness of epoxy coatings.[158–161] However, the challenge for neat EVO polymer coatings is to improve their mediocre mechanical and thermal performance, especially of polymer moduli and T_g values, which, to date, have prevented further market penetration. Co-polymerization with petroleum-based monomers of rigid structure and/or an application of inorganic compounds to form blended or nanocomposite coatings, respectively, are frequently applied strategies to enhance EVO coating properties.[46,92,162,163]

Because of their similar reactivity towards cationic polymerization and stiffer structure, cycloaliphatic epoxies such as 3,4-epoxycyclohexylmethyl carboxylate are commonly used as co-monomers with EVOs in coating applications. High-solid, cationically cured coatings based on cycloaliphatic epoxy resin, ESO and polyols were prepared by Raghavachar *et al.*[158] and had useful film properties as general-purpose coatings. A blend with 10 wt% ESO gave a coating with similar performance to the cycloaliphatic epoxy control. With further increases in the ESO content, the hardness of the coatings decreased, but this was regulated by the structure of the epoxy and polyol, or adjustment of the epoxy-to-polyol ratio.

Coatings derived from epoxidized *Mesua ferrea L.* seed oil and DGEBA were prepared by Das and Karak.[160] The results indicated that the EVO not only reduced the viscosity of the DGEBA but also enhanced the performance of the polymer. The performance of 50 wt% epoxidized seed oil was further enhanced by the formation of a nanocomposite using organically modified nanoclay. Adding 2.5 to 5 wt% clay improved the alkali resistance of the prepared coating.

Ultraviolet-initiated cationic polymerization of EVOs has been the subject of intensive research due to the convenience of curing at room temperature and the fast curing rate.[39,89,91,95] Bio-based coatings prepared by Thames and Yu[164] exhibited excellent adhesion, impact resistance, UV stability, gloss retention, and corrosion resistance properties. Vernonia oil or ESO blended with cycloaliphatic epoxy was used as an epoxy resin and cationically UV initiated. The incorporation level of EVOs was formulated by their compatibility with the other coating ingredients. Although both EVOs were compatible with cycloaliphatic epoxy at high concentrations, EVO epoxy blends were only partially compatible and formed hazy formulations with polyols or UV initiators. The pencil hardness and tensile strength of the coating films decreased, but the gloss retention increased. Optimum properties in terms of hardness, gloss and gloss retention were obtained at a 10 wt% of EVO in the coating.

Clear coatings containing ESO and cycloaliphatic epoxy resin were formulated by Gu *et al.*[165] using an onium tetrakis (pentafluorophenyl) gallate catalyst. The gallate catalyst showed better solubility and reactivity towards the non-polar monomers than the common UV-initiation catalysts diaryliodonium or triarylsulfonium salts. Up to 60 wt% ESO could be added to formulations without compromising the mechanical properties of the cured coatings.

Despite a current lack of optimum thermal and physical properties for commercial coating applications, high EVO-based or pure EVO coatings would be desirable because of their highly bio-renewable content. Goals for EVO development include improvement of T_g, hardness, moduli, and strength. Inorganic–organic hybrid films have been synthesized from EVO and titania or silicon-based or combined precursors.[166–171] These hybrid films generally showed improved properties, such as hardness, adhesion, chemical resistance, tensile strength, and T_g, but strongly depend on the type and concentration of inorganic content. Overloading the inorganic component led to decreased fracture toughness and elongation-at-break. A sharp transition from ductile to brittle material when loading with inorganic precursors has been observed.[171]

9.8 Composites from Epoxidized Vegetable Oils

Considerable attention has been focused on the development of VO-based composites due to their sustainable characteristics, greatly improved stiffness, modulus and strength.[172–175] A composite approach greatly expands the potential applications of VO-based polymeric materials, where some have been successfully commercialized and have behaved well as promising alternatives to petroleum-based materials in transport and construction applications.[176] Recent developments have been more toward high-performance bio-based materials and "green" composites for value-added and structural applications.[177–179] Based on the reinforcement type, EVO-based composites can be grossly divided into fiber-reinforced polymer composites (FRPs) and nanocomposites. The fibers, either natural or synthetic, can be continuous or chopped (short strand) and are on the macroscopic scale. The particle size and surface area per volume of nanocomposites provides reinforcement at the nanoscopic level.

9.8.1 Fiber-reinforced Composites

The mechanical strength of an FRP is dependent primarily upon the properties of the continuous phase reinforcement, while the matrix phase supports and binds the reinforcement together and distributes stress to the reinforcement. When designing a composite for a structural application, the polymer matrix must be strong enough to efficiently transfer stress to the reinforcement without initiation of cracks, *i.e.*, of sufficiently high cross-link density and of T_g higher than the temperature of its intended work environment. Since most pure EVO polymers generally show low cross-link density and T_g, even below room temperature, polymer blends of EVO with DGEBA have been frequently applied as polymer matrices for fiber reinforced composites (FRCs).

Some research has shown that EVOs are best limited to being minor components in blends, i.e., <30 wt% because EVO components cannot provide the mechanical and thermal properties desired for an FRC.[61,180] Pure EVO or high EVO content (e.g., >50 wt%) polymer matrices for high-performance composite applications are rare and better suited to non-structural applications.[181–184]

Glass fibers are one of the most widely used reinforcement materials in epoxy composites because of their availability, low cost, high modulus, and excellent adhesion to the matrix resin. Espinoza-Perez et al.[61] manufactured glass-fiber-reinforced composites using a hand lay-up method. A PACM-cured EVO commercial epoxy blend was used as the matrix. The 30 wt% EVO blended composite thermal and mechanical performance was slightly lower than the composites without EVO, but were comparable with those of anhydride-cured ones.

EAS, ESO, and EMS have been applied in bio-composite manufacturing using pultrusion processing but these epoxies were limited to being minor components in blends, e.g., ≤30 wt% of the epoxy blend.[185,186] Greater mechanical properties were demonstrated for EAS than for ESO or EMS due to an improved oxirane content and better reactivity. A further increase of EAS content, up to 50 wt%, was also attempted. While the T_g of the composite decreased from 78 to 52 °C, the impact strength improved.[187] The pulling force of the pultrusion manufacturing was significantly reduced due to good lubricity provided an oily bio-based component which apparently came from the saturated and unreactive components of EAS.

Using anhydride-cured pure EGS, a blend of EGS-DGEBA, or pure DGEBA as the polymer matrix, glass-fiber-reinforced composites were fabricated via vacuum-assisted resin transfer molding.[188] The EGS-based composite showed mechanical properties comparable to that of the DGEBA counterparts in terms of flexural strength/modulus and impact strength. Only a slightly reduced T_g and thermal stability were observed. This high-performance bio-based composite has good potential to replace petroleum-based epoxy resins as a value-added product from VOs.

Cellulosic fibers such as flax, hemp, or jute are also promising reinforcements for polymers composites due to their availability, high specific strength, low cost, and environmental friendliness.[189] VO-based polymer composites reinforced by cellulose are often called "green" composites, since both the matrix resin and reinforcement are from bio-renewable resources.[178] However, these composites tend to have lower mechanical strengths than similar composites reinforced with glass fibers. Due to the hydrophilic character of cellulosic fibers, surface modification is required to improve the adhesion or compatibility between the cellulose and the polymer matrix.[190,191]

Hemp-fiber-reinforced ELO composites were manufactured by Boquillon[192] using a hot pressing method. DMA results indicated the storage modulus in the rubbery region increased from 17 MPa for the neat resin to 850 MPa for 65 vol% fiber content composites; however, a reduced composite T_g at high fiber content was also observed. The adsorption of anhydride hardener on the hemp fiber surface leads to an off-stoichiometric reaction between the epoxy and anhydride. Reduction of T_g was also observed in flax-fiber-reinforced ELO composites by Fejős *et al.*,[193] which was ascribed to a chemical reaction between the hydroxyl groups of the fibers and the anhydride hardener.

Flax-fiber-reinforced composites were manufactured by Liu *et al.*[194] using a compression molding method. The polymer matrix was an amine-cured ESO and a 1,1,1-tris(*p*-hydroxyphenyl) ethane triglycidyl ether blend (THPE-GE). The flexural and tensile modulus increased with fiber content but decreased at higher fiber loadings due to the increased fiber–fiber interaction and dispersion problems. So the optimum fiber content was about 10 wt%. A high percentage of THPE-GE in the blend was essential to achieve high thermal and mechanical strength composites. Longer fiber composites had better mechanical properties than shorter fiber composites.

Manthey *et al.*[180] manufactured jute-fiber-reinforced bio-composites in which amine-cured EVO and DGEBA blends were used as the matrix. The epoxidized hemp oil (EHO) composites displayed marginally higher mechanical strengths than those of their ESO counterparts and both composites' mechanical performance decreased with increased ESO or EHO loading. A significant reduction in strength occurred above 30 wt% bio-resin concentration.

9.8.2   Nanocomposites

Polymer nanocomposites have attracted interest over the past few years for their ability to generate improved thermal and mechanical strength, low weight, and optical transparency at relatively low particle concentrations, *e.g.*, ≤5 wt%.[195] Various nanomaterials including nanoclay,[196,197] carbon nanotubes,[198] silica,[199] polyhedral oligomeric silsesquioxane[200] and alumina[201] have been used in EVO-based polymer nanocomposites. Among these, organomodified montmorillonite clay platelets (OMMT) are inexpensive but highly efficient reinforcement fillers for polymer nanocomposites. With the extremely large surface area and high aspect ratio of nanoclay platelets, strong interfacial interactions between the polymer and nanoclay play a key role in the confinement of polymer chain mobility under stress. Polymer properties can be substantially improved.[202,203] Therefore, the major challenge encountered during the preparation of polymer clay nanocomposites is proper dispersion of the clay in the polymer matrix on a nanometric scale to achieve exfoliated, intercalated or mixtures of these structures.

Wang and Schuman[204] reported nanocomposite morphologies and thermal and mechanical strengths as a function of the clay concentration and dispersion technique. A mechanical shear mixing method led to an intercalated structure of undisrupted tactoids. High-speed shear mixing combined with ultrasonication reduced the platelet tactoids toward much smaller scales and exfoliation, which in turn provided better properties compared to a shear mixing method alone. Compared to neat polymer, only 1 wt% of clay dispersed by ultrasonication improved the nanocomposite tensile strength and modulus by 22 and 13%, respectively. In other words, tensile modulus could be increased up to 34% by 6 wt% clay without any sacrifice of strength. The T_g was also increased by 4–6 °C depending on the OMMT concentration.

Tan et al.[205] studied anhydride-cured ESO nanocomposites. OMMT of 1 to 5 wt% concentration was dispersed into ESO by ultrasonication. The surface modifier, OMMT, and imidazole, co-catalyze the epoxy–anhydride curing reaction and, with an exfoliated structure, the tensile strength of the nanocomposite was increased with OMMT loading of up to 4 wt%. The tensile modulus, T_g and thermal stability of the nanocomposite were also increased but the fracture toughness and elongation-at-break were reduced due to higher stiffness and cross-link density. In a similar anhydride-cured ESO–clay nanocomposite system, Tanrattanakul and Saithai[112] indicated that exfoliation was prone to occur only at low OMMT content and higher clay concentrations led to intercalated structures with aggregations.

Miyagawa et al.[206,207] reported nanocomposites of anhydride-cured blends of DGEBF and ELO. Clay nanoplatelets were almost completely exfoliated and homogeneously dispersed in the epoxy network after ultrasonication dispersion. The resulting nanocomposites showed higher storage moduli than the neat polymer to offset a reduced storage modulus caused by re-placement of DGEBF by ELO. However, the Izod impact strength did not change after adding the clay, while the heat distortional temperature, and T_g were lower due to the plasticizing effect of the modifier (OMMT). The nano-composite was used as a matrix for carbon-fiber-reinforced polymer compos-ites,[208] and the results indicated that the interlaminar shear strength of the composite was improved after adding 5 wt% intercalated clay, but the exfoli-ated clay nanoplatelets were less effective in preventing crack propagation.

More significant improvements in strength have been observed for EVO-based nanocomposites of low T_g. Nanocomposite tensile strength and modulus were increased more than 300% for 8 wt% OMMT, as reported by Liu et al.[209] TETA was used as a curing agent for ESO, and the OMMT was dis-persed in ESO by ultrasonication to form an intercalated structure. The T_g was increased from 11.8 °C for the neat polymer to 20.7 °C with 5 wt% clay. Higher OMMT concentrations led to a reduction in properties, due to clay aggregation.

Shabeer et al.[210] synthesized nanocomposites using EAS and anhydride. Two types of dispersion technique, pneumatic and ultrasonication, were carried out to disperse the OMMT into EAS. The nanoclay was readily exfoliated into the resin due to a clay interaction and reaction with the anhydride. Tensile testing showed that the OMMT improved the tensile modulus and strength by 625% and 340%, respectively. These significant improvements in strength were explained by a strong interaction of epoxy with clay platelets, and the much higher modulus for clay platelets than flexible polymer chains. However, the T_g of the polymer, which was below room temperature, was further decreased with increased clay loading, and these polymers are best suited to non-structural applications.

9.9 Conclusions

The current major commercial application of EVO is as a stabilizer and plasticizer. EVO has already shown versatility as an epoxy monomer material resource and a variety of epoxy thermoset polymers ranging from flexible rubbers to rigid plastics have been synthesized from different EVOs by a number of different polymerization methods. Some of these thermoset polymers have possessed comparable properties to their petroleum-based counterparts and have shown promise as replacements or supplements to commercially available epoxy monomer materials.

However, the inherently fewer reactive internal epoxy groups and flexible carbon chain structures in EVOs have prevented their application as high-performance thermoset polymers for structural applications. A remaining opportunity is as a supplement for petroleum-based commercial epoxy monomers, as matrix materials for coatings, composites or as nanocomposites. The future trend in this area will be to increase the percentage of bio-based content whilst optimizing overall performance through structure–property studies. Developing new, VO-derived epoxy monomers with higher reactivity and oxirane functionality will provide an opportunity to expand the use of EVOs as green materials. As novel VO-based epoxy resins, EGSs have shown improved properties than other EVO structures but only when the saturated content is reduced. They are currently at an experimental stage towards commercialization.

The versatility of epoxy formulations not only depends on the epoxy monomer, but also on the combined effect of the curing agent, co-monomer, and polymerization conditions. Effective EVO curing systems with short polymerization times and lower curing temperatures are highly desired. EVO will be of continued interest with regard to environmental and renewable/sustainable efforts through industrial applications, but only if the materials meet customer performance requirements in terms of reactivity, compatibility and polymer mechanical, thermal, and environmental stability properties with minimal or no tradeoffs, and at a competitive cost. This chapter has summarized the issues of reactivity, compatibility, and properties as a function of EVO chemical structure. The continuing challenge is to create new cost-effective EVO-derived structures that improve upon existing performance levels.

References

1. H. Q. Pham and M. J. Marks, in *Encyclopedia of Polymer Science Technology*, ed. H. F. Mark, John Wiley & Sons, 2004, vol. 9, p. 678.

2. E. Santacesaria, R. Tesser, M. Di Serio, L. Casale and D. Verde, *Ind. Eng. Chem. Res.*, 2009, **49**, 964.

3. H. Kishi, Y. Akamatsu, M. Noguchi, A. Fujita, S. Matsuda and H. Nishida, *J. Appl. Polym. Sci.*, 2011, **120**, 745.

4. H. Kishi, A. Fujita, H. Miyazaki, S. Matsuda and A. Murakami, *J. Appl. Polym. Sci.*, 2006, **102**, 2285.

5. T. Koike, *Polym. Eng. Sci.*, 2012, **52**, 701.

6. G. Sun, H. Sun, Y. Liu, B. Zhao, N. Zhu and K. Hu, *Polymer*, 2007, **48**, 330.

7. A. M. Atta, R. Mansour, M. I. Abdou and A. M. El-Sayed, *J. Polym. Res.*, 2005, **12**, 127.

8. A. M. Atta, R. Mansour, M. I. Abdou and A. M. Sayed, *Polym. Adv. Technol.*, 2004, **15**, 514.

9. X. Q. Liu, W. Huang, Y. H. Jiang, J. Zhu and C. Z. Zhang, *eXPRESS Polym. Lett.*, 2011, **6**, 293.

10. K. Huang, J. Zhang, M. Li, J. Xia and Y. Zhou, *Ind. Crops Prod.*, 2013, **49**, 497.

11. X. Liu and J. Zhang, *Polym. Int.*, 2010, **59**, 607.

12. S. Carlotti, S. Caillol, C. Mantzaridis, A.-L. Brocas, G. Cendejas, A. Llevot, A. Remi and H. Cramail, *Green Chem.*, 2013, **15**, 3091.

13. H. Nouailhas, C. Aouf, C. Le Guerneve, S. Caillol, B. Boutevin and H. Fulcrand, *J. Polym. Sci., Part A: Polym. Chem.*, 2011, **49**, 2261.

14. S. Benyahya, C. Aouf, S. Caillol, B. Boutevin, J. P. Pascault and H. Fulcrand, *Ind. Crops Prod.*, 2014, **53**, 296.

15. J. Łukaszczyk, B. Janicki and M. Kaczmarek, *Eur. Polym. J.*, 2011, **47**, 1601.

16. X. Feng, A. J. East, W. B. Hammond, Y. Zhang and M. Jaffe, *Polym. Adv. Technol.*, 2011, **22**, 139.

17. F. Jaillet, E. Darroman, A. Ratsimihety, R. Auvergne, B. Boutevin and S. Caillol, *Eur. J. Lipid Sci. Technol.*, 2013, **116**, 63.

18. S. Ma, X. Liu, Y. Jiang, Z. Tang, C. Zhang and J. Zhu, *Green Chem.*, 2013, **15**, 245.

19. S. Ma, X. Liu, L. Fan, Y. Jiang, L. Cao, Z. Tang and J. Zhu, *ChemSusChem*, 2014, **7**, 555.

20. R. Auvergne, S. Caillol, G. David, B. Boutevin and J.-P. Pascault, *Chem. Rev.*, 2014, **114**, 1082.

21. L. Montero de Espinosa and M. A. R. Meier, *Eur. Polym. J.*, 2011, **47**, 837.

22. M. A. R. Meier, J. O. Metzger and U. S. Schubert, *Chem. Soc. Rev.*, 2007, **36**, 1788.

23. http://lipidlibrary.aocs.org/market/index.html.

24. N. Mann, S. Mendon, J. Rawlins and S. Thames, *J. Am. Oil Chem. Soc.*, 2008, **85**, 791.

25. P. Muturi, D. Wang and S. Dirlikov, *Prog. Org. Coat.*, 1994, **25**, 85.

26. J. V. Crivello and R. Narayan, *Chem. Mater.*, 1992, **4**, 692.

27. F. Gunstone, *The Chemistry of Oils and Fats: Sources, Composition, Properties and Uses*, CRC Press, Boca Raton, 2004.

28. D. Swern, *Chem. Rev.*, 1949, **45**, 1.
29. M. Rüsch gen Klaas and W. Siegfried, in *Recent Developments in the Synthesis of Fatty Acid Derivatives*, AOCS Publishing, 1999, p. 157.
30. V. V. Goud, A. V. Patwardhan, S. Dinda and N. C. Pradhan, *Chem. Eng. Sci.*, 2007, **62**, 4065.
31. S. Sinadinović-Fišer, M. Janković and Z. Petrović, *J. Am. Oil Chem. Soc.*, 2001, **78**, 725.
32. G. J. H. Buisman, A. Overeem and F. P. Cuperus, in *Recent Developments in the Synthesis of Fatty Acid Derivatives*, AOCS Publishing, 1999, p. 128.
33. T. Vlček and Z. Petrović, *J. Am. Oil Chem. Soc.*, 2006, **83**, 247.
34. A. Köckritz and A. Martin, *Eur. J. Lipid Sci. Technol.*, 2008, **110**, 812.
35. Z. S. Petrović, *Polym. Rev.*, 2008, **48**, 109.
36. J. Lu, S. Khot and R. P. Wool, *Polymer*, 2005, **46**, 71.
37. L. Gelb, W. Ault, W. Palm, L. Witnauer and W. Port, *J. Am. Oil Chem. Soc.*, 1959, **36**, 283.
38. L. Gelb, W. Ault, W. Palm, L. Witnauer and W. Port, *J. Am. Oil Chem. Soc.*, 1960, **37**, 81.
39. K. Zou and M. D. Soucek, *Macromol. Chem. Phys.*, 2005, **206**, 967.
40. J. Chen, M. D. Soucek, W. J. Simonsick and R. W. Celikay, *Polymer*, 2002, **43**, 5379.
41. Z. Zong, M. D. Soucek, Y. Liu and J. Hu, *J. Polym. Sci., Part A: Polym. Chem.*, 2003, **41**, 3440.
42. X. Pan, P. Sengupta and D. C. Webster, *Biomacromolecules*, 2011, **12**, 2416.
43. X. Pan, P. Sengupta and D. C. Webster, *Green Chem.*, 2011, **13**, 965.
44. T. Nelson, T. Galhenage and D. Webster, *J. Coat. Technol. Res.*, 2013, **10**, 589.
45. S. Alam, H. Kalita, A. Jayasooriya, S. Samanta, J. Bahr, A. Chernykh, M. Weisz and B. J. Chisholm, *Eur. J. Lipid Sci. Technol.*, 2014, **116**, 2.
46. S. Alam and B. Chisholm, *J. Coat. Technol. Res.*, 2011, **8**, 671.
47. G. Lligadas, J. C. Ronda, M. Galià and V. Cádiz, *J. Polym. Sci., Part A: Polym. Chem.*, 2006, **44**, 6717.
48. J. D. Earls, J. E. White, L. C. López, Z. Lysenko, M. L. Dettloff and M. J. Null, *Polymer*, 2007, **48**, 712.
49. J. Earls, J. White, M. Dettloff and M. Null, *J. Coat. Technol. Res.*, 2004, **1**, 243.
50. J. M. Fraile, J. I. García, D. Marco and J. A. Mayoral, *Appl. Catal., A*, 2001, **207**, 239.
51. K. Kamata, K. Sugahara, K. Yonehara, R. Ishimoto and N. Mizuno, *Chem.–Eur. J.*, 2011, **17**, 7549.
52. S. G. Yang, J. P. Hwang, M. Y. Park, K. Lee and Y. H. Kim, *Tetrahedron*, 2007, **63**, 5184.
53. *US Pat.*, 3 859 314 A, 1975.
54. *US Pat.*, 30 75 999 A, 1963.
55. *US Pat.*, 29 40 986 A, 1960.

56. K. Huang, P. Zhang, J. Zhang, S. Li, M. Li, J. Xia and Y. Zhou, *Green Chem.*, 2013, **15**, 2466.
57. R. Wang and T. Schuman, *eXPRESS Polym. Lett.*, 2013, **7**, 272.
58. H. Q. Pham and M. J. Marks, in *Ullmann's Encyclopedia of Industrial Chemistry*, ed. B. Elvers, Wiley-VCH, Weinheim, 2005.
59. P. Lu, J. O. Stoffer, R. A. Babcock and L. R. Dharani, presented at the 220th ACS National Meeting, Washington, DC, 2000.
60. J. Zhu, K. Chandrashekhara, V. Flanigan and S. Kapila, *J. Appl. Polym. Sci.*, 2004, **91**, 3513.
61. J. D. Espinoza-Perez, B. A. Nerenz, D. M. Haagenson, Z. Chen, C. A. Ulven and D. P. Wiesenborn, *Polym. Compos.*, 2011, **32**, 1806.
62. G. López Téllez, E. Vigueras-Santiago, S. Hernández-López and B. Bilyeu, *Des. Monomers Polym.*, 2008, **11**, 435.
63. A. Gandini, T. M. Lacerda and A. J. F. Carvalho, *Green Chem.*, 2013, **15**, 1514.
64. N. W. Manthey, F. Cardona, T. Aravinthan and T. Cooney, *J. Appl. Polym. Sci.*, 2011, **122**, 444.
65. N. W. Manthey, F. Cardona and T. Aravinthan, *J. Appl. Polym. Sci.*, 2012, **125**, E511.
66. V. del Río, M. P. Callao and M. S. Larrechi, , *Int. J. Anal. Chem.*, 2011, **2011**.
67. Z. Wang, X. Zhang, R. Wang, H. Kang, B. Qiao, J. Ma, L. Zhang and H. Wang, *Macromolecules*, 2012, **45**, 9010.
68. S. Miao, S. Zhang, Z. Su and P. Wang, *J. Appl. Polym. Sci.*, 2013, **127**, 1929.
69. P. Lu, PhD thesis, University of Missouri-Rolla, 2001.
70. A. M. Partansky, in *Epoxy Resins*, ed. H. Lee, ACS, Washington, DC, 1970, vol. 92, p. 29.
71. R. E. Harry-O'kuru, S. H. Gordon and A. Biswas, *J. Am. Oil Chem. Soc.*, 2005, **82**, 207.
72. N. Juangvanich, PhD thesis, University of Missouri-Rolla, 2003.
73. X. Fernàndez-Francos, X. Ramis and À. Serra, *J. Polym. Sci., Part A: Polym. Chem.*, 2014, **52**, 61.
74. F. I. Altuna, V. Pettarin and R. J. J. Williams, *Green Chem.*, 2013, **15**, 3360.
75. N. Supanchaiyamat, P. S. Shuttleworth, A. J. Hunt, J. H. Clark and A. S. Matharu, *Green Chem.*, 2012, **14**, 1759.
76. J.-P. Pascault, H. Sautereau, J. Verdu and R. J. J. Williams, *Thermosetting Polymers*, Marcel Dekker, New York, 2002.
77. S. G. Tan and W. S. Chow, *eXPRESS Polym. Lett.*, 2011, **5**, 480.
78. C. A. May, *Epoxy Resins: Chemistry and Technology*, Marcel Dekker, New York, 2nd edn, 1988.
79. N. Boquillon and C. Fringant, *Polymer*, 2000, **41**, 8603.
80. G. Liang and K. Chandrashekhara, *J. Appl. Polym. Sci.*, 2006, **102**, 3168.
81. S. G. Tan, Z. Ahmad and W. S. Chow, *Ind. Crops Prod.*, 2013, **43**, 378.

82. A. Mahendran, G. Wuzella, A. Kandelbauer and N. Aust, *J. Therm. Anal. Calorim.*, 2012, **107**, 989.
83. A. Nicolau, D. Samios, C. M. S. Piatnick, Q. B. Reiznautt, D. D. Martini and A. L. Chagas, *Eur. Polym. J.*, 2012, **48**, 1266.
84. A. J. Clark and S. S. Hoong, *Polym. Chem.*, 2014, **5**, 3238.
85. Z. Liu, K. M. Doll and R. A. Holser, *Green Chem.*, 2009, **11**, 1774.
86. Z. Liu and S. Z. Erhan, *J. Am. Oil Chem. Soc.*, 2009, **87**, 437.
87. Z. Liu and A. Biswas, *Appl. Catal., A*, 2013, **453**, 370.
88. B. K. Sharma, Z. Liu, A. Adhvaryu and S. Z. Erhan, *J. Agric. Food Chem.*, 2008, **56**, 3049.
89. L. Fertier, H. Koleilat, M. Stemmelen, O. Giani, C. Joly-Duhamel, V. Lapinte and J.-J. Robin, *Prog. Polym. Sci.*, 2013, **38**, 932.
90. M.-A. Tehfe, J. Lalevée, D. Gigmes and J. P. Fouassier, *Macromolecules*, 2010, **43**, 1364.
91. C. Decker, T. Nguyen Thi Viet and H. Pham Thi, *Polym. Int.*, 2001, **50**, 986.
92. W. D. Wan Rosli, R. N. Kumar, S. Mek Zah and M. M. Hilmi, *Eur. Polym. J.*, 2003, **39**, 593.
93. Y.-S. Li, M.-S. Li and F.-C. Chang, *J. Polym. Sci., Part A: Polym. Chem.*, 1999, **37**, 3614.
94. R. A. Ortiz, D. P. López, M. d. L. G. Cisneros, J. C. R. Valverde and J. V. Crivello, *Polymers*, 2005, **46**, 1535.
95. S. Chakrapani and J. V. Crivello, *J. Macromol. Sci., Part A*, 1998, **35**, 691.
96. M. Soucek and J. Chen, *J. Coat. Technol.*, 2003, **75**, 49.
97. A. Hartwig, K. Koschek and A. Lühring, in *Adhesion*, ed. W. Possart, Wiley-VCH, Weinheim, 2006, p. 205.
98. S.-J. Park, F.-L. Jin, J.-R. Lee and J.-S. Shin, *Eur. Polym. J.*, 2005, **41**, 231.
99. S.-J. Park, F.-L. Jin and J.-R. Lee, *Macromol. Rapid Commun.*, 2004, **25**, 724.
100. S.-J. Park, F.-L. Jin and J.-R. Lee, *Macromol. Chem. Phys.*, 2004, **205**, 2048.
101. A.-L. Brocas, C. Mantzaridis, D. Tunc and S. Carlotti, *Prog. Polym. Sci.*, 2013, **38**, 845.
102. I. E. Dell'Erba and R. J. J. Williams, *Polym. Eng. Sci.*, 2006, **46**, 351.
103. K. Boonkerd, B. K. Moon, M. C. Kim and J. K. Kim, *J. Elastomers Plast.*, 2013, DOI: 10.1177/0095244313483643.
104. C. Öztürk and S. H. Küsefoğlu, *J. Appl. Polym. Sci.*, 2011, **121**, 2976.
105. E. Del Rio, M. Galià, V. Cádiz, G. Lligadas and J. C. Ronda, *J. Polym. Sci., Part A: Polym. Chem.*, 2010, **48**, 4995.
106. C. H. Zhao, S. J. Wan, L. Wang, X. D. Liu and T. Endo, *J. Polym. Sci., Part A: Polym. Chem.*, 2014, **52**, 375.
107. J. La Scala and R. P. Wool, *Polymer*, 2005, **46**, 61.
108. A. Zlatanić, Z. S. Petrović and K. Dušek, *Biomacromolecules*, 2002, **3**, 1048.
109. Z. S. Petrović, W. Zhang and I. Javni, *Biomacromolecules*, 2005, **6**, 713.
110. M. Carme Coll Ferrer, D. Babb and A. J. Ryan, *Polymer*, 2008, **49**, 3279.

111. M. Samper, V. Fombuena, T. Boronat, D. García-Sanoguera and R. Balart, *J. Am. Oil Chem. Soc.*, 2012, **89**, 1521.
112. V. Tanrattanakul and P. Saithai, *J. Appl. Polym. Sci.*, 2009, **114**, 3057.
113. J. R. Kim and S. Sharma, *Ind. Crops Prod.*, 2012, **36**, 485.
114. Q. B. Reiznautt, I. T. S. Garcia and D. Samios, *Mater. Sci. Eng., C*, 2009, **29**, 2302.
115. A. Gerbase, C. Petzhold and A. Costa, *J. Am. Oil Chem. Soc.*, 2002, **79**, 797.
116. J. Rösch and R. Mülhaupt, *Polym. Bull.*, 1993, **31**, 679.
117. D. dos Santos Martini, B. A. Braga and D. Samios, *Polymer*, 2009, **50**, 2919.
118. A. P. Gupta, S. Ahmad and A. Dev, *Polym.-Plast. Technol. Eng.*, 2010, **49**, 657.
119. J. M. España, L. Sánchez-Nacher, T. Boronat, V. Fombuena and R. Balart, *J. Am. Oil Chem. Soc.*, 2012, **89**, 2067.
120. S. G. Tan and W. S. Chow, *J. Am. Oil Chem. Soc.*, 2011, **88**, 915.
121. N. Bouillon, J.-P. Pascault and L. Tighzert, *J. Appl. Polym. Sci.*, 1989, **38**, 2103.
122. M. J. Abdekhodaie, Z. Liu, S. Z. Erhan and X. Y. Wu, *Polym. Int.*, 2012, **61**, 1477.
123. S. Tan and W. Chow, *J. Am. Oil Chem. Soc.*, 2011, **88**, 915.
124. S. G. Tan and W. S. Chow, *Iran. Polym. J.*, 2012, **21**, 353.
125. S. Qureshi, J. A. Manson, J. C. Michel, R. W. Hertzberg and L. H. Sperling, In *Characterization of Highly Cross-linked Polymers*, ACS, Washington, DC, 1984, vol. 243, p. 109.
126. F. I. Altuna, L. H. Espósito, R. A. Ruseckaite and P. M. Stefani, *J. Appl. Polym. Sci.*, 2011, **120**, 789.
127. Y. Chen, L. Yang, J. Wu, L. Ma, D. Finlow, S. Lin and K. Song, *J. Therm. Anal. Calorim.*, 2013, **113**, 939.
128. J. Karger-Kocsis, S. Grishchuk, L. Sorochynska and M. Z. Rong, *Polym. Eng. Sci.*, 2014, **54**, 747.
129. R. Raghavachar, R. J. Letasi, P. V. Kola, Z. Chen and J. L. Massingill, *J. Am. Oil Chem. Soc.*, 1999, **76**, 511.
130. D. Ratna, *Polym. Int.*, 2001, **50**, 179.
131. A. Sarwono, Z. Man and M. A. Bustam, *J. Polym. Environ.*, 2012, **20**, 540.
132. I. Frischinger and S. Dirlikov, in *Interpenetrating Polymer Networks*, ACS, Washington, DC, 1994, vol. 239, p. 517.
133. F. R. Mustata, N. Tudorachi and I. Bicu, *Ind. Eng. Chem. Res.*, 2013, **52**, 17099.
134. F. Mustata, N. Tudorachi and D. Rosu, *Compos., Part B*, 2011, **42**, 1803.
135. A. P. Gupta, S. Ahmad and A. Dev, *Polym. Eng. Sci.*, 2011, **51**, 1087.
136. S. G. Tan and W. S. Chow, *J. Therm. Anal. Calorim.*, 2010, **101**, 1051.
137. H. Miyagawa, A. K. Mohanty, M. Misra and L. T. Drzal, *Macromol. Mater. Eng.*, 2004, **289**, 629.
138. H. Miyagawa, A. K. Mohanty, M. Misra and L. T. Drzal, *Macromol. Mater. Eng.*, 2004, **289**, 636.

139. P. Czub, *Macromol. Symp.*, 2006, **245**, 533.
140. P. Czub, *Macromol. Symp.*, 2006, **242**, 60.
141. E.-A.-C. Demengeot, I. Baliutaviciene, J. Ostrauskaite, L. Augulis, V. Grazuleviciene, L. Rageliene and J. V. Grazulevicius, *J. Appl. Polym. Sci.*, 2010, **115**, 2028.
142. A. Remeikyte, J. Ostrauskaite and V. Grazuleviciene, *J. Appl. Polym. Sci.*, 2013, **129**, 1290.
143. F.-L. Jin and S.-J. Park, *Polym. Int.*, 2008, **57**, 577.
144. R. Bagheri, B. T. Marouf and R. A. Pearson, *Polym. Rev.*, 2009, **49**, 201.
145. A. F. Yee and R. A. Pearson, *J. Mater. Sci.*, 1986, **21**, 2462.
146. J. C. Munoz, H. Ku, F. Cardona and D. Rogers, *J. Mater. Process. Technol.*, 2008, **202**, 486.
147. G. Zhan, L. Zhao, S. Hu, W. Gan, Y. Yu and X. Tang, *Polym. Eng. Sci.*, 2008, **48**, 1322.
148. S. G. Tan, Z. Ahmad and W. S. Chow, *Polym. Int.*, 2014, **63**, 273.
149. F. I. Altuna, V. Pettarin, L. Martin, A. Retegi, I. Mondragon, R. A. Ruseckaite and P. M. Stefani, *Polym. Eng. Sci.*, 2014, **54**, 569.
150. A. Shabeer, S. Sundararaman, K. Chandrashekhara and L. R. Dharani, *J. Appl. Polym. Sci.*, 2007, **105**, 656.
151. H. Miyagawa, M. Misra, L. T. Drzal and A. K. Mohanty, *Polym. Eng. Sci.*, 2005, **45**, 487.
152. S. G. Tan and W. S. Chow, *Polym.-Plast. Technol. Eng.*, 2010, **49**, 900.
153. F.-L. Jin and S.-J. Park, *Mater. Sci. Eng., A*, 2008, **478**, 402.
154. S.-J. Park, F.-L. Jin and J.-R. Lee, *Mater. Sci. Eng., A*, 2004, **374**, 109.
155. M. A. Sithique, S. Ramesh and M. Alagar, *Int. J. Polym. Mater.*, 2008, **57**, 480.
156. I. Frischinger and S. Dirlikov, in *Toughened Plastics I*, ed. C. K. Riew and A. J. Kinloch, ACS, Washington, DC, 1993, vol. 233, p. 451.
157. D. Ratna and A. K. Banthia, *J. Adhes. Sci. Technol.*, 2000, **14**, 15.
158. R. Raghavachar, G. Sarnecki, J. Baghdachi and J. Massingill, *J. Coat. Technol.*, 2000, **72**, 125.
159. M. Y. Shah and S. Ahmad, *Prog. Org. Coat.*, 2012, **75**, 248.
160. G. Das and N. Karak, *Prog. Org. Coat.*, 2009, **66**, 59.
161. S. Ahmad, F. Naqvi, E. Sharmin and K. L. Verma, *Prog. Org. Coat.*, 2006, **55**, 268.
162. M. D. Soucek, A. H. Johnson and J. M. Wegner, *Prog. Org. Coat.*, 2004, **51**, 300.
163. N. Jiratumnukul and R. Intarat, *J. Appl. Polym. Sci.*, 2008, **110**, 2164.
164. S. F. Thames and H. Yu, *Surf. Coat. Technol.*, 1999, **115**, 208.
165. H. Gu, K. Ren, D. Martin, T. Marino and D. Neckers, *J. Coat. Technol.*, 2002, **74**, 49.
166. D. M. Bechi, M. A. d. Luca, M. Martinelli and S. Mitidieri, *Prog. Org. Coat.*, 2013, **76**, 736.
167. M. Luca, M. Martinelli, M. Jacobi, P. Becker and M. Ferrão, *J. Am. Oil Chem. Soc.*, 2006, **83**, 147.

168. M. A. de Luca, M. Martinelli and C. C. T. Barbieri, *Prog. Org. Coat.*, 2009, **65**, 375.
169. D. Becchi, M. Luca, M. Martinelli and S. Mitidieri, *J. Am. Oil Chem. Soc.*, 2011, **88**, 101.
170. T. Tsujimoto, H. Uyama and S. Kobayashi, *Macromol. Rapid Commun.*, 2003, **24**, 711.
171. Z. Zong, J. He and M. D. Soucek, *Prog. Org. Coat.*, 2005, **53**, 83.
172. A. O'Donnell, M. A. Dweib and R. P. Wool, *Compos. Sci. Technol.*, 2004, **64**, 1135.
173. P. H. Henna, M. R. Kessler and R. C. Larock, *Macromol. Mater. Eng.*, 2008, **293**, 979.
174. Y. Lu and R. C. Larock, *Macromol. Mater. Eng.*, 2007, **292**, 1085.
175. R. L. Quirino, Y. Ma and R. C. Larock, *Green Chem.*, 2012, **14**, 1398.
176. R. P. Wool and X. S. Sun, *Bio-Based Polymers and Composites*, Elsevier, xxx, 2005.
177. M. A. Mosiewicki and M. I. Aranguren, *Eur. Polym. J.*, 2013, **49**, 1243.
178. E. Zini and M. Scandola, *Polym. Compos.*, 2011, **32**, 1905.
179. A. K. Mohanty, M. Misra and G. Hinrichsen, *Macromol. Mater. Eng.*, 2000, 276.
180. N. W. Manthey, F. Cardona, G. Francucci and T. Aravinthan, *J. Reinf. Plast. Compos.*, 2013, **32**, 1444.
181. J. V. Crivello, R. Narayan and S. S. Sternstein, *J. Appl. Polym. Sci.*, 1997, **64**, 2073.
182. Z. S. Liu, S. Z. Erhan, J. Xu and P. D. Calvert, *J. Appl. Polym. Sci.*, 2002, **85**, 2100.
183. A. Retegi, I. Algar, L. Martin, F. Altuna, P. Stefani, R. Zuluaga, P. Gañán and I. Mondragon, *Cellulose*, 2012, **19**, 103.
184. Z. S. Liu, S. Z. Erhan and P. D. Calvert, *Compos., Part A*, 2007, **38**, 87.
185. J. Zhu, K. Chandrashekhara, V. Flanigan and S. Kapila, *Compos., Part A*, 2004, **35**, 95.
186. K. Chandrashekhara, S. Sundararaman, V. Flanigan and S. Kapila, *Mater. Sci. Eng., A*, 2005, **412**, 2.
187. S. Sundararaman, A. Shabeer, K. Chandrashekhara and T. Schuman, *J. Biobased Mater. Bioenergy*, 2008, **2**, 71.
188. R. Wang and T. Schuman, presented at the 245th ACS National Meeting & Exposition, New Orleans, 2013.
189. F. P. La Mantia and M. Morreale, *Compos., Part A*, 2011, **42**, 579.
190. Z. Liu, S. Z. Erhan, D. E. Akin, F. E. Barton, C. Onwulata and T. A. McKeon, *Compos. Interfaces*, 2008, **15**, 207.
191. P. Tran, D. Graiver and R. Narayan, *J. Appl. Polym. Sci.*, 2006, **102**, 69.
192. N. Boquillon, *J. Appl. Polym. Sci.*, 2006, **101**, 4037.
193. M. Fejős, J. Karger-Kocsis and S. Grishchuk, *J. Reinf. Plast. Compos.*, 2013, **32**, 1879.
194. Z. Liu, S. Z. Erhan, D. E. Akin and F. E. Barton, *J. Agric. Food Chem.*, 2006, **54**, 2134.
195. S. Sinha Ray and M. Bousmina, *Prog. Mater. Sci.*, 2005, **50**, 962.

196. H. Uyama, M. Kuwabara, T. Tsujimoto, M. Nakano, A. Usuki and S. Kobayashi, *Macromol. Biosci.*, 2004, **4**, 354.
197. H. Uyama, M. Kuwabara, T. Tsujimoto, M. Nakano, A. Usuki and S. Kobayashi, *Chem. Mater.*, 2003, **15**, 2492.
198. H. Miyagawa, A. K. Mohanty, L. T. Drzal and M. Misra, *Nanotechnology*, 2005, **16**, 118.
199. G. Zhan, X. Tang, Y. Yu and S. Li, *Polym. Eng. Sci.*, 2011, **51**, 426.
200. G. Lligadas, J. C. Ronda, M. Galià and V. Cádiz, *Biomacromolecules*, 2006, **7**, 3521.
201. H. Miyagawa, A. Mohanty, L. T. Drzal and M. Misra, *Ind. Eng. Chem. Res.*, 2004, **43**, 7001.
202. J. Lu, C. K. Hong and R. P. Wool, *J. Polym. Sci., Part B: Polym. Phys.*, 2004, **42**, 1441.
203. Y. Lu and R. C. Larock, *Biomacromolecules*, 2006, **7**, 2692.
204. R. Wang, T. Schuman, R. R. Vuppalapati and K. Chandrashekhara, *Green Chem.*, 2014, **16**, 1871.
205. S. G. Tan, Z. Ahmad and W. S. Chow, *Appl. Clay Sci.*, 2014, **90**, 11.
206. H. Miyagawa, M. Misra, L. T. Drzal and A. K. Mohanty, *J. Polym. Environ.*, 2005, **13**, 87.
207. H. Miyagawa, M. Misra, L. T. Drzal and A. K. Mohanty, *Polymer*, 2005, **46**, 445.
208. H. Miyagawa, R. J. Jurek, A. K. Mohanty, M. Misra and L. T. Drzal, *Compos. Part A*, 2006, **37**, 54.
209. Z. Liu, S. Z. Erhan and J. Xu, *Polymer*, 2005, **46**, 10119.
210. A. Shabeer, K. Chandrashekhara and T. Schuman, *J. Compos. Mater.*, 2007, **41**, 1825.

Lubricity Characteristics of Seed Oils Modified by Acylation

ROGERS E. HARRY-O'KURU* AND GIRMA BIRESAW

USDA, ARS, National Center for Agricultural Utilization Research, Bio-Oils Research Unit, 1815 N. University Street, Peoria, IL 61604, USA[†]
*Email: rogers.harryokuru@ars.usda.gov

10.1 Introduction

Environmental stewardship among many residents of Earth in recent decades is triggering our recourse to explore bio-based materials in an effort to meet some of our industrial needs. In an analogous situation during World War II, when the floatation material supply, kapok from Indonesia was cut off, a search for replacement of that need led to the discovery of the lowly common milkweed (*Asclepias syriaca* L) as "a war strategic material".[1] The seeds of this plant contain a highly unsaturated oil that is 25–30% by weight of the dry seed. In an earlier study, the polyolefinic character of this seed oil was oxidized to polyoxirane derivatives,[2] the ring opening of which afforded the polyhydroxy triglycerides (PHMWO) of the oil. These derivatives provide synthetic platforms for the preparation of desirable and important industrial materials.[2–4] Stemming from the observed high viscosity of PHMWO, it was

---

[†]Mention of trade names or commercial products in this publication is solely for the purpose of providing specific information and does not imply recommendation or endorsement by the U.S. Department of Agriculture. USDA is an equal opportunity provider and employer.

RSC Green Chemistry No. 29
Green Materials from Plant Oils
Edited by Zengshe Liu and George Kraus
© The Royal Society of Chemistry 2015
Published by the Royal Society of Chemistry, www.rsc.org

conceived that esterification of the hydroxyl groups would reduce the high viscosity thus rendering the product more useful in lubricant applications. In this chapter, we describe the production of three polyesters: polyacetyl, polybutyroyl and polypentanoyl derivatives from PHMWO. Additionally, to explore the characteristics of shorter chain substituent polyesters, the parent oil was epoxidized for direct ring-opening with formic acid. The idea was to generate a traditional α-hydroxyl polyester. What was obtained instead was a vicinal acyl derivative of the milkweed oil, as analyzed by FT-IR and NMR spectroscopy. These polyesters were then evaluated for oxidative stability and suitability in bio-lubricant applications using a rotating pressurized vessel oxidation test (RPVOT), rheometry, elastohydrodynamic (EHD) film thickness under load at specified temperature and a four-ball anti-wear (4-ball AW) test. Most of the experimental data obtained for the polyesters of milkweed oil were essentially in excellent agreement with the predictive lubricity characteristics of lubricating oils.

10.1.1 Materials

Milkweed oil (MWO) was obtained from Natural Fibers Corporation (Ogallala, NE), whereas sodium chloride, sodium bicarbonate, ethyl acetate, hexane, acetic, butyric and valeric anhydrides were purchased from ACRŌS ORGANICS (Vineland, NJ); 4-dimethylammino pyridine (4-DMAP) and anhydrous magnesium sulfate were from Sigma–Aldrich (St. Louis, MO). PHMWO were synthesized from cold-pressed MWO as described previously.[2] All reagents were reagent-grade materials and were used as supplied.

Test balls used in 4-ball AW experiments were obtained from Falex Corporation (Aurora, IL) and had the following specifications: chrome-steel alloy made from AISI E52100 standard steel; 64–66 R_c hardness; 12.7 mm diameter; grade 25 extra polish. The test balls were degreased by two consecutive five-minute sonications in isopropyl alcohol and hexane solvents in an ultrasonic bath, before use in the 4-ball AW experiments.

10.1.2 Methods

10.1.2.1 Instrumentation

10.1.2.1.1 Fourier-transform Infrared Spectrometry. FT-IR spectra were measured on an Arid Zone FT-IR spectrometer (ABB MB-Series, Houston, TX) equipped with a DTGS detector. Liquid derivatives were pressed between two NaCl discs (25×5 mm) to give thin transparent oil films for analysis by FT-IR spectrometry. Absorbance spectra were acquired at 4 cm^{-1} resolution and signal-averaged over 32 scans. Interferograms were Fourier transformed using cosine apodization for optimum linear response. Spectra were baseline corrected, scaled for mass differences and normalized to the methylene peak at 2927 cm^{-1}.

10.1.2.1.2   NMR Spectroscopy. ^{1}H- and ^{13}C-NMR spectra were acquired on a Bruker AV-500 MHz spectrometer with a dual 5 mm proton/carbon probe (Bruker, Ballerica, MA). The internal standard used was tetramethylsilane.

10.1.2.1.3   Viscosity Measurements.   Measurements were determined on a Temp-Trol constant temperature bath (Precision Scientific, Chicago IL) using a Cannon–Fenske viscometer for transparent liquids (Cannon Instrument Company, State College, PA) in accordance with the American Oil Chemists Society (AOCS) Official Method Tq 1a-64.[5] The size of the Cannon–Fenske viscometer used was number 500. The cleaned dry tube was loaded at room temperature (RT) with the sample oil and placed in its holder in the constant temperature bath. The sample was allowed to equilibrate for 10 min at 40 °C or 15 min at 100 °C before being suctioned into the lower bulb until the meniscus just overshot the mark above the lower bulb. The suction was removed and the meniscus adjusted to the mark. The sample was allowed to flow at the same time the stop clock was started. The time (in seconds) it took the meniscus to reach the mark below the bulb, multiplied by the tube constant gave the viscosity of the sample. The measurements were replicated for reproducibility. All kinematic viscosity measurements were run in duplicate and the averaged values are reported.[6,7] Dynamic viscosity at 40 and 100 °C was directly measured on an ARES LS-2 Rheometer (TA Instruments Waters LLC New Castle, DE) as described elsewhere.[8] Viscosity indices (VIs) were calculated from the kinematic viscosity data at 40 and 100 °C following the procedure outlined in American Society for Testing and Materials (ASTM) D2270.[9]

10.1.2.1.4   Density Measurements.   Density measurements were performed using a pycnometer or Kimble Specific Gravity Bottle (10.0 mL) equipped with a thermometer (Kimble Glass Inc. Vineland, NJ). The pycnometer was cleaned with acetone and dried with an air jet at RT. The components were reassembled and weighed empty. Each sample was carefully introduced into the bottle until the liquid meniscus grazed the lower portion of the ground glass joint. The thermometer lid was carefully inserted allowing any excess sample to exit through the open side-arm. The completely filled pycnometer was carefully cleaned of excess sample, the side-arm cap replaced and the system was then weighed at the operating RT. The difference in mass from the empty pycnometer mass is the mass of 10.0 mL of sample. Two replicates for each sample were measured from which an average mass of sample was derived. Between samples, the pycnometer assembly was thoroughly rinsed with a mixture of hexane/acetone and dried before reuse.

10.2   Synthesis

10.2.1   Acetylation of Polyhydroxy Milkweed Oil

PHMWO (95.5 g, 92.3 mmol) was placed in a flame-dried round-bottomed flask (RBF) equipped with a magnetic stir bar. Acetic anhydride (94.23 g,

100 mL) and 4-dimethylamino pyridine (4-DMAP) (2.0 g) were added and stirred at RT for 24 h. The reaction mixture was then poured into a stirring saturated $NaHCO_3$ solution; stirring was continued until effervescence ceased. The aqueous mixture was extracted with EtOAc (150 mL×3); the combined extract was dried over anhydrous $MgSO_4$ and concentrated under reduced pressure at 57 °C. The concentrate was redissolved in ethyl acetate, heated close to boiling, activated charcoal added and then reheated to the boil for two min. After filtering, the filtrate was again reheated and 2.0 g of acid-washed clay added and heated to a boil. The mixture was filtered and concentrated at 57 °C under reduced pressure to yield 121.7 g (90.6%). The FT-IR spectrum of a sample of this derivative on a NaCl disc gave ν_{film} cm^{-1}: 2928, vs, 2857 s, 1741 vs, 1463 m, 1372 s, 1235 vs, 1171 s, 1026 s, 949 w, 725 w. Its density at 22 °C was 1.13 g·cm^{-3} and its kinematic viscosity at 40 °C was 1733.6 cSt, and 78 cSt at 100 °C, *i.e.*, a VI of 105.0. ^{1}H-NMR (CDCl$_3$) δ: 4.3 m (3H), 4.25 m (5H), 2.30 t ($J = 7.3$ Hz, 7.4 Hz, 8H), 2.05 s (15H), 1.55 m (22H), 1.30 m (68H), 0.84 t ($J = 6.8$ Hz, 5.0 Hz, 12H). ^{13}C-NMR (CDCl$_3$) δ: 173.23, 173.20, 173.14, 172.74, 171.08, 170.55, 170.51, 170.46, 170.36 (C=O), 160.51 (C=C of 4-DMAP); 81.73, 81.66, 81.19, 81.09, 81.02, 80.92, 77.64, 75.10, 75.04 (–CHO– on the triglyceride chains); 68.88 (–CHO– backbone methine carbon on glycerol), 63.71 (–CH$_2$O–); 62.77 (–CH$_2$O–), 60.83 (–CH$_2$O–); 35.49, 34.29, 34.08, 34.00, 33.97, 33.92 (–CH$_2$s proximal to the carbonyl groups), 31.88, 31.78, 31.62, 31.53, 31.17, 27.82, 26.59, 26.30, 26.12, 25.86, 25.79, 22.45(2) (–CH$_2$s–); 21.12, 21.03, 20.95, 20.93, 20.80, 20.69, 20.55 (–CH$_3$ of acetate); 14.21, 14.12, 14.10, 14.03, 13.99 (terminal CH$_3$ of triglycerides).

10.2.2 Butyroylation of PHMWO

Butyric anhydride (310 g, 1.941 mol) and 4-DMAP (2.0 g) were added to PHMWO (203.0 g, 196.2 mmol) in a 500 mL RBF. The reaction mixture was stirred and gently heated to 45 °C. The reaction was monitored by TLC on a pre-coated silica gel (20×5 cm) plate using hexane/ethyl acetate/acetic acid (7:5:1) as the solvent. The reaction was stopped after 26 h when TLC showed complete consumption of the starting material. The product showed a more defined $R_f = 0.85$–0.96 compared to the smeared trace of the HMWO (hydroxy milkweed oil or polyhydroxy milkweed oil) starting material. The product mixture was poured into a saturated $NaHCO_3$ solution and stirred until effervescence stopped. The aqueous mixture was then extracted with EtOAc (300 mL×3) and the combined dark-amber-colored extract was heated to near boiling; activated carbon was added and allowed to boil. The mixture was hot filtered and the filter cake rinsed with more EtOAc. This filtrate was again heated with 2.5 g of clay (Nevergreen). The mixture was filtered and dried over Na_2SO_4 and concentrated at 57 °C under reduced pressure to yield 250.7 g (73.6% of the theoretical yield), of a viscous oil. The FT-IR spectrum of a sample of this material on a NaCl disc gave ν_{film} cm^{-1}: 2926 vs, 2858 s, 1818 w, 1737 vs, 1461 m-s, 1373 m-s, 1251 s, 1170 vs, 1096 s, 1035 m-s, 725 w.

Its kinematic viscosity was 926.2 cSt at 40 °C and 63.2 cSt at 100 °C, *i.e.*, a VI of 131.0. The density at 24 °C was 1.11 g·cm^{-3}; the specific gravity 1.1103. ^{1}H-NMR (CDCl$_3$) δ (ppm): 5.2 (H–CO– glyceryl unit), 4.36 m (6H), 4.3 m (3H), 3.9 m (2H), 2.3 m (24H), 1.66 m (44H), 1.29 bs (86H), 0.95 (overlapping triplets of terminal methyl groups of butyroyl substituents, 21H) and 0.86 m (overlapping triplets of methyl groups of triglycerides, 15H). ^{13}C-NMR (CDCl$_3$) δ: 173.24, 173.20, 173.14, 173.10, 172.74, 171.08 (C=Os); 160.0 (–HCO$_2$ small amount of formate ester); 74.99, 74.90, 74.78, 74.76, 74.69, 74.66, 74.02, 73.75, 73.54, 73.51, 73.15, 73.11, 70.50 (CHO– of the hydroxylated triglyceride chains); 68.92 (–CHO– of the glyceride backbone); 63.76, 62.09, 60.39 (–CH$_2$O– of the glycerol backbone); 33.97, 33.93, 31.88, 31.78, 31.58, 30.76, 30.66 (–CH$_2$s– next to carbonyls); 29.66, 29.62, 29.43, 29.36, 29.32, 29.30, 29.23, 29.13, 29.08, 29.04, 29.00, 28.95, 28.93; 25.09, 24.79, 24.73, 20.98 (–CH$_2$–); 18.57, 18.55, 18.52, 18.46, 18.41, 17.74 (–CH$_2$s– contiguous to terminal –CH$_3$s of butyrate substituents); 14.16, 14.07, 14.04 (–CH$_3$s of the triglycerides); 13.98, 13.95, 13.92, 13.66, 13.64, 13.57 (–CH$_3$s of substituents).

10.2.3   Valeroylation of PHMWO

Valeric anhydride (98%, 200.0 g, 1.052 mol) and 4-DMAP (1.0 g) were added to HMWO (173.9 g, 168.1 mmol). The mixture was stirred and warmed to 40 °C until the hydroxylated groups on the oil had been completely esterified, *ca.* 24 h. The product mixture was then poured into a saturated NaHCO$_3$ solution and stirred until neutralized. The neutralized aqueous mixture was extracted with ethyl acetate (300 mL×3) and the extract was treated with a mixture of charcoal and clay as above, filtered and the filtrate dried over Na$_2$SO$_4$. The solution was filtered and concentrated under reduced pressure to give 224.0 g (78.1% of theoretical). The FT-IR spectrum of this product ν_{film} on a NaCl disc cm^{-1}: 2957 vs, 2931 vs, 2859 s, 1819 w, 1739 vs, 1465 m, 1378 m, 1244 m-s, 1172 s, 1036 s; its kinematic viscosity was 489.4 cSt at 40 °C and 42.2 cSt at 100 °C, *i.e.*, a VI of 136. Its density at 22 °C was 1.096 g·cm^{-3} and its specific gravity 1.099. ^{1}H-NMR (CDCl$_3$) δ (ppm): 4.30 m (3H), 4.10 m (6H), 3.90 m (2H), 2.44 t (J = 7.4 Hz, 7.5 Hz, 3H), 2.30 m (23H), 1.55 m (45H), 1.30 m (100H), 0.90 (overlapping triplets of terminal methyls of valeroyl substituents, 24 H), 0.85 (overlapping terminal methyl groups of triglycerides, 15H). ^{13}C-NMR (CDCl$_3$) δ: 178.22, 173.30, 173.26, 173.22, 173.19, 173.13, 172.72, 171.06, 168.54 (C=Os); 160.51 (HCO$_2$ of formate side-product); 77.27, 74.72 (–CHO–); 68.87 (–CHO– of the triglyceride backbone); 63.68 (–CH$_2$O–), 62.04, 60.33 (–CH$_2$O–); 34.93, 34.28, 34.20, 34.16, 34.14, 34.12, 34.07, 34.04, 34.00 (–α-CH$_2$– to carbonyls); 33.96, 33.92, 33.84, 33.54 (–CH$_2$s– at C11 between hydroxylated C10 and C12 of the linoleyl chains); 31.14, 31.10, 30.73, 30.64, 29.65, 29.61, 29.57, 29.45, 29.43, 29.35, 29.31, 29.29, 29.22, 29.14, 29.12, 29.07, 29.03, 28.99, 28.94, 28.92, 28.84, 28.78, 27.14, 27.10, 27.05, 27.00, 26.93, 26.84, 26.76, 26.53, 26.22, 25.76, 25.28, 25.25, 25.07, 25.05, 25.03, 24.95, 24.82, 24.78, 24.75, 24.72, 22.64, 22.60, 22.47, 22.44, 22.24, 22.15, 21.96, 20.96 (–CH$_2$–); 14.15, 14.06, 14.04,

13.97, 13.95, 13.91 (terminal methyl groups of triglycerides); 13.68, 13.64, 13.59 (terminal methyl groups of valeroyl substituents).

10.2.4 Synthesis of Milkweed Polyformate Esters

In a typical experiment, milkweed epoxy oil, 215.20 g (0.225 mol) previously prepared and characterized[3] as above, was placed in a 1.0 L dry RBF containing a magnetic stirrer bar. Into this RBF was added 124.35 g (2.702 mol, 102 mL) formic acid followed by stirring. The reaction mixture was then heated to gentle reflux and monitored every 40 min by sampling the contents for progress in the disappearance of the epoxy band at 820–840 cm^{-1} using FT-IR spectroscopy. The reflux was allowed to run overnight after which FT-IR indicated complete disappearance of the oxirane absorption band. The heat source was then removed and the system allowed to cool to RT after which the mixture was transferred into a beaker of saturated NaHCO$_3$ to neutralize the excess formic acid. The solution was stirred and more NaHCO$_3$ added until effervescence ceased. The organic phase was separated and saved while the aqueous layer was extracted twice with ethyl acetate (100 mL×2). The combined organic phase was dried over Na$_2$SO$_4$ and concentrated under reduced pressure at 56 °C followed by further drying with a vacuum pump to yield 294.6 g (98%) of the performyl ester of MWO. The product density was 1.09 g cm^{-3} (24 °C); viscosity 388.46 cSt (40 °C) and 44.60 cSt (100 °C), *i.e.*, a VI of 172. The second derivative of the FT-IR spectrum (film on a NaCl disc) ν_{NaCl} cm^{-1}: 2958 s (–CH$_3$ asym stretch), 2927s (–CH$_2$– asym stretch), 2875 s (–CH$_3$ sym stretch), 2857 vs (–CH$_2$– sym stretch), 1748 s (C=O triglyceride ester), 1727 vs (C=O formyl ester), 1469 m (–CH$_2$– deform.), 1380 m-w (–CH$_3$ deform), 1170 vs (–CO$_2$ stretch), 724 w (–CH$_2$– wag). ^{1}H-NMR (CDCl$_3$) δ: 8.05–8.2 (s methine H–CO$_2$, 10H), 5.18 (methine H–C–O on the glyceride backbone), 4.4–4.0 (–CH$_2$O– on the glyceride backbone), 2.3 m, 2.05 s, 1.5 m, 1.3 m, 0.9 t (6.7 Hz, 7.0 Hz); ^{13}C-NMR (CDCl$_3$) δ: 178.83, 171.16 (C=O on the triglyceride backbone); 161.01, 160.43, 160.30, 160.26 (C=O formyl esters); 81.75, 81.67, 81.09, 80.99, 77.90, 77.79, 77.74, 77.39, 77.26, 74.91, 74.85, 74.80, 73.89, 73.29, 73.24 (–HC–O–); 68.87, 68.57, 68.49 (–CH–O– glyceride backbone); 62.07, 61.90, 61.82, 61.67, 61.42 (–CH$_2$–O– glyceride backbone), 35.50, 35.42, 34.08, 34.01, 33.83, 31.89, 31.76, 31.52, 30.51, 29.62, 29.55, 29.40, 29.32, 28.89, 28.85, 28.65, 26.03, 25.78, 25.72, 25.10, 25.01, 24.93, 24.83, 24.68, 24.64, 24.56, 22.65, 22.59, 22.46, 22.43, 22.41, 22.35; 20.99, 14.16, 14.08, 14.05, 13.97, 13.95, 13.92 (–CH$_3$).

10.2.5 Cold Flow and Stability Measurements

10.2.5.1 Pour Point

Pour Point (PP) was measured following ASTM method D97-96a[9] to an accuracy of ± 3 °C. The procedure involves placing a test jar with 50 mL of the sample into a cylinder submerged in a cooling medium. The sample temperature was measured in 3 °C increments at the top of the sample until the material stopped pouring. The sample no longer poured when the material in the test jar did not flow when held in a horizontal position for 5 s. The temperature of the cooling medium was chosen based on the expected PP of the material. Samples with PPs that ranged from +9 to −6, −6 to −24, and −24 to −42 °C were placed in baths of temperatures from −18, −33, and −51 °C, respectively. PP is defined as the coldest temperature at which the sample still pours. All PPs were run in duplicate and averaged values are reported.

10.2.5.2 Cloud Point

Cloud Point (CP) was determined following ASTM method D2500-99[10,11] to an accuracy of ±1 °C. The method involves placing a test jar with 50 mL of the sample into a cylinder submerged in a cooling medium. The sample temperature was measured in 1 °C increments at the bottom of the sample until any cloudiness was observed at the bottom of the test jar. The temperature of the cooling medium was chosen based on the expected CP of the material. Samples with CPs that ranged from RT to 10, 9 to −6, −6 to −24 and −24 to −42 °C were placed in baths of temperatures from 0 −18, −33, and −51 °C, respectively. All CP measurements were conducted in duplicate and averaged values are reported.

10.2.5.3 Oxidative Stability

10.2.5.3.1 Rotating Pressurized Vessel Oxidation Test. RPVOT experiments were conducted on a RPVOT apparatus manufactured by Koehler (Bohemia, NY) using ASTM method D2272-98.[12] A commercial anti-oxidant package, Lubrizol™ 7652 (The Lubrizol Corporation, Wickliffe, OH), was blended into the test oil samples on a wt% basis. Formulated samples were stirred for 24 h before use in RPVOT experiments. All RPVOT experiments were conducted at 150 °C. The copper catalyst used in the experiments was 3 m long and sanded with 220 grit silicon carbide sandpaper produced by Abrasive Leaders and Innovators (Fairborn, OH) and was used immediately. The wire was wound to have an outside diameter of 44–48 mm, a height of 40–42 mm, and a weight of 55.6 ± 0.3 g. In a typical experiment, 50.0 ± 0.5 g of blended samples were weighed into the pressure vessel, to which 5.0 mL of reagent-grade water and the freshly prepared catalyst copper wire were added. The vessel was then assembled and slowly purged with reagent-grade oxygen twice. The pressure vessel was charged with 90.0 ± 0.5 psi (620 kPa) of reagent-grade oxygen, and then checked for oxygen leaks by immersing the vessel in water. The test began after confirming no leaks, and was stopped and the time recorded when the pressure in the vessel dropped by 175 kPa from the maximum pressure. All samples were run in duplicate and the averaged RPVOT times in min are reported.

10.2.6 Friction and Wear Measurements

Friction and wear tests were conducted using a 4-ball tribometer (model KTR-30L, Koehler Instruments, Bohemia, NY) following the 4-ball AW test method in ASTM D4172-94[12] under the following test conditions: load, 392 ± 2 N (40 ± 0.2 kgf, the force in kilograms exerted on the top rotating ball); speed, 1200 ± 40 rpm; lubricant temperature, 75 ± 2 °C; test duration, 60 ± 1 min. During the test, the torque required to overcome the friction opposing the rotation of the ball was continuously measured and recorded along with the spindle speed, load and lubricant temperature. The measured frictional torque was converted into the coefficient of friction (COF) following the procedure outlined in ASTM D5183-95.[13] At the end of the test period, the wear scar diameter on each of the three bottom balls was measured along and transverse to the wear direction using an automated wear scar measurement system, comprising hardware and software (Scar View software version 2005, Koehler Instruments). The instrument load and torque cells were properly calibrated prior to use, and freshly degreased test balls were used in each test. In a typical experiment, the rotating ball was first secured in the collect of the top spindle. The three stationery balls were secured in the pot using a wrench set to the recommended pressure. The test lubricant was then poured into the mixture until the balls were completely submerged. The assembled mixture was then placed in the tribometer and the load applied carefully so the top ball contacted all three bottom balls. The test parameters (lubricant, speed, load temp, duration) were entered into the computer and the heater turned on. The test started automatically as soon as the test temperature was reached, and continued until the test duration expired. At the end of the test, the mixture was removed, the oil poured out, rinsed with hexanes and used in wear scar diameter measurement. For each test lubricant, two AW tests were conducted using fresh balls and test lubricants. The COF and wear scar diameter from the two tests were averaged and used in further analysis.

10.2.7 Film Thickness

Lubricant film thickness was measured using the optical interferometry method between a glass disc and a steel ball. Details regarding the basic principles of the optical interferometry method can be found elsewhere.[14] Film thickness measurement was conducted on an EHL Ultra Thin Film Measurement System (PCS Instruments, UK). The glass disc used with this instrument is coated with a semi-reflecting chrome layer, which is coated with a silica spacer layer of pre-determined thickness to allow for instrument calibration as well as for measuring very thin (ultra-thin) lubricant films. The system has several components including: a mechanical unit; electronic unit; optical system; and a computer with the appropriate software.

A detailed description of the main components has been given previously.[15] The specifications of the instruments are: disc, 100 mm diameter by 12 mm thick float glass, coated with a ~ 20 nm semi-reflecting chrome layer, which in turn is coated with a ~ 500 nm thick silica spacer layer; precision steel ball, super finished 3/4 inch diameter, G10 carbon Cr steel (AISI 52100); measured film thickness, $1-1000 \pm 1$ nm; rolling speed, $0-3$ m s^{-1}; slide-to-roll ratio, 0–200%; applied load, 0–50 N; maximum contact pressure, 0.7–1.1 GPa; lubricant temperature, ambient to 150 °C. In this work, film thickness test parameters were varied as follows: speed, 0.02–3.0 m s^{-1} in 20% increments; load, 10–40 N in 10 N increments; temperature, 30, 40, 70, 100 °C. A detailed description of the procedure for film thickness measurement this system, including instrument set-up/calibration and data acquisition/analysis, has been given previously.[15]

10.3   Results and Discussion

The PHMWO used as the starting material in this investigation Figure 10.1(a) was a pale yellow, very viscous substance as described previously.[2] Table 10.1 contains a comparative listing of the fatty acid composition and unsaturation indices of some other vegetable oils. The densities of different milkweed derivatives are summarized in Table 10.2.

The kinematic and dynamic viscosities of milkweed acyl derivatives are summarized in Tables 10.3 and 10.4. VIs were calculated from viscosities at 40 °C and 100 °C. An earlier constitutive modeling study of the rheology of PHMWO had indicated this oil exhibited non-Newtonian viscoelastic behavior which could be explained by the presence of the many hydroxyl groups in its structure. These would naturally form both inter- and intramolecular hydrogen bonds and thus show unusual flow behavior at lower temperatures.[16] The FT-IR spectrum of PHMWO [Figure 10.1(b)] is characterized by a prominent broad band centered at 3462 cm^{-1} which is attributable to hydrogen-bonded O–H stretching frequencies as a consequence of the multihydroxyl group content of this starting material. In contrast, one of the first noticeable FT-IR spectral features observed in each peracylated derivative is the disappearance of the hydroxyl stretching mode (3600 cm^{-1} to 3200 cm^{-1}). This is expected from substitution of the hydroxyl protons of the secondary alcohols of the triglyceride chains by the esterifying substituent acyl moieties. The other notable features of the FT-IR spectra are the prominence of the alkyl and carbonyl frequencies around 2940–2850 cm^{-1} and 1740 cm^{-1}, respectively, resulting from the substituent moieties' contribution to the spectrum. The NMR spectra of these derivatives are shown in Figures 10.2–10.4.

Again the proton spectra confirmed the absence of any residual hydroxyl functionalities in all the derivatives. The most interesting information was obtained from stacking the distortion-less enhancement proton transfer experiments, DEPT, of the ^{13}C-NMR spectra [Figure 10.3(a)] of the three

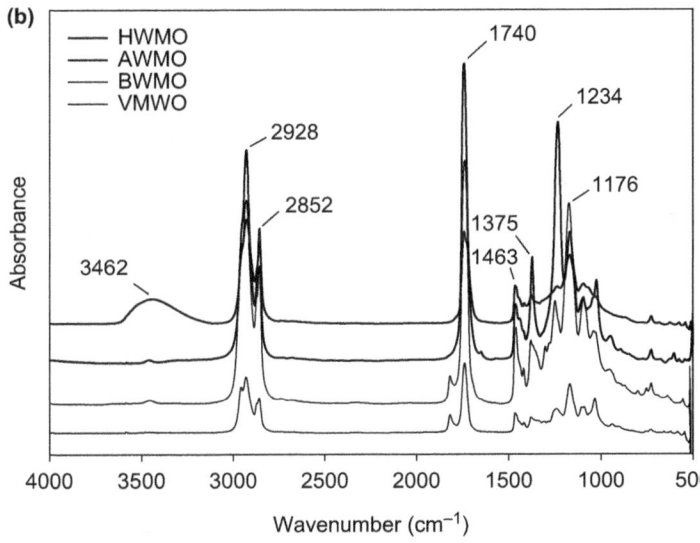

Figure 10.1 (a) Generation of milkweed polyhydroxy triglyceride and its acyl derivatives from milkweed oil oxiranes. (b) FT-IR spectral overlay of three peracyl derivatives of milkweed polyhydroxy triglycerides with the starting HMWO.

Table 10.1　Fatty acid composition and unsaturation index (UI) of milkweed and other selected commodity oils.

Cn:DB	LINO[b]	SBO[b]	MWO[a]	CANO[b]	SESO[b]	CASTO[b]	PalmO[b]
C16:0	5.4	10.6	5.9	4.1	9.3	—	44.4
C16:1, 16:2	—	0.1	9.6	0.3	0.1	—	0.2
C18:0	3.5	4.0	2.3	1.8	5.4	1.8	4.1
C18:1	19.9	23.2	31.0	60.9	39.8	92.7	39.3
C18:2	17.9	53.7	50.5	21	43.7	4.2	10.0
C18:3	51.2	7.6	1.2	8.8	0.4	—	0.4
C20:0	0.2	0.3	0.2	0.7	0.2	—	0.3
C20:1	0.3	—	—	1	0.2	—	—
UI[c]	209.6	153.5	142.6	130.6	128.7	101.1	60.7

[a]Ref. 4.
[b]Ref. 22.
[c]Ref. 7.

Table 10.2　Densities of some milkweed oil derivatives.

Sample	Density/g·cm^{-3} (measured at 21.8 °C unless otherwise stated)[a]
MWO	<1
EMWO	1.05 (23 °C)
HMWO	1.14 (24.2 °C)
AMWO	1.14
BMWO	1.11
VMWO	1.10
MWF	1.09 (24 °C)

[a]Measured in a pycnometer.

Table 10.3　Kinematic viscosities and VIs of selected seed oils, milkweed and derivatives (cSt).

Oil[a]	Kinematic viscosity at 25 °C	Kinematic viscosity at 40 °C[b]	Kinematic viscosity at 100 °C[b]	VI[c]
MWO	—	33.8	7.3	210
SBO[d]	—	31	7.6	227
CANO[d]	—	34	7.8	215
ESBO	420[e]	—	—	—
EMWO	—	164.4	81.3	142
HMWO	—	2332	75.5	85
AMWO	—	1733	78	105
BMWO	—	926.2	63.2	131
VMWO	—	489.4	42.2	136
MWF	—	388.3	44.6	172

[a]All data from this work except data from the literature which are indicated with a different symbol.
[b]ASTM D2270, using a Cannon–Fenske viscometer according to AOCS Official Method Tq 1a-64.
[c]VI measured according to ASTM D2770.
[d]Data for soybean oil (SBO) and canola oil (CANO) obtained using ASTM D2770 (ref. 23).
[e]Technical Information, Vikoflex® 7170 Epoxidized Soybean Oil, Arkema Inc, 2010 from www.arkema-inc.com.

Table 10.4 Dynamic viscosities of selected seed oils *vs.* milkweed derivatives (m Pa · s).

Oil[a]	Dynamic viscosity at 40 °C[a]	Dynamic viscosity at 70 °C[a]	Dynamic viscosity at 100 °C[a]
MWO	33.8	—	7.3
SBO[b]	30.3	—	7.3
CANO[b]	33.3	—	7.0
HMWO	31 986[c]	494.6 ± 13.7	103.1 ± 5.5
HMWO	—	482.3 ± 29.6	105.6 ± 5
AMWO	1360.6 ± 14.1	308.1 ± 15.1	96.8 ± 5.8
AMWO	—	300.1 ± 10.8	97.4 ± 5.3

[a]Measured on a TA Instruments ARES LS1 controlled strain rheometer at $0.1–100 \text{ s}^{-1}$ shear rate using a 50 mm cone and plate geometry, cone angle of 0.0401 radians and a gap of 0.046 mm.
[b]Dynamic viscosity of soybean oil (SBO) and canola oil (CANO) (ref. 15).
[c]The sample was non-Newtonian at 40 °C, and the viscosity was obtained by analyzing the data using the Cross model.

derivatives. This allowed us to unambiguously distinguish the substituent terminal methyl groups from those of the parent triglyceride in each derivative. For instance, in Figure 10.3, entry A is the polyacetyl triglyceride (AMWO) in which the methyl groups of the substituent acetyl units are chemically shifted downfield to around 21 ppm compared to the terminal methyl groups of the parent triglycerides which resonate at around 14 ppm. This is understandable from the deshielding effect of the contiguous carboxyl C=O on the acetate methyl carbon. In contrast, the terminal methyl groups of the longer chain substituents which are farther removed from this influence in the butyroyl and pentanoyl substituents, resonate at a higher field position (around 13 ppm) from the terminal methyl groups of the parent triglyceride observed consistently at around 14 ppm in all derivatives. An additional noticeable point in the butyroyl derivative, entry B, is that the methylene groups next to the terminal carbons resonate at a higher field at around 18 ppm compared to the equivalent methylene groups in the pentanoyl derivative. This observation is in agreement with spectra of butyroyl esters and bears mention here. Another feature to re-emphasize is that neither the ^{1}H-NMR spectra nor the FT-IR spectra of these acyl derivatives had any indication of the presence of residual hydroxyl groups in the polyesters.

For the polyformyl esters, the principal diagnostic feature of the oxirane starting material for the formyl ester synthesis is the $820–845 \text{ cm}^{-1}$ doublet in the FT-IR spectrum, Figure 10.5.

This is the epoxy –C–O–C– asymmetric stretching absorbance of the three-membered ring. The corresponding ^{13}C-NMR resonances for the oxirane spectrum are the 10 carbon resonance lines between 57 and 53.8 ppm, shown in Figure 10.6; this region of the spectrum is transparent in the triglyceride oil and in the subsequent formyl ester. Conceptually, from the characteristically low pH (2.3) of formic acid, one would expect a facile epoxy

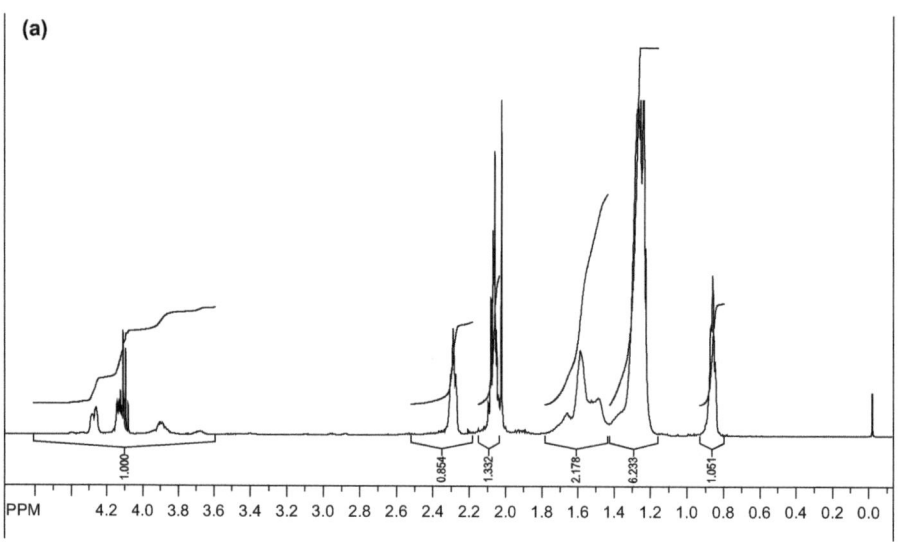

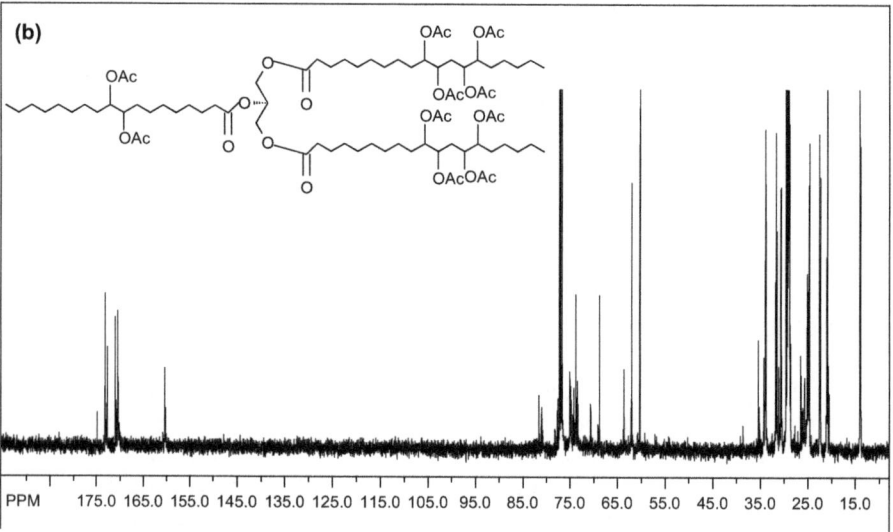

Figure 10.2 (a) ^{1}H-NMR spectrum of the peracetyl polyhydroxy triglyceride derivative of milkweed oil. (b) ^{13}C-NMR spectrum of the peracetylated polyhydroxy triglyceride derivative of milkweed oil.

ring-opening with generation of an α-hydroxyl formate ester with or without a strong acid catalyst. We opted not to use any catalyst. What was actually observed in the reaction process under gentle reflux, as the reaction progressed and was sampled for FT-IR spectroscopy, was the non-appearance of the expected hydroxyl group absorbance in its characteristic 3500–3300 cm^{-1} region as the epoxy ring absorption (820–840 cm^{-1}) diminished. The

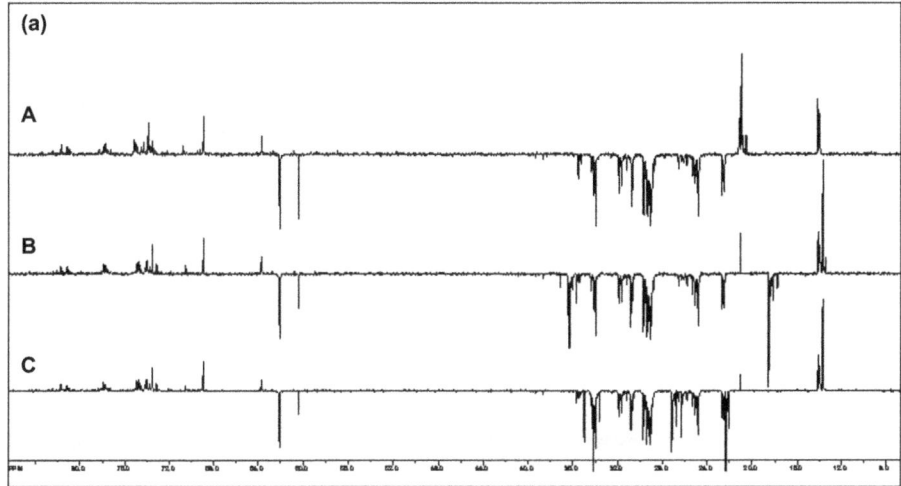

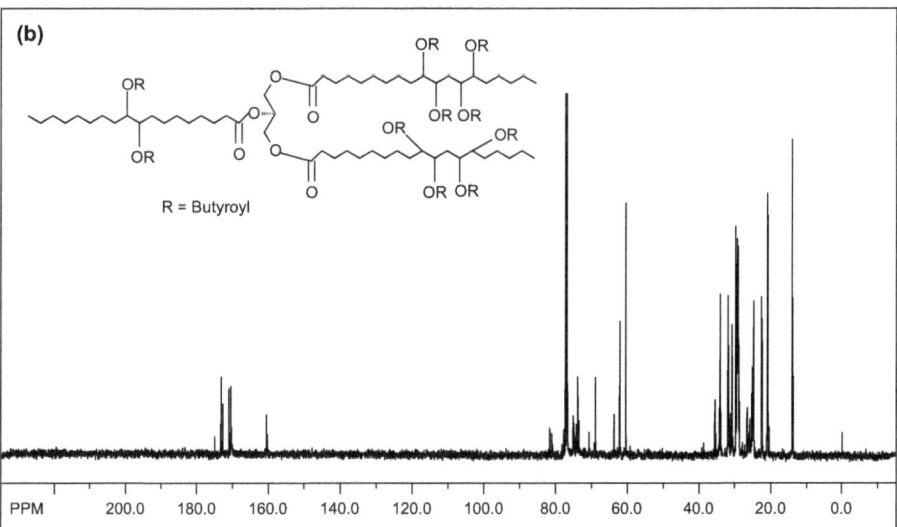

Figure 10.3 (a) ¹³C-NMR DEPT spectra of the: peracetyl (A), perbutyroyl (B) and pervaleroyl (C) polyhydroxy triglyceride derivatives of milkweed oil. (b) ¹³C-NMR spectrum of the perbutyroyl polyhydroxy triglyceride derivative of milkweed oil.

inference would be a condensation step following the initial ring-opening reaction of the epoxide. To confirm this postulate, toluene was added to the reaction mixture in a subsequent ring-opening reaction in order to azeotrope off any condensed water formed into a Dean–Stark trap. This was confirmed by the quantity of water collected in the trap during the reaction. Thus the resulting ring-opening of the oxirane gave a vicinal diformate ester for each oxirane unit of the substrate. The FT-IR spectrum of the product shows a

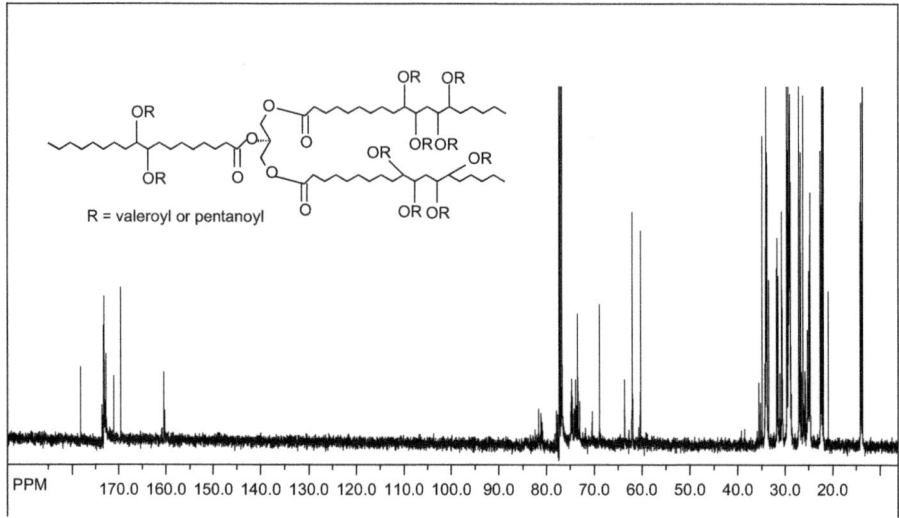

Figure 10.4 ^{13}C-NMR spectrum of the pervaleroyl polyhydroxy triglyceride derivative of milkweed oil.

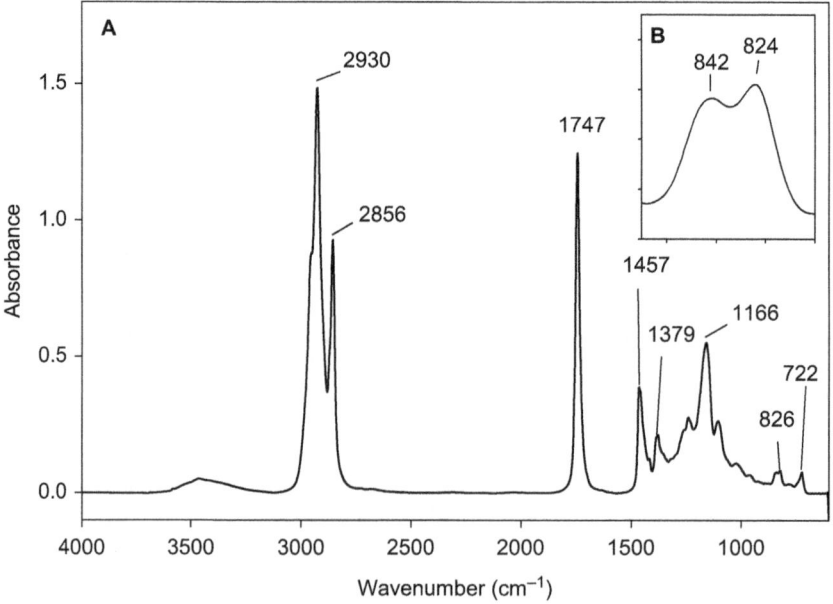

Figure 10.5   FT-IR spectrum of milkweed polyepoxy triglyceride. A is the whole FTIR spectrum of the polyepoxy material and B, the insert, is an expansion of the epoxy spectral band region.

strong carbonyl band whose 2nd derivative indicates the presence of two carbonyl types, *i.e.*, the parent triglyceride (C=O) with a weaker carbonyl band intensity at 1748 cm^{-1} and a very strong band at 1724 cm^{-1} for the H–C=O ester, Figure 10.7. The intensity of the 1724 cm^{-1} band derives from

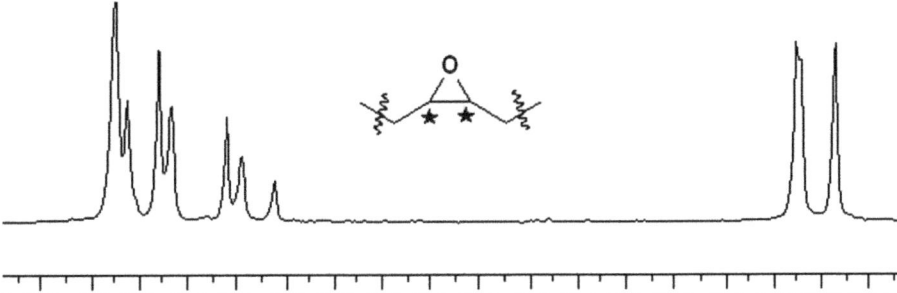

Figure 10.6 ^{13}C-NMR spectrum of the epoxy resonances of the derivatized triglyceride.

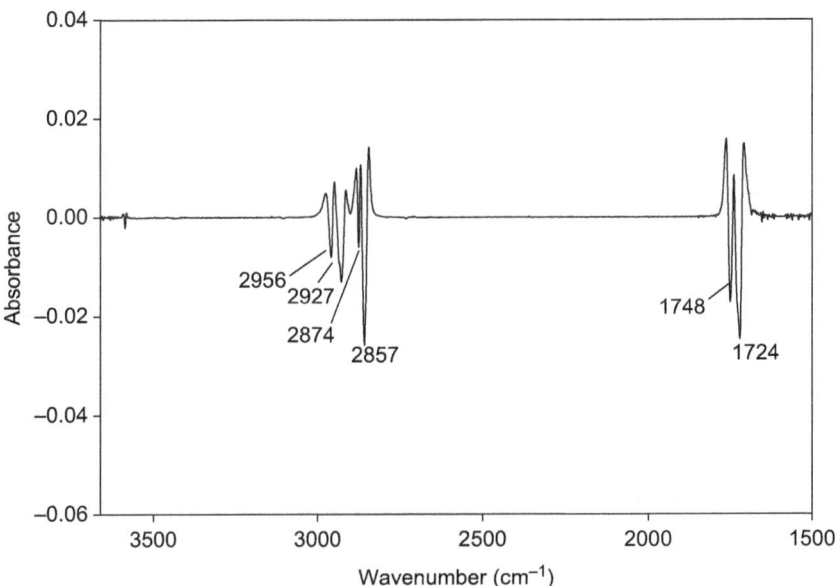

Figure 10.7 Second derivative of the FT-IR spectrum of the polyformyl derivative of milkweed oil.

the mass effect of the contribution of the many methine carbonyls compared to the relatively fewer parent triglyceride carbonyl esters present. A moderately weak band generally observed around 3500 cm^{-1} is attributable to the overtone of the very intense carbonyl absorption band at 1728 cm^{-1}. The ^{1}H-NMR spectrum of the performyl ester of milkweed oil shows characteristic resonance frequencies of the formate methine protons at 8.0–8.20 ppm, Figure 10.8.

Expansion of this absorption region reveals some 10 singlet resonances (inset), whereas the ^{13}C-NMR spectrum of the polyformate ester shown in Figure 10.9 has methine carbonyls of the formate ester occurring up field around 160–161 ppm relative to the parent triglyceride ester carbonyls that

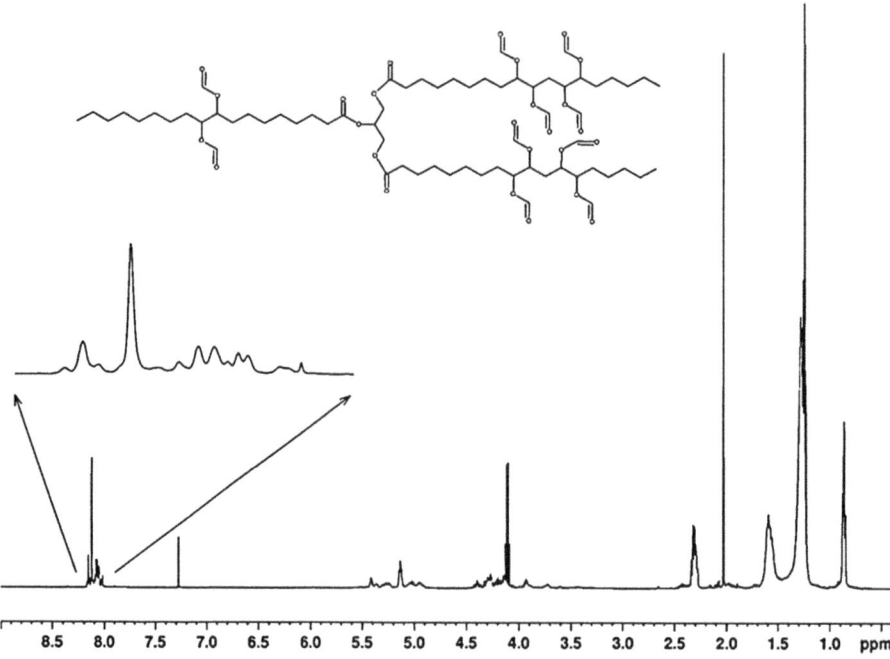

Figure 10.8 ^{1}H-NMR spectrum of the polyformate of milkweed oil.

are farther downfield at around 170–174 ppm. Mechanistically, it can be inferred that the α-hydroxyl group of the formate ester putatively formed, rapidly loses a H_2O molecule following protonation by a second molecule of formic acid as shown in Figure 10.10. The conjugate base (formate anion) then displaces the conjugate acid $(-OH_2)^+$ in a presumably concerted reaction step thus giving rise to the vicinal diformate ester per oxirane unit of the starting material. The reaction is a general one for vegetable oil oxiranes as both soybean and pennycress oil oxiranes gave the analogous diformate esters under these reaction conditions.

10.3.1 Viscosity and other Physical Measurements

10.3.1.1 Kinematic Viscosity

The kinematic viscosities of MWO and its acyl derivatives at 40 and 100 °C are summarized in Table 10.3. For purposes of comparison, literature viscosity and VI values for soybean (SB) and canola (CAN) oils and derivatives are also included in Table 10.3. The viscosities of some milkweed derivatives were obtained on two different instruments using ASTM D2270 methods.[10] MWO displayed an average kinematic viscosity of 33.8 and 7.3 cSt at 40 and 100 °C, respectively. The viscosity value for the MWO derivative is slightly higher than that of the SBO derivative, Table 10.3. This may be to do with differences in the degree of purification between these three oils. Epoxidation resulted in a

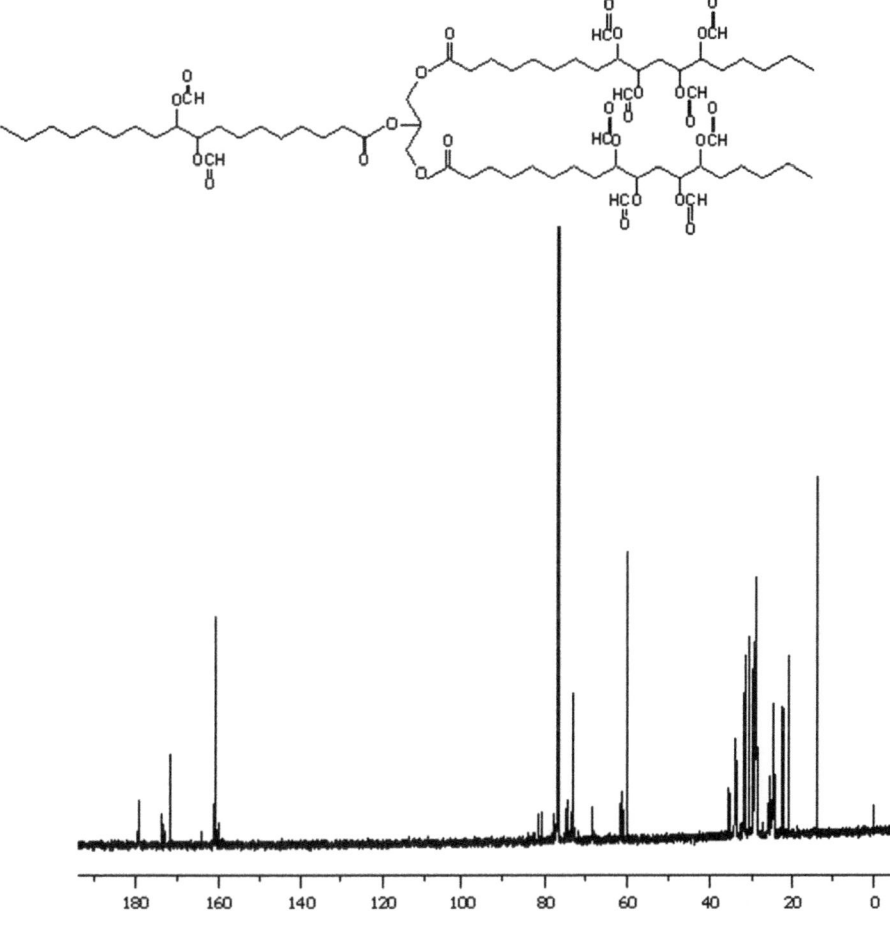

Figure 10.9 ^{13}C-NMR spectrum of the polyformate of milkweed oil.

large increase in the kinematic viscosity of EMWO. An even greater, almost 50-fold, increase in kinematic viscosity was observed for the PHMWO derivative. This increase in viscosity of HMWO is attributable to a high degree of intermolecular hydrogen-bonding between HMWO molecules. Esterification of PHMWO, however, resulted in a reduction of kinematic viscosity, Table 10.3. The degree of attenuation in viscosity of esterified PHMWO was a function of the substituent group chain length.[17]

The VI of unmodified MWO is 210, which is slightly lower than the values for SB and CAN oils, but within the range expected of seed oils.[18] Epoxidation and esterification resulted in a dramatic decrease in the VI of milkweed. For example, the VI of PHMWO was 85, which improved slightly with esterification. The VI of esterified PHMWO increased with increasing chain length of the substituent group.

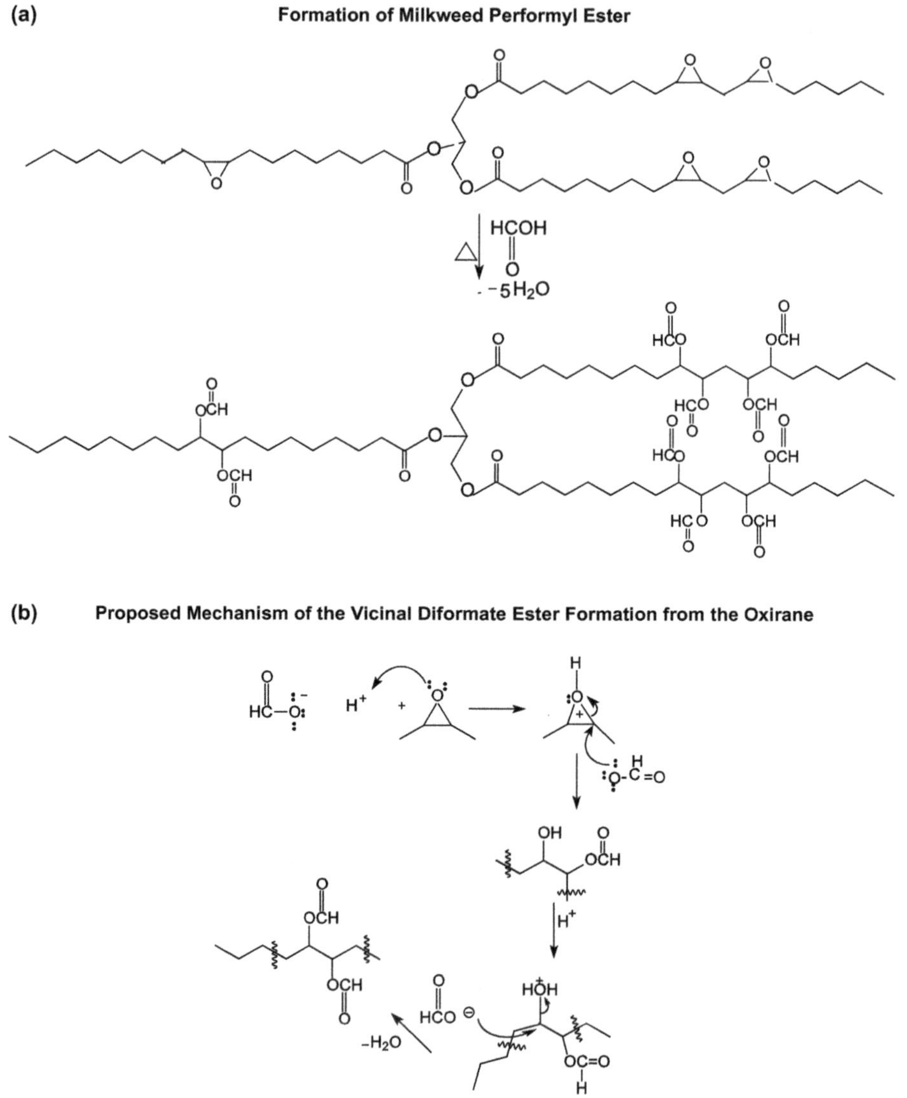

(a)   **Formation of Milkweed Performyl Ester**

(b)   **Proposed Mechanism of the Vicinal Diformate Ester Formation from the Oxirane**

Figure 10.10   (a) Formation of milkweed performyl ester. (b) Proposed mechanism of vicinal diformate ester formation from the oxirane.

10.3.1.2 Dynamic Viscosity

The dynamic viscosities of MWO and its derivatives were measured using two different instruments at 40, 70 and 100 °C and are summarized in Table 10.4. The instruments used were a controlled-strain cone and plate rheometer (TA Instruments), and the A P SVM 3000 viscometer. Data for SB and CAN oils have been included in Table 10.4 for comparative purposes. The dynamic viscosity of MWO at 40 and 100 °C was 33.8 and 7.3, respectively, which are within the range reported for SB, CAN and other seed oils.[19] PHMWO displayed a very high dynamic viscosity ($\sim$1000-fold relative to MWO) and non-Newtonian behavior at 40 °C. At 70 °C or higher temperatures, however, it displayed Newtonian behavior.[16] The latter behavior confirms the strong intermolecular hydrogen-bonding resulting in non-Newtonian behavior of this compound at lower temperatures. Higher temperatures therefore, thermally disrupt these intermolecular hydrogen-bonds freeing the molecules to behave as a Newtonian fluid. Esterification similarly disrupts hydrogen-bonding, hence, the dynamic viscosities of the esterified PHMWOs decreased with increasing ester group chain length as: PHMWO$\gg$AMWO$>$BMWO$>$VMWO. This trend was observed for data generated by both viscometers and at all temperatures. Analogous observations of the effect of branching on viscosity have been reported in branched poly-saccharides,[17] and branched synthetic polymers.[20–23]

10.3.2 Cold Flow

The PPs and CPs of MWO and its derivatives are summarized in Table 10.5 with PP values for SBO and CANO, and similar derivatives of SBO included. PHMWO displayed a similar PP to hydroxyl soybean oil (HSBO), which showed an improvement over SB. AMWO displayed a similar PP to ASBO, which was worse than HSBO, PHMWO, and even SBO. BMWO and VMWO displayed similar PP values to PHMWO. However, the PP of BSBO (−3 °C) was identical to those of ASBO and AMWO but much higher than that of BMWO (−18 °C). This result is clearly anomalous since almost all SBO and MWO derivatives displayed similar PP values except for BMWO and VMWO. The CP of PHMWO displayed a four-fold improvement upon esterification. There were no effects of the ester group chemical structure on the CP. The esterified PHMWO derivatives had CPs that were in the range −25 to −28 °C.

Table 10.5 Low-temperature flow properties of soybean (SB) and milkweed (MW) derivatives.

Oil	Pour Point/°C	Cloud Point/°C	Ref.
SBO	−9	—	22
CANO	−21	—	22
ESBO	3	—	25
HSBO	−18	—	25
HMWO	−15	−7	*a*
ASBO	−3	—	25
AMWO	−3	−25	*a*
BSBO	−3	—	25
BMWO	−18	−30	*a*
VMWO	−18	−28	*a*

*a*This work.

10.3.3 Oxidative Stability

The oxidative stability (OS) data for MWO and its derivatives, which were obtained using the RPVOT method, are summarized in Table 10.6. Oxidation stability evaluations were conducted on samples with (3.5 wt%) and without blended commercial anti-oxidant additives. For comparison, similar literature data for CANO and SBOs are included in Table 10.6. As can be seen, MWO derivatives (PHMWO, *etc.*) without anti-oxidant additives displayed slight improvements in OS compared to unmodified SB or CAN oils. On the other hand, samples containing the commercial anti-oxidant additives displayed significant improvements in OS. However, the improvements due to added anti-oxidant varied and were dependent on the nature of the oil being tested. For PHMWO, the oxidation stability improved by a third (from 15 to 20 min). For the polyesters (AMWO, BMWO and VMWO), however, the changes were more significant – up to a four-fold increase in RPVOT was observed. A closer examination of the OS data indicated that, for the MWO polyesters, the oxidative stability decreases with increasing substituent chain length.

10.3.4 EHD Film Thickness

Film thickness was investigated using an optical interferometry method. Film thickness was measured as a function of lubricant entrainment speed, temperature and applied load. In these studies, only acetylated (AMWO) and valerylated (VMWO) milkweed esters were investigated. These two oils were chosen because they had the highest and lowest viscosities, respectively, among the three milkweed esters investigated in this work (Tables 10.3 and 10.4). Also, PHMWO was not investigated because it was too viscous for the measurement system. A typical data set from film thickness measurements by optical interferometry is illustrated in Figure 10.11. The data in Figure 10.11 is for VMWO, but similar data were also obtained for AMWO. AMWO and VMWO oils displayed the familiar entrainment speed *vs.* film thickness profiles displayed by lubricating oils. The film thickness increased

Table 10.6 RPVOT oxidative stability (min) results for soybean (SB) and milkweed (MW) derivatives, with and without a blended commercial anti-oxidant.

Oil	No anti-oxidant	3.5 wt% anti-oxidant[a]	Ref.
SBO	13	28	22
CANO	10	43	22
HMWO	15	20	[b]
AMWO	—	74	[b]
BMWO	16	67	[b]
VMWO	—	52	[b]

[a]LZ 7652 from Lubrizol Corporation.
[b]This work.

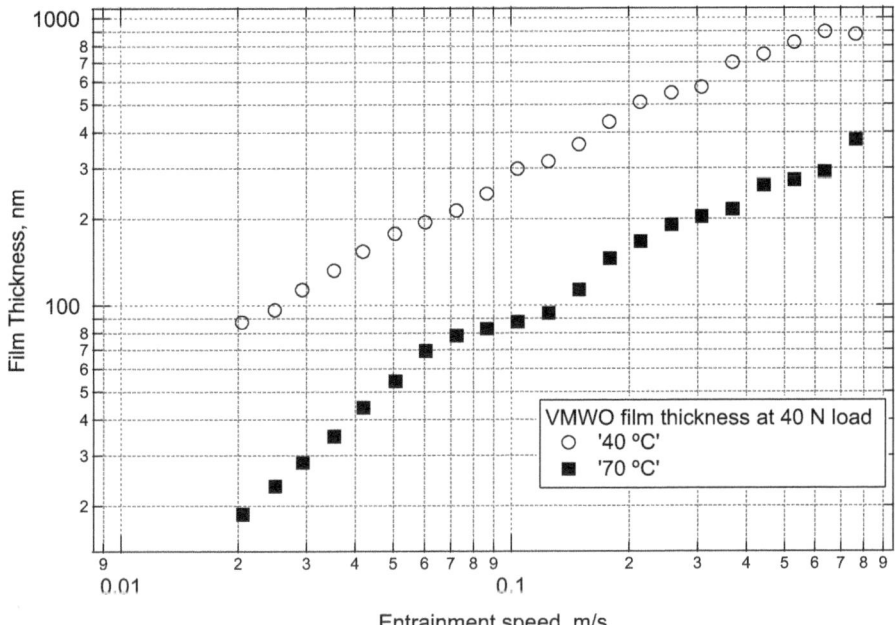

Figure 10.11   Effect of entrainment speed and temperature on film thickness of valeroyl milkweed oil.

with increasing entrainment speed and decreasing temperature. This is illustrated by the data in Figure 10.11 which compares the film thickness of VMWO as a function of entrainment speed at 40 and 70 °C. The data in Figure 10.11 was obtained at a load of 40 N. As can be seen, the film thickness of VMWO increased with increasing entrainment speed and decreasing temperature. Similar profiles were obtained for the film thickness of AMWO as a function of entrainment speed and temperature.

The effect of load on film thickness is illustrated in Figure 10.12. The obtained data is for VMWO measured at 70 °C, at four different loads: 10, 20, 30, and 40 N. The data clearly show that the film thickness of VMWO is independent of the applied load. Similar results were obtained for AMWO. This observation is consistent with those generally observed for lubricating oils, which display little or no change with varying load.[15]

The effect of chemical structure on film thickness is illustrated in Figure 10.13, where film thickness data for milkweed AMWO and VMWO are compared. The points in Figure 10.13 are measured film thickness data at 20 N of load, at 70 °C. Examination of the data in Figure 10.13 clearly shows that at both temperatures, AMWO produces a much thicker film than VMWO. Similar results were obtained for these two oils at other temperatures.

The difference in the film-forming properties between these two oils is related to their difference in chemical structure. These two milkweed

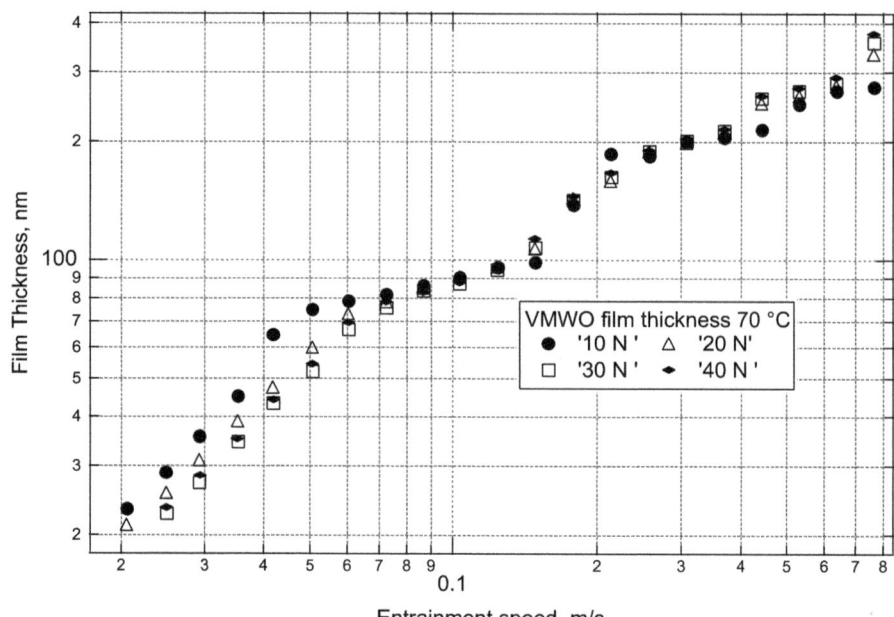

Figure 10.12   Effect of load on film thickness of valeroyl milkweed oil.

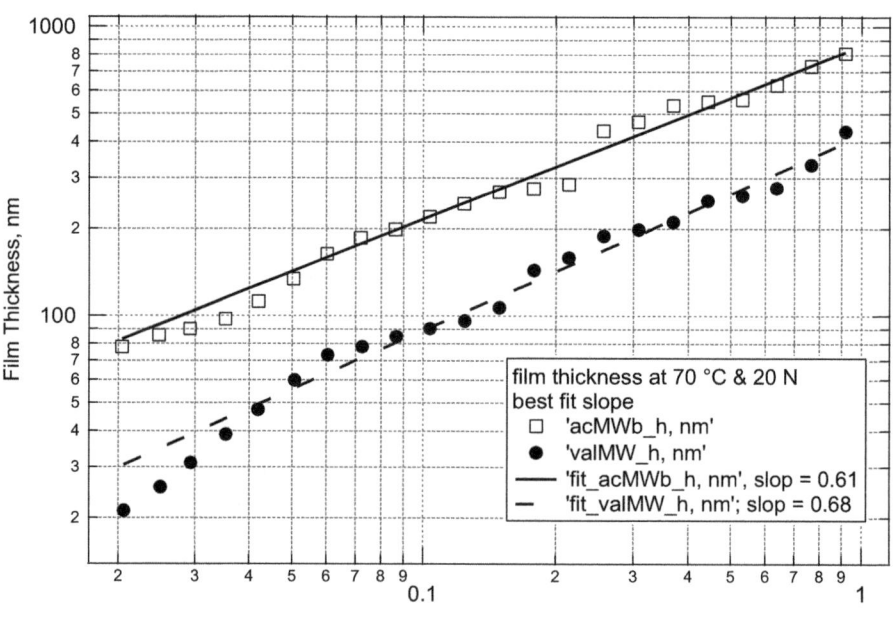

Figure 10.13   Effect of milkweed polyhydroxy ester chemical structure on measured *vs.* predicted elastohydrodynamic film thickness. The open squares are data for AcMWO whereas filled circles are data for VaMWO; the solid and dashed lines represent the predicted Hamrock-Dawson for these derivatives.

derivatives differ in the chain length of the carboxylic acid substituents on the PHMWO. The longer carboxylate groups in the polyvalerates relative to those in the polyacetates leads to a dramatic difference in the density and viscosity of these two materials (Tables 10.2–10.4). As shown in Tables 10.2–10.4, both the viscosity and the density of HMWO decreased due to ester-ification of the hydroxyl groups. In addition, both the viscosity and the density of the esterified MWOs were functions of the chain length of the substituent groups. Both properties also decreased with increasing chain length, but the decrease in viscosity was the more dramatic. Viscosity, along with entrainment speed, are the two major parameters affecting lubricant oil film thickness, and can be used to predict film thickness using the Hamrock–Dowson (H–D) relationship.[24] Predicted film thicknesses of AMWO and VMWO oils using the H–D equation is illustrated by the lines in Figure 10.13. Examination of the measured *vs.* predicted film thickness data in Figure 10.13 reveals a number of interesting features. For AMWO at 70 °C, the measured and predicted values showed excellent agreement at all en-trainment speeds. For VMWO at 70 °C, an excellent agreement between ex-perimental and predicted film thickness was observed mainly in the high entrainment speed region, which also corresponds to high film thicknesses. In the low entrainment speed region, however, VMWO displayed a negative deviation, *i.e.*, the measured film thickness was smaller than that predicted by H–D theory. Such deviations between measured and H–D predicted EHD film thicknesses, in thin and ultrathin film regions, have been reported for a number of oils.[15,25–28] Various mechanisms have been proposed to explain these deviations but more work is needed to develop a comprehensive theory.

10.3.5 Friction and Wear

The friction properties of these fluids were investigated using a 4-ball trib-ometer configured for AW evaluation using ASTM D4172.[12] In this method, a steel ball is loaded (392 N) against three balls and rotated (1200 rpm) while completely immersed in the test lubricant maintained at 75 °C. During the test, which lasts 60 min, the frictional torque, temperature, load, and speed are continuously monitored and recorded. At the end of the test, the wear scar diameters (WSDs) on the three balls are measured parallel and trans-verse to the wear direction, and the six measurements averaged. The re-corded frictional torque and load were used to calculate the COF using the procedure outlined in ASTM D5183-95.[13] Two tests were conducted on each test lubricant and the resulting COF and WSD values from each measure-ment averaged and used in further analysis. An example of COF *vs.* time data from a repeat measurement on a test lubricant is illustrated Figure 10.14. The data in Figure 10.14 is for neat VMWO. As shown in both measurements, the COF displayed an initial sharp increase and decrease. This was followed by a more or less steady-state value from about 500 s until the end of the test. The average and standard deviation of the COF values in the steady-state region of the repeat measurements for this lubricant were 0.051 ± 0.008 and

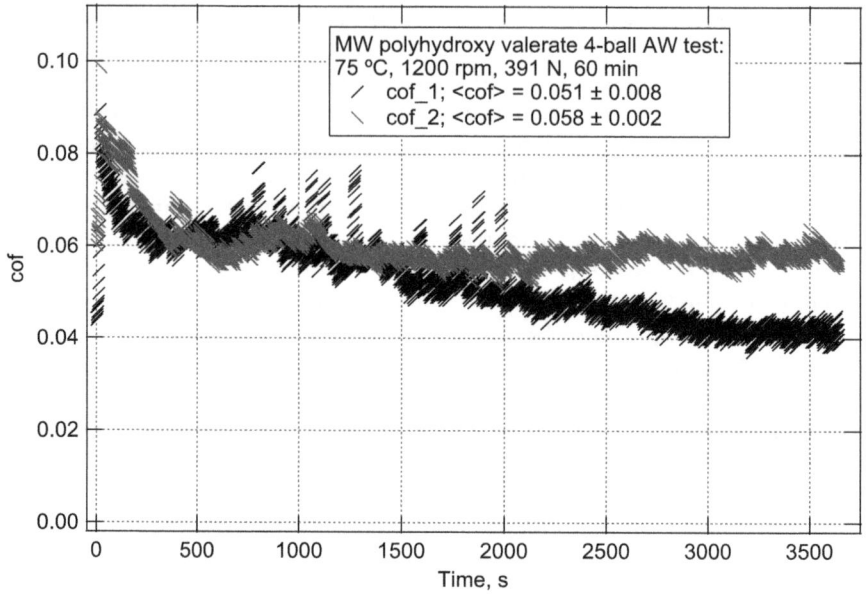

Figure 10.14 Time *vs.* COF data for VMWO from repeat four-ball anti-wear measurements.

Table 10.7 COF and WSD values of milkweed polyhydroxy and polyformyl esters from four-ball anti-wear tests.

Oil	COF ± standard deviation	WSD ± standard deviation/mm
AMWO	0.059 ± 0.008	0.576 ± 0.030
BMWO	0.064 ± 0.014	0.622 ± 0.067
VMWO	0.055 ± 0.005	0.778 ± 0.058
MWF	0.06 ± 0.00	0.38 ± 0.01

0.058 ± 0.002. These values were further averaged and used in data analysis. The corresponding WSD data from each of the two tests (six measurements per test) on VMWO gave the following average and standard deviation WSD values: 0.737 ± 0.058 and 0.819 ± 0.033. These average WSD values as well as those from the other milkweed ester oils were identical within one standard deviation so they were further averaged and used in analysis. The average and standard deviation COF and WSD values of the milkweed polyhydroxy esters investigated here are summarized in Table 10.7. Examination of the COF results indicates that the three polyesters displayed similar COF values within one standard deviation. These values were within the range of 0.04 to 0.11 reported for unformulated seed oils.[24] However, close comparison indicates that the COF values of the milkweed polyesters are lower than those of commodity oils that have similar structures to milkweed (*e.g.*, SBO, 0.08; CANO, 0.07). In fact, the COF of the milkweed esters is close to the lower end of the literature range, which is populated by very few commodity oils such as castor oil (COF = 0.04).[22] The relatively lower COF of milkweed polyesters may be to do with their high polarity because of the presence of multiple ester groups in their structure. Ester groups are known to promote adsorption of oil molecules to metallic surfaces resulting in lower COFs.[28] The presence of multiple ester groups in molecules, as in the case of the oils studied in this work, has been found to lead to stronger adsorption and further lowering of COF.[29] This is attributed to possible binding of the oil molecules to multiple sites using more than one ester group. The relatively low COF results shown in Table 10.7 for the oils studied in this work are consistent with expectations based on multiple binding of these molecules to the friction surfaces. It is interesting to note that, even though these oils had similar degrees of esterification, they differ in the structure of the ester group, particularly the chain length. One might expect, based on steric or other considerations that the three esters would have to overcome varying barriers to attain multiple bonding on the metal surface. The fact that they all had similar COF might be an indication that partial multiple bonding is responsible for the observed COF values. Such partial multiple bonding can be attained by all three esters to an equal degree and does not require steric or other barriers to be overcome. If complete multiple bonding was responsible, the COF values would have been different as a function of structural or other barriers to adsorption.

Examination of the WSD data in Table 10.7 shows that the values for the milkweed esters are within the reported range for commodity oils (0.51–0.87 mm)[22] except for the anomalous polyformyl ester (MWF) (0.38 ± 0.01). Comparison of the values for the three longer chain substituents indicates that there was no difference in WSD values between AMWO and BMWO within one standard deviation. However, the WSD value for VMWO was much higher than the other two, whereas the lowest wear was exhibited by the polyformyl ester (0.38 ± 0.01). VMWO has the lowest viscosity among the three longer chain substituents which might have contributed to its relatively higher WSD. The polyformyl derivative, on the other hand, with the shortest substituent chain, appears to have the most improved lubricating character of the four derivatives.

10.4 Conclusions

To summarize this work, the combination of favorable oxidative stability, cold flow character, EDH film thicknesses observed under load, low COF and the consequent low WSD for these polyesters of MWO, shows that appropriately modified MWO (or indeed many other vegetable oils) would be a good bio-based industrial base material for lubrication.

References

1. B. Berkman, *Econ. Bot.*, 1949, **3**, 223.
2. R. E. Harry-O'kuru, R. A. Holser, T. P. Abbott and D. Weisleder, *Ind. Crops Prod.*, 2002, **15**, 51.
3. R. E. Harry-O'kuru, S. H. Gordon and A. Biswas, *J. Am. Oil Chem. Soc.*, 2005, **82**, 207.
4. R. E. Harry-O'kuru, *US Pat.*, 7 351 403, 2008.
5. *Official Methods and Recommended Practices of the AOCS*, AOCS Press, Champaign, 5th edn, 1997.
6. *ASTM D445-97*, ASTM, Philadelphia, 1997.
7. G. Biresaw and G. Bantchev, *J. Synth. Lubr.*, 2008, **25**, 159.
8. *ASTM D2270-93*, ASTM, Philadelphia, 1998.
9. *ASTM D97-96a*, ASTM, Philadelphia, 1996.
10. *ASTM D2500-99*, ASTM, Philadelphia, 1999.
11. *ASTM D2272-98*, ASTM, Philadelphia, 1998.
12. *ASTM D4172-94*, ASTM, Philadelphia, 2002.
13. *ASTM D5183-95*, ASTM, Philadelphia, 2002.
14. G. J. Johnston, R. Wayte and H. A. Spikes, *Tribol. Trans.*, 1991, **34**, 187.
15. G. Biresaw, *J. Am. Oil Chem. Soc.*, 2006, **83**, 559.
16. R. E. Harry-O'kuru and C. J. Carriere, *J. Agric. Food Chem.*, 2002, **50**, 3214.
17. R. L. Whistler and J. N. BeMiller, in *Industrial Gums, Polysaccharides and Their Derivatives*, ed. R. L. Whistler, Academic Press, New York, 1959.
18. S. S. Lawate, K. Lal and C. Huang, in *Tribology Data Handbook*, ed. E. R. Booser, CRC Press, New York, 1997, p. 103.
19. A. Adhvaryu, Z. Liu and S. Z. Erhan, *Ind. Crops Prod.*, 2005, **21**, 113.
20. U. J. Jonsson, *Wear*, 1999, **232**, 185.
21. M. L. Gee, P. M. McGuiggan and J. N. Israelachvili, *J. Chem. Phys.*, 1990, **93**, 1895.
22. I. Sendijarevic and A. J. McHugh, *Macromolecules*, 2000, **33**, 590.
23. P. Wood-Adams, *J. Rheol.*, 2001, **45**, 203.
24. B. T. Hamrock and D. Dowson, *The Elastohydrodynamics of Elliptical Contacts*, John Wiley & Sons, New York, 1981.
25. G. Guangteng and H. A. Spikes, *Tribol. Trans.*, 1997, **40**, 461.
26. H. A. Spikes, *Langmuir*, 1996, **12**, 4567.
27. H. A. Spikes, *R. Soc.-Unilever Indo-UK Forum Mater. Sci. Eng.*, 3rd, 1996, 334.
28. S. Jahanmir and M. Beltzer, *J. Tribol.*, 1986, **108**, 109.
29. G. Biresaw, A. Adharyu, S. Z. Erhan and C. J. Carriere, *J. Am. Oil Chem. Soc.*, 2002, **79**, 53.

CHAPTER 11

The Potential of Vegetable Oils for Lubricants

BRAJENDRA K. SHARMA,*[a] ZENGSHE LIU[b] AND
SEVIM Z. ERHAN[c]

[a] Illinois Sustainable Technology Center, UIUC, 1 Hazelwood Drive,
Champaign, IL 61820, USA; [b] USDA, ARS, National Center for Agricultural
Utilization Research, Bio-Oils Research Unit, 1815 N. University Street,
Peoria, IL 61604, USA[†]; [c] USDA, ARS, Eastern Regional Research Center,
600 E. Mermaid Lane, Wyndmoor, PA 19038, USA[†]
*Email: bksharma@illinois.edu

11.1 Introduction

Environmental concerns and petroleum shortages have encouraged exten-
sive research into bio-lubricants. Because of their biodegradability, low
ecotoxicity and excellent tribological properties, vegetable oil-based lubricants, such as soybean oil find applications in many areas from greases
to hydraulic oils which offer lower coefficients of friction, improved
wear characteristics, a higher viscosity index, lower volatility, and lower
flashpoints than mineral-based oils. The global lubricant demand was
38.7 million metric tons (MMT) in 2012 and is forecast to reach 42.1 MMT in
2017, growing at less than 2% per year.[1,2] The lubricant growth rates are
expected to be stagnant for Europe (0.6% per year) and North America

---

[†]Mention of trade names or commercial products in this publication is solely for the purpose of
providing specific information and does not imply recommendation or endorsement by the
U.S. Department of Agriculture. USDA is an equal opportunity provider and employer.

RSC Green Chemistry No. 29
Green Materials from Plant Oils
Edited by Zengshe Liu and George Kraus
© The Royal Society of Chemistry 2015
Published by the Royal Society of Chemistry, www.rsc.org

(0.4% annual growth), while there will be a decent demand in other regions, such as the Asia-Pacific (2.7%), South America (2.4%), and Africa and the Middle East (1.4%). The global lubricant market was worth US$44.0 billion and the value of global lubricants will grow even faster than the volume, rising at a CAGR (compound annual growth rate) of 5.5% from 2012 to 2018 and expected to reach US$65.2 billion in 2018. This will be caused by the estimated increase in the price of lubricants from US$960 per MT in 2005, to US$1330 per MT by 2015.

The USA has remained the leader with just under 22% of the global market, but lubricant volume demand in the USA is declining and is expected to grow merely 0.4% per year, a reversal of the declining production from 2006 to 2009. The USA's production of paraffinic and naphthenic base oils has been on an increasing trend from 55 million barrels to 60.2 million barrels in 2010, and 62.0 million barrels in 2011, but overall production is still lower than the 2006 level (66.8 million barrels).[3] In automotive lubricants (47%), multigrade engine oils constituted 38% of total USA lubricant sales in 2011 followed by automotive transmission and hydraulic fluids (6%), while in industrial lubricants (50%), process oils have the largest share (23%) followed by general industrial oils (14%), industrial engine oils (7%), and metalworking oils (6%). The remaining 2% of total lubricant sales is grease. Currently, only 5–10% of lubricants are made using non-conventional basestocks, including polyalphaolefins synthetic esters, rerefined oils, naphthenic oils, white oils, vegetable oils and animal fats. Clearly, there is room in this market for the expansion of bio-based lubricants.

Although many vegetable oils have excellent lubricity, they often have poor oxidation stability due to a high degree of multiple unsaturations in the fatty acid (FA) chains of vegetable oils. This poor thermal and oxidative stability confines their use as lubricants to a modest range of temperatures. Several reports in the literature[4–6] claim the use of vegetable oils such as rapeseed oil and sunflower oil as substitutes for petroleum-based lubricating oils and synthetic esters. Attempts have been made to improve the oxidative stability by transesterification of trimethylolpropane and rapeseed oil methyl ester[7] and by selective hydrogenation of the polyunsaturated bonds of the FA chains.[8] Reports have discussed the use of epoxidized unsaturated FAs as metalworking fluids[9] and epoxy oils as lubricating additives to eliminate corrosion from chlorine-containing compounds.[10] Esters of dicarboxylic acids with branching have been used as lubricants and hydraulic fluids over a wide range of temperatures.[11,12]

Moringa oleifera (referred to as moringa in this study), a member of the Moringaceae family is a multipurpose plant native to sub-Himalayan regions of Northwest India, Africa, Arabia, Southeast Asia, the Pacific and Caribbean islands, and South America. It has also been distributed in many other regions such as the Philippines, Cambodia, and Central and North America.[13] The fully matured, dried seeds of this plant are round or triangular shaped, and the kernel is surrounded by a lightly wooded shell with three thin flexible wings.[14,15] Moringa seeds contain between 33 and 41% (w/w) vegetable oil.[15] Several authors have investigated its composition including its FA profile[15–18] and found its oil to be high in oleic acid (>70%). Its oil is commercially known as "ben oil" or "behen oil" due to its content of behenic (docosanoic) acid. It possesses significant resistance to oxidative degradation,[19] and has been extensively used in the enfleurage process.[14] A recent survey conducted on 75 indigenous (India) plant-derived non-traditional oils concluded that moringa oil, among others, has good potential for bio-diesel production,[20] and is also a good candidate as a lubricant base oil.[21]

In this chapter, we explore the effectiveness of using epoxidized soybean oil (ESBO), high-oleic soybean oil (HOSBO) obtained by genetic modification of SBO, chemically modified soybean oil (Ace-SBO, But-SBO, Isobut-SBO and Hex-SBO) and moringa oil in lubricant applications. The base oil properties for ESBO, HOSBO, chemically modified SBO and moringa oil are compared with soybean oil (SBO).

11.2 Experimental

11.2.1 Materials

SBO was obtained from Pioneer High Bred International (Des Moines, IA) HOSBO from Optimum Inc. (Urbandale, IA) and ESBO from Elf Atochem (Blooming Prairie, MN). The oils were used as received commercially without any further purification and processing. Moringa oil was obtained after crushing *M. oleifera* seeds (University of Agriculture, Faisalabad, Pakistan), extraction using hexane in a Soxhlet extractor, and finally solvent removal at 45 °C under vacuum using a rotary evaporator (Eyela, N-N Series, Rikakikai Co. Ltd, Tokyo). Acetic anhydride (99.5%), propionic anhydride (97%), butyric anhydride (≥97.5%), isobutyric anhydride (97%), valeric anhydride (97%), hexanoic anhydride (97%), hexanes, acetone, methyl oleate, methyl linoleate, monoolein, diolein, and triolein were obtained from Sigma–Aldrich (St. Louis, MO), while heptanoic anhydride (99%), sodium chloride and sodium bicarbonate were obtained from Fisher Scientific (Fairlawn, NJ), and used as received. Polystyrene standards with molecular weights of 1700, 2450, 5050, 7000, 9200, and 10 665 Da, were obtained from Polymer Laboratories Ltd (Amherst, MA).

11.2.2 Preparation of Chemically Modified SBO

11.2.2.1 Ring-opening Reaction

The reaction was carried out by refluxing a 2.5 L aqueous solution of 127.4 g epoxidized soybean oil at 100 °C for 48 h in a three-neck 5 L round-bottomed flask. Perchloric acid (26.05 g) was added drop-wise to the reaction mixture as it was constantly agitated by a mechanical stirrer. After the reaction was complete, the mixture was cooled to room temperature and the organic phase was extracted with chloroform and washed three times with water to remove any traces of acid remaining in the reaction mixture. The solvent was removed under reduced pressure at 80 °C and the product (the dihydroxy derivative of soybean oil DiOH-SBO) was stored under dry vacuum overnight.

11.2.2.2 Esterification of the Hydroxyl Groups in the Dihydroxy Derivative

The dihydroxy derivative of SBO (50 g) was added to 44.67 g of acetic anhydride in a 1:2 ratio (1 epoxy ring and 2 moles of anhydride), and then 21.5 g of pyridine in an equimolar ratio was further added to the reaction mixture. The mixture was stirred with a mechanical stirrer in a 500 mL glass round-bottomed flask for 48 h at room temperature. Then the reaction mixture was cooled by pouring onto ice cubes in a beaker and again stirred for 12 h. The reaction mixture was extracted three times with 100 mL of dichloromethane (chloroform or diethyl ether can also be used). Then the organic phase was washed with 75 mL of 3% HCl, 5% $NaHCO_3$, and water (each three times) and finally dried over 100 g anhydrous $MgSO_4$ for 24 h. The solvent was removed using a rotavapor and unreacted anhydride was removed using Kugelrohr's distillation under reduced pressure (0.2 Torr vacuum) at 80–100 °C. The product (Ace-SBO) was then stored under vacuum. The above procedure was repeated using butyric, isobutyric and hexanoic anhydride for preparation of the diester derivatives But-SBO, Isobut-SBO and Hex-SBO respectively.

11.2.3 Characterization

^1H- and ^{13}C-NMR spectra were recorded using a Bruker (Boston, MA) Avance 500 NMR operating at a frequency of 500.13 and 125.77 MHz, respectively, using a 5 mm broadband inverse Z-gradient probe in $CDCl_3$ (Cambridge Isotope Laboratories, Andover, MA) and the Bruker Icon NMR software was used. Peaks were referenced to sodium 3-trimethylsilylpropionate-2,2,3,3-d_4 (TSP) at 0.0000 ppm. Each spectrum was Fourier transformed, phase corrected, and integrated using ACD spectrum manager. The integration values in the ^1H-NMR spectra were referenced to 4.00 between the range of 4.1 and 4.4 ppm. Simulations of ^{13}C-NMR spectra were performed by ACD/Labs 6.00 ACD/CNMR predictor software.

FT-IR spectra were recorded on a Thermo Nicolet (Madison, WI) Nexus 470 FT-IR system in a scanning range of 650–4000 cm^{-1} for 32 scans at a spectral resolution of 4 cm^{-1} with a pair of KBr crystals in the thin film. Data was collected and processed using the Omnic 6.2 software.

GPC profiles were obtained on a PL-GPC 120 high-temperature chromatograph (Polymer Laboratories, Amherst, MA) equipped with a column, autosampler and built-in differential refractive index detector. The starting material and products were dissolved in tetrahydrofuran (THF). THF was used as the eluent with flow rate of 1.00 mL min^{-1} at 40 °C. The injection volume was 100 μL. Two PL gel 3 μm mixed E columns (300×7.5 mm) were used in series. The GPC was calibrated using a mixture of polystyrene (MW 1700, 2450, 5050, 7000, 9200, and 10 665 Da), methyl oleate (296.48), methyl linoleate (294.48), monoolein (353), diolein (619.2), and triolein (885.4) in THF at 40 °C.

11.2.4 Performance Properties

The kinematic viscosity of vegetable oils was measured using a Cannon–Fenske calibrated viscometer (Cannon Instrument Co., State College, PA) in a Cannon temperature bath (CT-1000) at 40 and 100 °C according to ASTM (American Society for Testing and Materials) standard method D445-95. The viscosities obtained are average values of 2–3 determinations and the precision is within the limits of the ASTM method specification. To compare base oils with respect to viscosity variations with temperature, the ASTM method D2270 provides a means to calculate a viscosity index (VI). VI values were therefore calculated using the kinematic viscosity at 40 and 100 °C.

Pour Points were measured by following the ASTM D-5949 method using a Phase Technology Analyzer, Model 70X (Phase technology, Hammersmith Gate, Canada). The Pour Point is defined as the temperature in °C when the sample still pours when the jar is tilted. Statistically the method has shown quite good consistency for determining the low-temperature flow properties of fluids.

The free FA content, iodine value and peroxide value for the vegetable oils were determined as per AOCS (American Oil Chemists' Society) official methods Ca 5a-40, Cd 1-25 and Cd 3-25, respectively.

The anti-oxidant used for chemically modified soybean oils (CMSBO) is an alkylated phenolic compound (Lubrizol™ 7652) from Lubrizol, Wickliffe, OH. It was blended into the CMSBO at 40 °C in various concentrations (0.5, 1.0, 1.5 and 2.0 wt%).

11.2.4.1 Thin Film Micro Oxidation

A small amount of oil (25 μL) was oxidized as a thin film on a freshly polished high-carbon steel catalyst surface with a steady flow (20 cm^3 min^{-1}) of dry air. Oxidation tests were done at various temperatures (175, 200, 225, 250 and 275 °C) and time lengths (30, 60, 90, 120 and 150 min) inside a bottom-less glass reactor. The temperature was maintained at ±1 °C with a heated aluminum slab placed on top of a hot plate. This arrangement eliminates the temperature gradient across the aluminum surface and transfers heat to the catalysts placed on the slab. The constant airflow ensured removal of volatile oxidation products. The test was designed to eliminate any gas diffusion limitation.

After oxidation, the catalyst containing the oxidized oil sample was removed from the oxidation chamber and cooled rapidly under a steady flow of dry N$_2$ and transferred to a desiccator for temperature equilibration. After approximately 2 h, the catalyst containing the oxidized oil was weighed to determine the volatile loss (or gain) due to oxidation and then soaked (30 min) in THF to dissolve the soluble portion of the oxidized oil. After dissolving the soluble portion of the oxidized oil, the catalyst was dried and weighed to determine the remaining insoluble deposit.

11.2.4.2 Pressurized Differential Scanning Calorimetry

The experiments were carried out using a PC-controlled DSC 2910 thermal analyzer from TA Instruments (New Castle, DE). The instrument has a maximum sensitivity of 5 mV cm^{-1} and a temperature sensitivity of 0.2 mV cm^{-1}. A 1.5–2.0 mg sample was placed in a hermetically sealed type aluminum pan with a pinhole lid for interaction of the sample with the reactant gas (dry air). The controlled diffusion of the gas through the hole greatly restricts the volatilization of the oil while still allowing for saturation of the liquid phase with air. A film thickness of less than 1 mm was required to ensure proper oil–air interaction and to eliminate any discrepancy in the result due to gas diffusion limitations. The module was first temperature calibrated using the melting point of indium metal (156.6 °C) at a 10 °C min^{-1} heating rate. Dry air was pressurized in the module at a constant pressure of 3450 kPa and a scanning rate of 10 °C min^{-1} was used throughout the experiment. The signal maximum (SM) and onset (T_{onset}) temperatures were calculated from the exotherm in each case. The induction time (I_t) was measured for oils containing different additive (phenolic anti-oxidant) concentrations (0.5–2.0 wt%) using an isothermal scanning rate.

11.2.4.3 Friction Measurement by the Ball-on-disk Method

The boundary lubrication properties of SBO, HOSBO and ESBO were studied using a multispecimen friction measurement apparatus (FALEX, Sugar Grove, IL). Ball-on-disk experiments (1018 steel disk, R_c 15–25) were carried out under low speed, 6.22 mm s^{-1} (5 rpm), and high load 181.44 kg at 25 °C using test oils diluted to different concentrations with hexadecane. Measurements of the coefficient of friction (COF) and torque were made in each case. The COF values reported are averages of two or three independent experiments and the standard deviation observed was ± 0.02.

11.3 Results and Discussion

11.3.1 Physical Properties of the Vegetable Oils Investigated

The physicochemical properties of SBO, HOSBO, ESBO and moringa oil are presented in Table 11.1. The viscosity of ESBO at 40 °C is significantly larger than the other oils. The higher molecular weight and more polar structure compared with SBO and HOSBO results in stronger intermolecular interactions in ESBO. This property of ESBO would translate into enhanced lubricity

Table 11.1 Physical properties of vegetable oils. (Reprinted from ref. 46 with permission from Elsevier.)

Property	Vegetable oil			
	SBO	HOSBO	ESO	Moringa
Appearance	Light yellow	Pale yellow	Colorless	Light yellow
Kinematic viscosity/cSt^a at 40 °C	32.93	41.34	170.85	27.1
Kinematic viscosity/cSt^a at 100 °C	8.08	9.02	20.41	7.0
Acid value/mg KOH g^{-1} (AOCS Ca 5a-40)^b	0.16	0.12	0.09	0.32
Peroxide value/meq · kg^{-1} (AOCS Cd 8-3)^b	9.76	4.78	0.0	1.65
Iodine value/mg I$_2$ g^{-1} (AOCS Cd 1-25)^b	144.8	85.9	9.11	70
Fatty acid composition by GC/% (AACC 58-18)^c				
C16:0	6.0	6.0	70.0	6.65
C18:0	5.5	3.0	30.0^d	6.09
C18:1	22.0	85.0	0.0	73.85
C18:2	66.0	4.0	0.0	0.99
C18:3	0.5	2.0	0.0	0.0
C20:0	0.0	0.0	0.0	3.98
C20:1	0.0	0.0	0.0	1.99
C22:0	0.0	0.0	0.0	5.85

^{a}*ASTM D-445*, ASTM, Philadelphia, 2000.
^{b}*Official Methods and Recommended Practices of the American Oil Chemists' Society*, AOCS, Urbana, 5th edn, 1998.
^{c}*Approved Methods of the American Association of Cereal Chemists*, AACC, St. Paul, 10th edn, vol. 2, 2000.
^{d}Obtained from the epoxidation of C18:1, C18:2 and C18:3.

in a dynamic system. The data indicate that nearly all the unsaturations in the FA chains have been converted to epoxy groups. These properties will cumulatively influence the thermal and oxidative behavior of ESBO.

11.3.2 Chemically Modified SBO

Many nucleophilic reagents are known to add to oxirane rings, resulting in ring opening. These ring-opening reactions in ESBO could result in branching at the oxirane carbons (earlier sites of unsaturation in SBO). The hypothesis is that appropriate branching groups would interfere with the formation of macrocrystalline structures during low-temperature applications and would provide enhanced fluidity to vegetable oils. Triacylglycerols that are hydrogenated to eliminate polyunsaturation will solidify at room temperature due to alignment and stacking of adjacent molecules. For this reason, it is important that there should be at least one unsaturation site available for functionalization that will generate two branching points on the chain. The ester branching groups are quite effective for attaining the desired molecular spacing. An ester of six-carbon chain length has been observed to deliver the most desired Pour Point properties for these oils.[22] These modified vegetable oils with chain branching are reported to have superior performance properties and are promising as biodegradable lubricants in applications such as hydraulic fluids, metalworking fluids, crankcase oils, drilling fluids, two-cycle engine oils, wear-resistant fluids, and greases. A similar strategy was employed for the synthesis of CMSBO with branching groups at the epoxy ring carbons.

The nucleophilic attack by water on the oxirane ring of ESBO in the presence of perchloric acid results in the ring-opened product DiOH-SBO, as

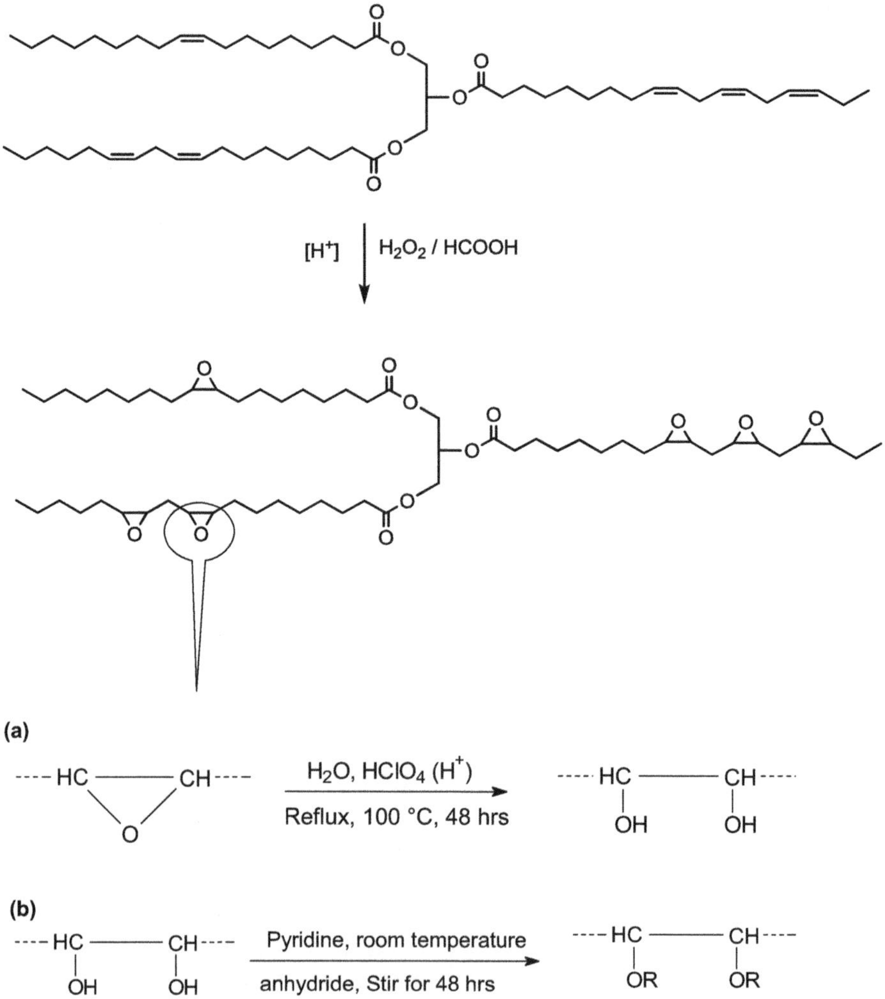

Scheme 11.1 (a) Epoxy ring opening with formation of a dihydroxy derivative. (b) Reaction with anhydride to form diester derivatives: Ace-SBO if R = –COCH$_3$; But-SBO if R = –COCH$_2$CH$_2$CH$_3$ (R1); Isobut-SBO if R = –COCH(CH$_3$)$_2$ (R2); Hex-SBO if R = –COCH$_2$CH$_2$CH$_2$CH$_2$CH$_3$ (R3). (Reproduced from ref. 47 with kind permission from Springer Science and Business Media from Springer and the AOCS Press.)

shown in Scheme 11.1(a). The extent of reaction was monitored by FT-IR spectroscopy of small aliquots taken at 4 h intervals. As the ring-opening reaction progressed, the oxirane C–O twin bands at 823 and 842 cm^{-1} decreased and disappeared, while a broad hump appeared and increased in intensity at 3700–3100 cm^{-1}. This hump is assigned to the H-bonded O–H stretching vibrations of alcohols (Figure 11.1).

The presence of hydrogen-bonded OH groups results in an increased viscosity for the DiOH-SBO derivative over ESBO. The reaction completion was further confirmed by their ^{1}H-NMR spectra (Figure 11.2), where the peaks at 3.2–2.8 ppm (–CH– protons of the epoxy ring) completely disappeared and additional peaks appeared in the range 4.1–3.5 ppm (protons attached to the carbon of the –CHOH group) and 3.5–3.0 ppm (protons attached to oxygen of group –CHOH). In addition, the disappearance of epoxy carbon peaks in the range 54–57 ppm and the appearance of –CHOH peaks in the range 73–75 ppm in the ^{13}C-NMR spectra of DiOH-SBO confirmed the conversion of epoxy groups to dihydroxy derivatives. Ring opening of ESBO with water was done at 100 °C with other catalysts such as sulfuric acid, and HCl. Unfortunately, these catalysts resulted in hydrolysis of ester linkages. In perchloric acid, ring opening is the major reaction with minimal ester hydrolysis. This is observed from the retention of the NMR peaks of the backbone glycerol structure in the ring-opened DiOH-SBO derivative. These are the NMR peaks at 5.2–5.3 ppm and 4.1–4.4 ppm for the CH and CH$_2$ protons of the –CH$_2$–CH–CH$_2$– glycerol backbone, respectively. The retention of the triacylglycerol backbone is important for maintaining high biodegradability. The triacylglycerol structure remains unaffected at temperatures lower than 100 °C. Other ^{1}H-NMR peaks which are common in ESBO and DiOH-SBO are 2.25–2.5 ppm for CH$_2$ protons α to >C=O, 1.68–1.85 ppm for CH$_2$ protons in between two epoxies, 1.58–1.68 ppm for CH$_2$ protons β to >C=O, 1.15–1.58 ppm for all other CH$_2$ protons and 0.8–1.0 ppm for terminal CH$_3$ protons. Table 11.1 shows quantitative data obtained from ^{1}H-NMR spectra of different products. These values are computed using a reference value of four protons for peaks at 4.1–4.4 ppm corresponding to four CH$_2$ protons of –CH$_2$–CH–CH$_2$– in the glycerol backbone of the triacylglycerol structure. It shows the same number of CH$_3$ and CH$_2$ protons for both DiOH-SBO and ESBO molecules. Additional protons in other ranges are due to protons of –CHOH. A combination of time and temperature is selected for this step to ensure that the reaction goes to almost completion (at least 90%).

In the second step, the dihydroxy derivative is reacted with acetic, butyric, isobutyric and hexanoic anhydride to yield the diesters Ace-SBO, But-SBO, Isobut-SBO and Hex-SBO respectively [Scheme 11.1(b)]. In this step, the reaction was monitored using FT-IR for the disappearance of the O–H stretching band in the region 3700–3100 cm^{-1} and the change in relative intensities of the 1465 and 1377 cm^{-1} bands in the fingerprint region (Figure 11.1). The epoxy absorption band (823 and 842 cm^{-1}), which was present in ESBO, no longer exists in the final diester products. The completion of the reaction was also confirmed by their respective ^{1}H-NMR

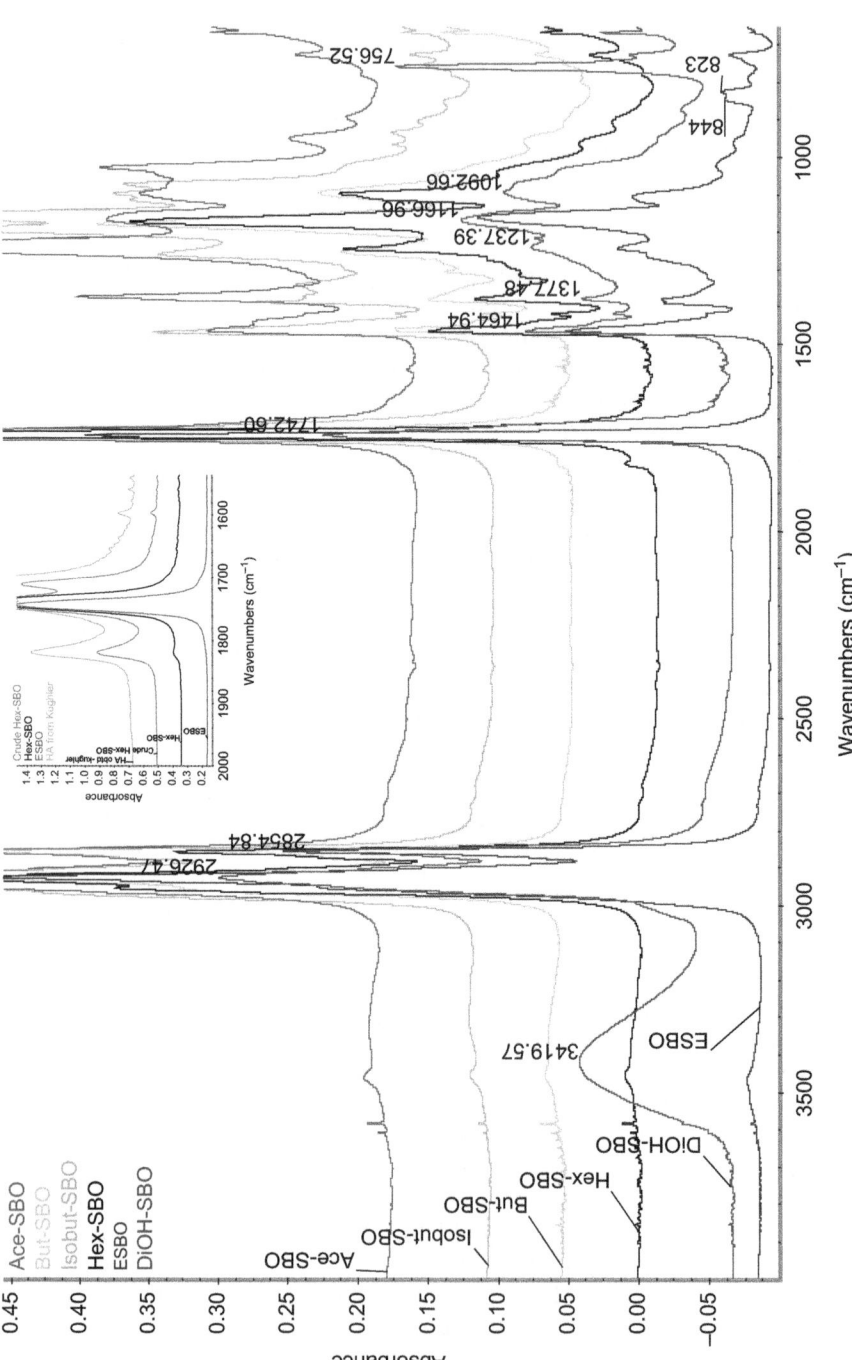

Figure 11.1 FT-IR spectra of ESBO, DiOH-SBO, Ace-SBO, But-SBO, Isobut-SBO and Hex-SBO. Inset shows the FT-IR spectra of ESBO, crude Hex-SBO (before Kugelrohr distillation), Hex-SBO (after Kugelrohr distillation), and hexanoic anhydride (HA) obtained as the Kugelrohr distillate from crude Hex-SBO in the region 2000–1500 cm^{-1}. (Reproduced from ref. 47 with kind permission from Springer Science and Business Media from Springer and the AOCS Press.)

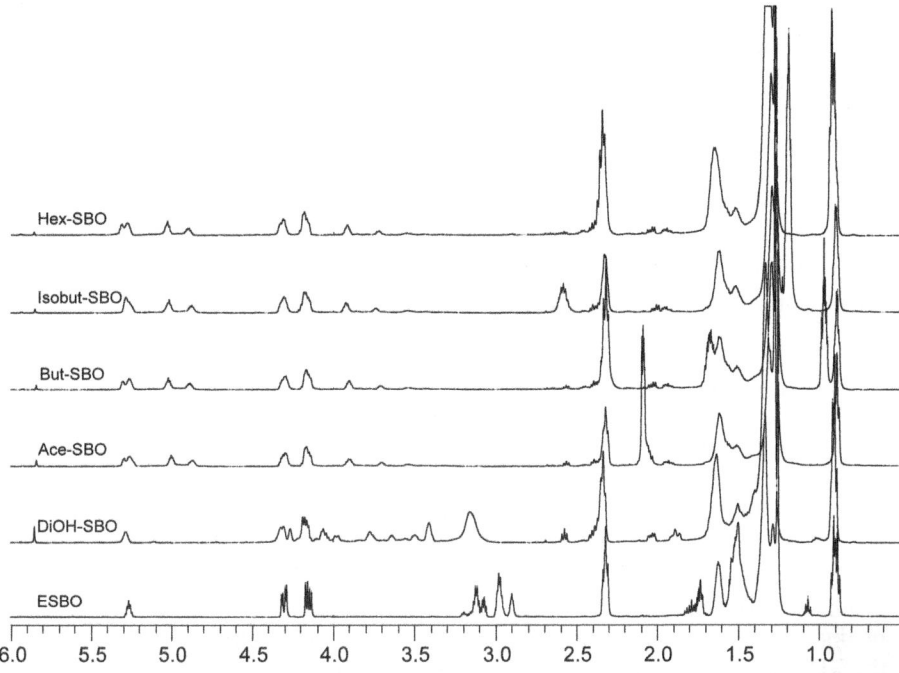

Figure 11.2 ¹H-NMR spectra of ESBO, DiOH-SBO, Ace-SBO, But-SBO, Isobut-SBO, and
Hex-SBO.
(Reproduced from ref. 47 with kind permission from Springer Science
and Business Media from Springer and the AOCS Press.)

spectra (Figure 11.2). Peaks in the region 3.5–3.0 ppm (protons attached to
oxygen of group –CHOH) have disappeared completely and new peaks at
4.8–5.4 ppm (protons at the point of substitution –CH–O–CO–R in FA chains)
appeared in all diester products. The presence of ¹³C-NMR peaks at
173.1 ppm due to the carbonyl carbon of triacylglycerols, 68.9 and 62 ppm
respectively for the CH and CH$_2$ carbons of the –CH$_2$–CH–CH$_2$– glycerol
backbone and the corresponding protons at 5.2–5.3 ppm and 4.1–4.4 ppm in
the ¹H-NMR spectra of all diester products confirm that the glycerol–FA
linkage is intact and did not undergo any hydrolysis.

11.3.2.1 FT-IR Characterization of Diester Products

The FT-IR spectra (Figure 11.1) of all these products have some common
peaks at 2926 and 2855 cm^{-1} (methylene asymmetric stretching), 1743 cm^{-1}
(triglyceride carbonyl stretching), 1465 cm^{-1} (CH$_2$ bending vibration),
1377 cm^{-1} (CH$_3$ symmetrical bending vibration), 724 cm^{-1} (CH$_2$ rocking
vibration), and additional peaks at 1242, 1160 and 1104 cm^{-1} due to
stretching vibrations of the C–O group in esters.[23] Formation of the Ace-SBO
product was also confirmed by the increase in peak intensity ratio

$1465 : 1377$ cm^{-1}. In ESBO and DiOH-SBO, the 1377 cm^{-1} peak is almost half the intensity of the 1465 cm^{-1} peak, while in Ace-SBO, due to the addition of methyls (–COCH$_3$) in the structure, the peak at 1377 cm^{-1} (CH$_3$ bending vibration) almost doubled compared to the one at 1465 cm^{-1} (CH$_2$ bending vibration). Also, the intensities of peaks such as 1743 cm^{-1} (ester carbonyl) and 1160 cm^{-1} (C–O stretching of esters) relative to the 2926 cm^{-1} peak increased in Ace-SBO compared to ESBO and DiOH-SBO. This is due to an increase in ester functionalities in the structure. A similar trend was observed in the spectra of But-SBO, Isobut-SBO and Hex-SBO confirming the formation of the final diester products.

11.3.2.2   NMR Characterization of Diester Products

Figure 11.2 shows the ^{1}H-NMR spectra of ESBO, DiOH-SBO, Ace-SBO, But-SBO, Isobut-SBO and Hex-SBO in CDCl$_3$, while their quantitative ^{1}H-NMR spectral data is shown in Table 11.2. It can be seen that peaks at 2.8–3.0 ppm due to epoxy groups are not present, except for in ESBO. The peaks at 3.0–3.2 ppm in DiOH-SBO due to the protons of the hydroxyl in the –CHOH group are also not present in the diester products. The presence of additional peaks at 1.85–2.18 ppm in the ^{1}H-NMR spectra, due to the methyl protons of the –COCH$_3$ group, and the corresponding methyl carbons at 21–22.6 ppm and carbonyl carbons at 171 ppm in the ^{13}C-NMR spectra confirms the

Table 11.2 ^{1}H-NMR data of ESBO and dihydroxy and diester derivatives of SBO. (Reproduced from ref. 47 with kind permission from Springer Science and Business Media from Springer and the AOCS Press.)

NMR range/ ppm	Assignment	ESBO	DiOH-SBO	Ace-SBO	Isobut-SBO	But-SBO	Hex-SBO
5.2–5.4	CH backbone + CH–O–R[a]	1	1	1.7	1.7	1.7	1.9
4.8–5.1	CH–O–R[a]	—	—	1.9	1.8	1.8	2.1
4.08–4.3	CH$_2$ backbone	4	5.4	4	4	4	4
3.47–4.05	–CHOH dihydroxy	—	6.1	1.4	1.4	1.3	1.5
3.0–3.47	–CHOH dihydroxy	—	7.7	—	—	—	—
2.83–3.2	CH epoxy	8.6	—	—	—	—	0.4
2.5–2.8	CH isobutyric	—	0.8	—	3.3	0.4	
2.23–2.5	CH$_2$ α to CO	6.2	7.5	6.1	6.3	13.5	14.1
1.83–2.2	CH$_3$ acetyl	—	1.4	11.7	1.1	1.6	3.1
1.16–1.9	CH$_2$ other	73.2	73.3	62.4	61.7	70.1	83.4
1.1–1.2	CH$_3$ isobutyric	—	—	—	21.8	—	—
0.95–1.1	CH$_3$ butyric	0.8	0.6	—	—	10.5	
0.72–0.95	CH$_3$	8.8	9.1	8.5	8.7	8.1	18.6
	N Epoxy rings[b]	4.3	0	0	0	0	0.2
	N Substituents[c]	—	—	3.9	3.6	3.5	3.2

[a]R = acetyl and other branching groups.
[b]Number of epoxy rings (CH epoxy protons/2).
[c]Number of substituents on SBO estimated as: for acetyl, CH$_3$ protons (1.83–2.25 ppm)/3; for isobutyric CH$_3$ protons (1.1–1.2 ppm)/6; for butyric CH$_3$ protons (0.95–1.1 ppm)/3; for hexanoic [CH$_3$ protons (0.72–1.02 ppm) − 9]/3.

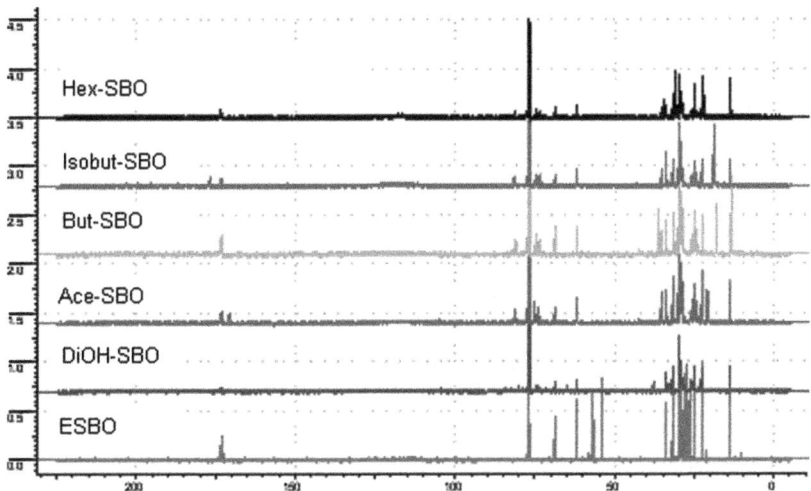

Figure 11.3 ^{13}C-NMR spectra of ESBO, DiOH-SBO, Ace-SBO, But-SBO, Isobut-SBO, and Hex-SBO products.
(Reproduced from supplementary information for ref. 47 with kind permission from Springer Science and Business Media from Springer and the AOCS Press.)

formation of the Ace-SBO product. The peaks at 73.8–75.2 ppm can be assigned to the fatty acid carbons of the branching site –CH–O–Ac (Figure 11.3). The formation of the But-SBO product was confirmed by the presence of a new branching group (R1 = –COCH$_2$CH$_2$CH$_3$) in its structure. This is also shown by the increase in intensity of the peaks at 2.26–2.46 ppm due to CH$_2$ protons α to the carbonyl (Table 11.1), and the appearance of new peaks at 0.95–1.06 ppm due to methyl protons in the ^{1}H-NMR spectra. The carbonyl peak of the R1 group appears at 173.1 ppm, thus increasing the intensity of the already existing carbonyl peak of the base molecule. Additional peaks appeared at 36.0–36.5 ppm due to CH$_2$ carbons α to carbonyl groups, and at 18.4–18.5 ppm due to CH$_2$ carbons α to the methyl of the R1 group in the But-SBO molecule. The addition of the R2 group [–COCH(CH$_3$)$_2$] in DiOH-SBO to give Isobut-SBO can be confirmed by the following observations. In the ^{1}H-NMR spectra two additional peaks appeared, one at 2.5–2.68 ppm due to a CH proton α to the carbonyl, and the other at 1.10–1.23 ppm due to the CH$_3$ protons of the R2 group. In addition, the appearance of new ^{13}C-NMR peaks at 176.6 ppm due to the carbonyl carbons, and 18.8–19.0 ppm due to the methyl carbons of the R2 group, further confirmed the formation of the Isobut-SBO derivative (Figure 11.3). The peak due to the CH carbon of the R2 group appeared at 33.8 ppm and it merged there with other peaks. There are no additional peaks in the ^{1}H-NMR spectra of Hex-SBO due to the addition of the branching group R3 (–COCH$_2$CH$_2$CH$_2$CH$_2$CH$_3$). The peaks for the CH$_3$ and CH$_2$ protons α to the carbonyl of the R3 group resulted in an intensity increase of signal at 0.83–1.01 and 2.25–2.44 ppm respectively (Table 11.2). In the ^{13}C-NMR spectra of Hex-SBO, the carbonyl carbon peak of the R3 group overlapped with the carbonyl carbon peak of the base molecule (173.4 ppm), and an additional peak appeared at 35.6 ppm due to CH_2 carbons α to the carbonyl of the R3 group. These observations confirm the formation of diester products from DiOH-SBO and various anhydrides.

Percentage conversions calculated from ^{1}H-NMR data for signals in the range 2.95–3.45 ppm (hydroxyl proton of the –CHOH group) yield values >95%. If there is only one substitution on each epoxy carbon (others still have an –OH group), then the peak position for the remaining hydroxyl proton may change, and this approach may not give accurate conversion values. Using specific ^{1}H-NMR peaks arising from branching groups (–COCH$_3$, R1, R2 and R3), the number of these groups attached to the triacylglycerol structure can be calculated as shown in Table 11.2. In an average triacylglycerol molecule, there are 8.6 sites available for substitution, $\sim$4 sites are substituted by these branching groups, while the others retain hydroxyl groups. Steric hindrance may be a possible reason for there being only one substituent on each epoxy carbon. This explanation seems valid, as the average number of substituents in diester derivatives decreases with the bulkiness (chain length) of the branching group.

The presence of hydroxyl and polar ester groups in diester products may cause enhanced intermolecular interactions[24] resulting in highly viscous products. GPC was performed on ESBO, dihydroxy and diester products to check the extent of internal polymerization. It was found that oligomerization occurred at some stage during the reaction. Apart from the main peak of the expected products, there were approximately 20% oligomerized products (with higher molecular weights) in the DiOH-SBO and diester products. The molecular weights (M_w) of the main product peaks for the starting and final products are: ESBO 901 Da; DiOH-SBO 1107 Da; Ace-SBO 1193 Da; But-SBO 1286 Da; Isobut-SBO 1277 Da; and Hex-SBO 1653 Da as estimated by using a calibration mixture of polystyrene, fatty acid methyl ester, and mono-, di-, and triolein.

11.3.2.3   *Anhydride Determination*

FT-IR and TGA were used to check for unreacted anhydride in the final diester products. It was found that in Hex-SBO, and But-SBO unreacted anhydride remained in the product even after vacuum distillation. The amount of unreacted anhydride was quantified using TGA. For the analysis, the sample was heated in the presence of nitrogen at 20 °C min^{-1} to 500 °C in high-resolution mode. Hexanoic anhydride (HA) has the highest boiling of all the anhydrides used, and was found to evaporate before 200 °C. No evaporation loss below 200 °C was detected in ESBO. On using a solution of 20% HA in ESBO, at 200 °C, $\sim$20% weight loss was detected, which can be assigned to anhydride present in that mixture. Using TGA, the unreacted anhydride in the crude Hex-SBO obtained after vacuum distillation, was found to be 28% (Figure 11.4). Kugelrohr's distillation was then used at 100 °C under 0.2 Torr vacuum to remove the unreacted anhydride. Figure 11.4 shows the TGA thermograms of ESBO, the 20% HA in ESBO,

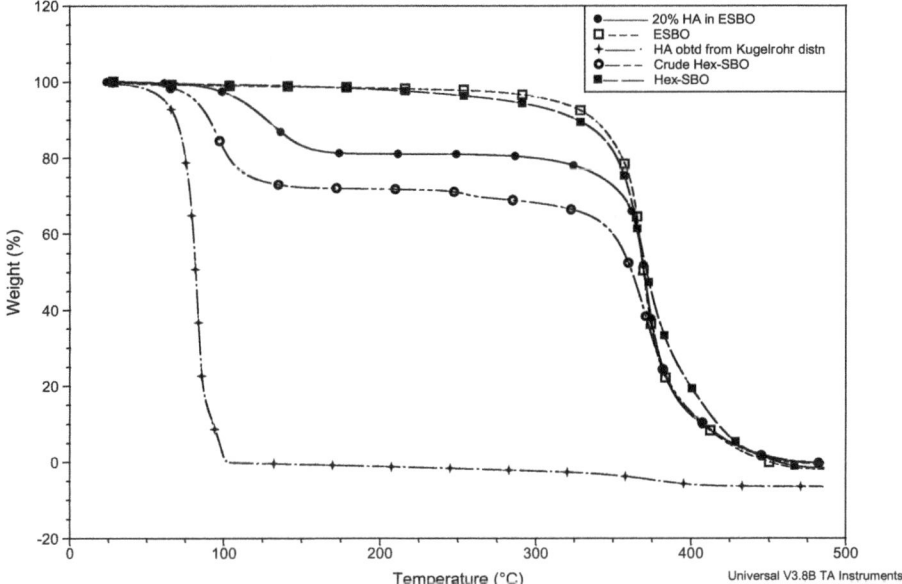

Figure 11.4 TGA thermograms of ESBO, the 20% hexanoic anhydride (HA) in ESBO, crude Hex-SBO, Hex-SBO and HA obtained from the Kugelrohr distillation of crude Hex-SBO, showing weight (%) loss with temperature. (Reproduced from ref. 47 with kind permission from Springer Science and Business Media from Springer and the AOCS Press.)

crude Hex-SBO, and HA obtained from the Kugelrohr distillation of crude Hex-SBO and the final diester product Hex-SBO. The percentages of anhydride obtained using the TGA method for the final diester products were Ace-SBO 2%; Isobut-SBO 1.9%; But-SBO 1.7%; and Hex-SBO 2%.

Similarly, FT-IR was used to monitor the removal of unreacted anhydride. The peak for the anhydride carbonyl group appears at 1818 cm^{-1} and is easily distinguishable from the ester carbonyl peak at 1743 cm^{-1}. In Figure 11.1, a peak at 1818 cm^{-1} can be seen in a crude Hex-SBO. Complete removal of unreacted anhydrides using Kugelrohr distillation of the final diester products (Ace-SBO, But-SBO, and Isobut-SBO) was confirmed by the absence of the 1818 cm^{-1} peak in their spectra (Figure 11.1).

11.3.3 Oxidation and Thermal Stability

The TGA profiles show that the diesters are as thermally stable as ESBO, even at high temperatures (Figure 11.4). Noack volatilities were also determined as per the ASTM D6375 method using TGA and are shown in Table 11.3. Noack volatilities for all the samples were less than 3%, which are excellent compared to the 10–30% for mineral oils and synthetic esters. These diester derivatives are therefore thermally stable. PDSC (pressure differential scanning calorimetry) experiments were used to measure the oxidation stability of CMSBO samples. This test is widely used in the lubricant industry and

Table 11.3   TGA Noack volatilities, PDSC data (at 10 °C min^{-1} in air under 200 psi pressure), and PP values of dihydroxy and diester products, and vegetable oils. T_{onset} and SM values are averages of three experiments with a standard error of ±1 °C. (Reproduced from ref. 47 with kind permission from Springer Science and Business Media from Springer and the AOCS Press.)

Test fluid	TGA Noack volatility/%	Onset temperature (T_{onset})/°C	Signal maximum temperature (SM)/°C	Pour point (PP)/°C
SBO	2.6	174	186	−9
DiOH-SBO	3.6	163	213	—
Ace-SBO	2.7	174	213	−3
But-SBO	2.5	169	208	−15
Isobut-SBO	3	157	241	—
Hex-SBO	2.4	161	234	−21
ESO	—	188	215	0
HOSBO	—	186	201	−21
MO	—	191	217	4

considered reliable for base oils as well as finished lubricants.[25] The PDSC results, the onset temperature (T_{onset}) and the signal maximum temperature (SM) for the diester samples are shown in Table 11.3. The diester samples showed higher oxidation stability than the dihydroxy derivatives and also SBO. This shows that oxidation stability is considerably improved by the removal of unsaturation. The T_{onset} decreases with the chain length of the branching group. Ace-SBO had the highest value (174 °C) of T_{onset}, while Hex-SBO had the lowest (161 °C). Longer side-chains are more prone to oxidative cleavage than small compact ones. This observation is supported by another study on synthetic esters,[26] where they mentioned that short-chain acids are more stable than long-chain acids.

Table 11.3 shows the PDSC T_{onset} data which is higher for moringa oil (MO) than for the soybean oils. Similar results were obtained for the oxidative stability index (OSI) measured using the Rancimat method EN 14112. The PDSC technique uses a microgram amount of the sample, while Rancimat uses 3 g of the sample. PDSC experiments were run in triplicate and the values shown are an average of three measurements. Moringa oil showed a higher coefficient of variation (8.4%) than the other oils (0.3–2.3%). This shows that oxidative stability data generated using PDSC is more repeatable than OSI. It has been shown in an earlier study that higher stability oils like moringa oil show poor repeatability in Rancimat tests,[27] while the PDSC method can be used for low, as well as very high oxidative stability oils. The PDSC T_{onset} shows good correlation with the Rancimat OSI values (R^2 value of 0.91). The higher oxidative stability shown by moringa oil is ascribed to its low unsaturation number (78) compared to other oils and the presence of naturally occurring anti-oxidants.[19] This low unsaturation number is a result of the higher amounts of monounsaturated fatty acids (MUFAs) and saturated fatty acids (SFAs) present in it. The unsaturation number correlates very well with the

T_{onset} ($R^2 = 0.95$) and OSI data ($R^2 = 0.87$). The highest unsaturation number of soybean oil makes it the least oxidatively stable, with the low T_{onset} value.

The thermal stability of the oils is important in defining the operability conditions for the lubricants prepared using such vegetable oils.[28-32] Hence, the thermal stability of moringa and soybean oils was measured using TGA under an inert atmosphere of nitrogen. Both oils show a temperature range of a particular length before any mass loss occurs. This induction period is followed by a steep fall in the curve with varying slope, which indicate losses due to either evaporation or cracking. The slope and behavior of the curve do not show any indication as to whether a distillation mass loss is continuing or the sample has already been thermally cracked. T_{onset} marks the onset temperature of the thermal mass loss transition. Apart from thermal stability of the oil, T_{onset} also provides an idea of the initial boiling point of the oil. The thermal stability, as measured using T_{onset}, does not correspond to the oxidative stability data, although moringa oil, with the lowest unsaturation and the highest oxidative stability, showed the highest T_{onset} value (347 °C) compared to soybean oil (343 °C). This is because of the different atmospheres used in these analyses. The mass loss temperatures closely correspond to the distillation curves and can be used to determine the boiling range of the oils.[33]

The deposit values were obtained in static mode at 175 °C using thin film micro-oxidation (TFMO), under a steady flow of air (20 mL min^{-1}) in a bottomless glass reactor. Reactions were terminated and analyzed at the indicated times. After the stipulated test time, the oxidized oil was washed with THF and the residue left on the coupon is termed the "insoluble deposit". During the oxidation process, several primary oxidation products are formed, which later in the presence of excess O_2 undergo a further oxy-polymerization reaction to form an insoluble deposit.[34] The tendency to form such a deposit is the main detrimental factor associated with vegetable oil for use in high-temperature lubricants. The data shows that ESBO remains fairly stable. No significant oxidative degradation occurs until after 60 min into the oxidation. At 60 min, the percentage of insoluble deposits is less than 3% for ESBO, while for HOSBO and SBO it is 45 and 65% respectively. After 60 min, there is a sharp increase in the deposit formation, suggesting a rapid breakdown of the epoxy groups leading to oxidative polymerization through oxygen bonding. The induction time for deposit formation of ESBO is roughly twice that of HOSBO (~ 30 min) under similar oxidation temperatures (175 °C). The percentage of insoluble deposits becomes stable after 2 h of oxidation with ESBO (50%) remaining lower than the other oils (HOSBO 65%, SBO 75%). The larger insoluble deposits observed for SBO and HOSBO at the different time periods compared with ESBO, are mainly due to the linoleic and oleic content in their FA chains (see Table 11.1). SBO shows a greater deposit than HOSBO primarily due to the presence of polyunsaturation in its FA chains relative to HOSBO.

Anti-oxidant additives were used to further improve the oxidation stability of the diester samples. Different anti-oxidants were tried with varying concentrations in SBO. The best additive combination was then used 4% (w/w) in the

Table 11.4   Results of the PDSC temperature ramp at 10 °C min^{-1} in air under static 200 psi pressure, and isothermal experiments at 200 °C (SBO at 170 °C) under 200 psi pressure with 20 mL min^{-1} flow of air, for diester SBO derivatives with additives (4% w/w). (Reproduced from ref. 47 with kind permission from Springer Science and Business Media from Springer and the AOCS Press.)

Test fluid	Onset temperature (T_{onset})/°C	Oxidation induction time (OIT)/min
SBO	192	25.7[a]
Ace-SBO	256	75.3
But-SBO	247	51.0
Isobut-SBO	255	56.2
Hex-SBO	254	55.6

[a]Isothermal experiment at 170 °C.

diester samples. The effect of these anti-oxidants was studied using PDSC temperature ramping and isothermal experiments.[25] The results obtained are shown in Table 11.4. It was found that the anti-oxidant additive responded very well to these diester samples and increased their T_{onset} by 80–100 °C. The isothermal PDSC experiment was done to determine the oxidation induction time at 200 °C under 200 psi at a constant flow of 20 mL min^{-1} of air. These results also suggested that the anti-oxidant additive could prevent the start of oxidation for at least 50 min at 200 °C. Both the T_{onset} and OIT of these diester samples are higher compared to SBO. Even mineral-oil-based formulations have an OIT of 20 min at 200 °C.[35] This shows that chemical modification of SBOs along with a suitable anti-oxidant additive can improve their oxidation properties to be comparable with mineral oils. The effect of chain length on the oxidation stability of additive-enhanced diester samples is similar to that for non-enhanced samples, *i.e.*, decreasing oxidation stability was observed with increasing chain length of the substituents on the CMSBO. Using an optimized set of additives (2% zinc diamyl dithiocarbamate and 2% antimony dialkyldithiocarbamate), the oxidation stability can be improved further. With this combination, the T_{onset} for SBO increased to 215 °C, while that for HOSBO increased to 254 °C. The OIT values also increased to 49 min for HOSBO and 1 min for SBO at 200 °C.

11.3.4   Low-temperature Properties

Triacylglycerols that are completely hydrogenated for the purpose of eliminating sites of unsaturation will tend to harden at room temperature due to alignment and stacking of adjacent molecules. Therefore, it is important that there should be at least one site of unsaturation available for derivatization that will yield branching sites. This approach is used here to improve the low-temperature-flow behavior of SBO by attaching ester branching at the double-bond sites.

The PP measurements provide a good estimate of the low-temperature fluidity of the lubricants. PP values for SO, ESO, and CMSBO derivatives are given in Table 11.3. The PP of SBO is −9 °C, while that of ESO is 0 °C. The resultant modified SBO products are characterized by lower PP values, as demonstrated by the superior PP values for Hex-SBO (-21 °C). The PP values decreased with increasing chain length of ester branching. This can be rationalized by the presence of branching groups attached to the sites of unsaturation, which does not allow individual molecules to come close for easy stacking, due to steric interactions, and thus inhibits crystallization, resulting in a lower PP. Unmodified vegetable oils have a tendency to form macrocrystalline structures *via* a uniform stacking of the 'bent' triacylglycerol backbone at low temperatures. The branching groups of optimum length at the epoxy carbons, not only serve to eliminate the sites of unsaturation, but also impose spacing from other triacylglycerol molecules, thereby interfering with the formation of macrocrystalline structures. The ester branching groups with chain lengths of at least six carbons, were found to be the most effective for imposing the desired molecular spacing and thus imparting the most desirable PP properties.

Further improvement in the low-temperature fluidity of Hex-SBO was attained by using a Pour Point depressant (PPD) additive (L7671-A from Lubrizol Corporation, Wickliffe, OH). Blending was carried out by stirring the CMSBO derivatives with an optimized additive dose at room temperature for 2 h. The purpose of the PPD additives is to sterically hinder crystallization of triacylglycerol molecules at low temperatures by disrupting the stacking mechanism. In general, PPDs can lower PP by 30 °C in mineral oils at concentrations of only 0.1–0.4% (w/w), while in vegetable oils, a PP reduction of only 9–12 °C can be achieved with a higher treat rate of 1%. Since the additive response of PPD in vegetable oils is not as good as in mineral oils, a higher amount of PPD is needed to lower the PP significantly. An optimum PPD additive concentration of 1% in the Hex-SBO enabled a PP of -30 °C. Further addition of PPD additives made no significant improvement in the PP. In order to further improve the low-temperature properties of the formulation, diluents such as polyalphaolefins, dibutyl adipate and high-oleic vegetable oils can be used. Here we used a biodegradable synthetic ester, dibutyl adipate (96% purity), that was uniformly blended into the formulation as a diluent at several concentrations. The final optimized formulation was Hex-SBO + 1% PPD + diluent (70 : 30 oil-to-diluent ratio) and had a PP of -42 °C.

The results thus showed that use of a suitable anti-oxidant additive can bring the oxidation performance of CMSBO derivatives on a par with mineral oils. The fully formulated lubricant using CMSBO as a base fluid, dibutyl adipate as a diluent, PPD and an anti-oxidant additive has excellent oxidation stability (PDSC T_{onset} of 248 °C and an OIT of 28.5 min). This approach of chemical modification significantly improves the thermo-oxidative stability of vegetable oils. Their cold-flow properties were also improved using a combination of PPD and suitable diluents. The basic vegetable oil structure is retained even after chemical modification of SBO, thus maintaining excellent biodegradability.

Moringa oil showed a higher PP than SBO and HOSBO despite the fact that it has more UFAs. This high PP is due to the presence of $\sim$23% SFAs compared to 11.5% for SBO and 9% for HOSBO. It appears that low-temperature-flow properties are influenced by the amount of FAs and not by their chain lengths.[36] Therefore a low amount of SFAs with a good combination of MUFAs and polyunsaturated fatty acids (PUFAs) favors low PP and CP values.

The crystallization behavior of moringa oil and other vegetable oils was also studied using Cryo-DSC.[37] The long-chain compounds (FA chains) of vegetable oils in the solid state exist in more than one crystalline form and thus have multiple melting points. The melting points of triacylglycerols depend on the chain length, the nature of the unsaturation (*cis*- or *trans*-olefins), and the number and position of double bonds.[38,39] Monoacid triacylglycerols show three distinct melting points corresponding to the three crystalline forms α, β′, and β.[40,41] Since most vegetable oils are mixed triacylglycerols, no β-form is present, and they take the highest melting β′-form.[42] When the oil is cooled quickly it solidifies in the, lowest melting, α-form. When heated slowly, the monoacid triacylglycerols melt, and, held just above the α melting point, resolidify into the β′ crystalline form. The β-form is the highest melting form, in the case of monoacid triacylgly-cerols and is produced by crystallization from solvents. Since crystallization kinetics are very sensitive to temperature fluctuations, cooling rate and thermal history, the Cryo-DSC method is relatively simple, reproducible and robust enough to provide good control of temperature fluctuations, cooling rate *etc.*

Cryo-DSC has already been utilized for the identification of various crys-tallization forms of pure single-acid triacylglycerols.[41,43,44] Since vegetable oils are mixtures, the situation becomes complex. Multiple transitions have been observed during the cooling and heating cycle of DSC experiments. Three measurements (of each heating and cooling cycle) were carried out for each oil. It was found that more variation in results was observed in the cooling cycle than the heating cycle. In the case of the cooling cycle, 8 out of 16 measurements (50%) showed standard deviations higher than 1 °C runs and varied from 1.2 to 14.2 °C. For the heating cycle, only 1 out of 20 (5%) showed a standard deviation of 1.5 °C. A previous study[45] also showed that in a heating experiment, DSC thermograms were minimally affected by thermal history and variations in cooling rate, while in cooling experiments, these factors can influence the shape of the peak and cause more variation in the results. Therefore, only the results obtained from the heating cycle were used to analyze the influence of FA composition on the cold-flow properties of vegetable oils. The wax disappearance temperatures (WDTs) and SM values of various peaks were determined. Moringa oil shows a single sharp peak at a SM of −0.7 °C and a WDT of −11.7 °C because of its predominant oleic acids. Moringa oils have almost the same amounts of SFAs and UFAs, but the makeup of its UFAs is different. In moringa oil, MUFAs make up 76% and PUFAs make up 1%. This change in composition of UFAs, *i.e.*, lower amounts of PUFAs in moringa oil results in higher SM and WDT values compared to other oils (sunflower, cotton, jatropha and canola). More SFAs in an oil results in close-packing of the triacylglycerol molecules during cooling, leading to gel-like structures which entrap low-melting-point molecules, thus resulting in higher PP, CP, WDT and SM values. PUFAs in their bent configurations prevent the close-packing of triacylglycerol molecules during cooling more than MUFAs, thus high amounts of PUFAs in oils (such as in jatropha and sunflower oils) result in lower PP, CP, WDT and SM values.

SM values obtained using Cryo-DSC were found to be a better measure of PP as shown by a better correlation ($R^2 = 0.90$) than WDT ($R^2 = 0.70$). Similarly, CP also showed better correlation with SM values ($R^2 = 0.78$) than WDT ($R^2 = 0.64$). In all these correlations, the major outlier point was jatropha oil. On removing this point, the R^2 value for the PP–SM correlation increased to 0.99, for PP–WDT to 0.88, for CP–SMT to 0.92, and for CP–WDT to 0.96. Testing more oil samples will help to develop better correlations between these two methods.

11.3.5 Lubricity Data

An important property of lubricants is their ability to maintain a stable lubricating film at the metal contact zone. Triacylglycerols of vegetable oils are known to provide excellent lubricity due to their ester functionalities. This is because, the polar head of the triacylglycerol molecule, *i.e.*, the glycerol end, attaches to the metal surface and allows a monolayer film to form, with the non-polar end of the FA chains sticking away from the metal surface. This prevents metal-to-metal direct contact by providing a sliding surface. Without a good sliding surface, the two metals at the contact zones of moving parts come in direct contact with each other, and this results in an increase in temperature causing adhesion, scuffing or even welding. The ester structures in triacylglycerols offer active oxygen sites that trigger binding to the metal surface forming a protective film. This protective film builds further with time to reduce friction. During this rubbing process under lubricated conditions, at high load and low speed, bond cleavage of FA molecules might take place. Under such conditions, the epoxy groups of ESBO offer active oxygen sites that trigger polymerization on the metal surface forming a protective film. This protective film builds further with time to reduce friction. The COF data from the ball-on-disk experiments on vegetable oils blended into hexadecane at a 0.003 molar concentration are 0.269 for SBO, 0.248 for HOSBO, and 0.232 for ESBO, while at 10 times higher concentrations (0.03 molar), the COF values are 0.135 for SBO, 0.132 for HOSBO, and 0.104 for ESBO. Under a high load of 181.44 kg and a low speed of 6.22 mm s^{-1} (5 rpm), all of the vegetable oils show an excellent reduction in COF at low molar concentrations.

The anti-wear and friction-reducing properties of moringa oil and other vegetable oils were evaluated using a high-frequency reciprocating rig (HFRR) lubricity tester. The HFRR method determines the lubricity or the ability of a fluid to affect friction and wear between the surfaces in relative motion under load. The average ball scar diameter, disk wear scar width and length on the *x*-axis, film percentage, and coefficient of friction (COF) for

Table 11.5   HFRR data showing the average ball wear scar diameter, the disk wear
scar width and length on the *x*-axis, the film percentage, and the
coefficient of friction (COF) for moringa and other vegetable oils.
(Reproduced with kind permission from Springer Science and
Business Media from Springer / *Journal of Thermal Analysis and
Calorimetry*, **96**(3), 2009, 999–1008. Lubricant properties of Moringa oil
using thermal and tribological techniques. B.K. Sharma, U. Rashid,
F. Anwar, and S.Z. Erhan, Table 5.)

Sample	Ball wear scar diameter/µm	Disk wear scar width on the *x*-axis/µm	Disk wear scar length on the *x*-axis/µm	Film/%	COF
Moringa oil	156	217	1144	97	0.092
Jatropha oil	146	204	1168	97	0.096
Cottonseed oil	202	253	1142	95	0.097
Canola oil	149	173	1202	94	0.076
Sunflower oil	151	192	1134	95	0.069

moringa and other vegetable oils are shown in Table 11.5. With moringa oil,
the ball scar diameter was 156 µm, which is similar to other vegetable oils.
This high wear scar diameter for cottonseed oil may be due to the low
amount of free fatty acids (FFAs) present in it. Jatropha oil provided the
lowest wear scar diameter, which may be due to the presence of a higher
amount of FFAs present in it. The disk *x*-scar results were also similar, with
the biggest scar width being for cottonseed oil. The average COF was higher
(0.09) for moringa, jatropha and cottonseed oils, followed by canola oil
(0.076) and was the least for sunflower oil (0.07). The lower COFs in sunflower and canola oil may be due to their higher viscosities compared to
other oils. Despite the higher COF for moringa oil, no decrease was observed
in the average lubricant film percentage. An attempt was made to see the
effect of various groups of FAs, *i.e.*, SFAs, UFAs, MUFAs, and PUFAs on the
lubricity properties of these oils. The COF was found to have a good correlation with the amount of SFA ($R^2 = 0.85$) and UFA ($R^2 = 0.89$). This correlation shows that a low amount of SFAs and high amount of UFAs results
in a low COF. A high amount of UFAs and a low amount of SFAs may be the
reason for the lower COFs in the case of canola and sunflower oil. More
studies are needed to confirm this observation. These results show that the
lubricity properties of moringa oil are on a par with those of other
vegetable oils.

11.4   Conclusions

In this chapter we have discussed the properties of soybean oil and how its
oxidative stability and cold-flow properties can be improved either using
genetically modified oils (HOSBO), or chemically modifying soybean oils to
ESO and CMSBOs and then using these as basestocks for lubricant formulations. Use of chemical additives made further improvements in important

lubricant properties, such as oxidative stability, low- temperature-flow properties, and the friction and wear behavior of these oils. Among these oils, the best ones were high-oleic vegetable oils, which when formulated with chemical additives, provide lubricants exhibiting improved low-temperature properties, and superior oxidative stabilities and wear properties.

References

1. G. Gill, *Lubes'n'Greases*, 2014, **20**, 37.
2. *Transparency Market Research Report on Technology Trends in Lubricants (Mineral, Synthetic, and Bio-based) Market for Turbine Oil, Compressor Oil, Gear Oil, Hydraulic Oil, Bearing Oil and Heat Transfer Fluid Lubricant Applications - Global Industry Analysis, Size, Share, Growth, Trends and Forecast, 2012–2018*, 2013, **1**, pp. 1–150, http://www.transparencymarketresearch. com/technology-trends-lubricants-market.html.
3. N. DeMarco, *Lubricant Industry Factbook*, LNG Publishing, Falls Church, 2012, vol. 1, p. 6.
4. S. J. Randles and M. Wright, *J. Synth. Lubr.*, 1992, **9**, 145.
5. S. Asadauskas, J. M. Perez and J. L. Duda, *Lubr. Eng.*, 1996, **52**, 877.
6. X. Wu, X. Zhang, S. Yang, H. Chen and D. Wang, *J. Am. Oil Chem. Soc.*, 2000, **77**, 561.
7. E. Uosukainen, Y.-Y. Linko, M. Lämsä, T. Tervakangas and P. Linko, *J. Am. Oil Chem. Soc.*, 1998, **75**, 1557.
8. L. E. Johansson and S. T. Lundin, *J. Am. Oil Chem. Soc.*, 1979, **56**, 974.
9. S. Watanabe, T. Fujita and M. Sakamota, *J. Am. Oil Chem. Soc.*, 1988, **65**, 1311.
10. D. H. Tao, H. L. Zhu and Z. M. Hu, *Lubr. Sci.*, 1996, **8**, 397.
11. R. G. Kadesch, *J. Am. Oil Chem. Soc.*, 1979, **56**, 845A–849A.
12. P. Bondioli, L. Della Bella and A. Manglaviti, *Ol., Corps Gras, Lipides*, 2003, **10**, 150.
13. J. F. Morton, *Econ. Bot.*, 1991, **45**, 318.
14. *The Wealth of India: Raw Materials*, Council of Scientific and Industrial Research, New Delhi, 1962, vol. 6, p. L425.
15. A. Sengupta and M. P. Gupta, *Fette, Seifen, Anstrichm.*, 1970, **L72**, 6.
16. F. Anwar, S. Latif, M. Ashraf and A. H. Gilani, *Phytother. Res.*, 2007, **21**, 17.
17. F. Anwar and M. I. Bhanger, *J. Agric. Food Chem.*, 2003, **51**, 6558.
18. M. A. Somali, M. A. Bajnedi and S. S. Al-Fhaimani, *J. Am. Oil Chem. Soc.*, 1984, **61**, 85.
19. S. Lalas and J. Tsaknis, *J. Am. Oil Chem. Soc.*, 2002, **79**, 677.
20. M. M. Azam, A. Waris and N. M. Nahar, *Biomass Bioenergy*, 2005, **29**, 293.
21. B. K. Sharma, U. Rashid, F. Anwar and S. Z. Erhan, *J. Therm. Anal. Calorim.*, 2009, **96**, 999.
22. H. S. Hwang and S. Z. Erhan, *J. Am. Oil Chem. Soc*, 2001, **78**, 1179.
23. E. T. Akintayo, O. Olaofe, S. O. Adefemi and C. O. Akintayo, *Int. J. Chem.*, 2002, **72**, 151.

24. B. Tamami, S. Sohn and G. L. Wilkes, *J. Appl. Polym. Sci.*, 2004, **92**, 883.
25. B. K. Sharma and A. J. Stipanovic, *Thermochim. Acta*, 2003, **402**, 1.
26. S. J. Randles, in *Synthetic Lubricants and High-Performance Functional Fluids*, ed. L. R. Rudnick and R. L. Shubkin, CRC Press, Boca Raton, 1999, p. 63.
27. D. R. Kodali, *J. Agric. Food Chem.*, 2005, **53**, 7649.
28. J. C. O. Santos, L. N. Lima, I. M. G. Santos and A. G. Souza, *J. Therm. Anal. Calorim.*, 2007, **87**, 639.
29. J. C. O. Santos, A. D. Oliveira, C. C. Silva, J. D. S. Silva, A. G. Souza and L. N. Lima, *J. Therm. Anal. Calorim.*, 2007, **87**, 823.
30. C. C. Garcia, P. I. B. M. Franco, T. O. Zuppa, N. R. A. Filho and M. I. G. Leles, *J. Therm. Anal. Calorim.*, 2007, **87**, 645.
31. S. Vecchio, L. Campanella, A. Nuccilli and M. Tomassetti, *J. Therm. Anal. Calorim.*, 2008, **91**, 51.
32. B. Lin, L. Yang, H. Dai, Q. Hou and L. Zhang, *J. Therm. Anal. Calorim.*, 2009, **95**, 977.
33. M. L. A. Gonçalves, D. A. Ribeiro, D. A. P. da Mota, A. M. R. F. Teixeira and M. A. G. Teixeira, *Fuel*, 2006, **85**, 1151.
34. A. Adhvaryu, S. Z. Erhan, Z. S. Liu and J. M. Perez, *Thermochim. Acta*, 2000, **364**, 87.
35. B. L. Papke, W. Song, W. J. Heilmann, A. R. De Kraker, Y. H. Jois, L. M. Morrison and M. Pozebanchuk, presented at the 14th International Colloquium of Tribology: Tribology and Lubrication Engineering, 2004.
36. H. Imahara, E. Minami and S. Saka, *Fuel*, 2006, **85**, 1666.
37. L. Yang, H. Dai, A. Yi, B. Lin and G. Li, *J. Therm. Anal. Calorim.*, 2008, **93**, 875.
38. F. D. Gunstone, *Fatty Acid and Lipid Chemistry*, Aspen Publishers, p. 101.
39. M. Noordin and L. Chung, *J. Therm. Anal. Calorim.*, 2009, **95**, 891.
40. E. S. Lutton and A. J. Fehl, *Lipids*, 1970, **5**, 90.
41. J. W. Hagemann and J. A. Rothfus, *J. Am. Oil Chem. Soc.*, 1983, **60**, 1123.
42. K. Sato, *Chem. Eng. Sci.*, 2001, **56**, 2255.
43. D. J. Cebula and K. W. Smith, *J. Am. Oil Chem. Soc.*, 1991, **68**, 591.
44. J. W. Hagemann, W. H. Tallent and K. E. Kolb, *J. Am. Oil Chem. Soc.*, 1972, **49**, 118.
45. A. K. Aboul-Gheit, T. Abd-el-Moghny and M. M. Al-Eseimi, *Thermochim. Acta*, 1997, **306**, 127.
46. A. Adhvaryu and S. Z. Erhan, *Ind. Crops Prod.*, 2002, **15**(3), 247–254.
47. B. K. Sharma, A. Adhvaryu, Z. S. Liu and S. Z. Erhan, *J. Am. Oil Chem. Soc.*, 2006, **83**(2), 129–136.

CHAPTER 12

Utilization of Green Materials for Coating Applications

SHAILESH N. SHAH,*[a] SHARATHKUMAR K. MENDON[b] AND
SHELBY F. THAMES[b]

[a] Bio-Energy Research Group, Chemistry Department, Faculty of Science,
The Maharaja Sayajirao University of Baroda, Vadodara 390002, India;
[b] School of Polymers and High Performance Materials, 118 College Drive
#5217, Hattiesburg, MS 39406, USA
*Email: shilshilp@hotmail.com

12.1 Introduction

A coating is a covering applied to the surface of an object (substrate) for the purpose of decoration (*e.g.*, gloss, appearance, color) and/or protection (*e.g.*, corrosion resistance, wear resistance, scratch resistance). The coating process involves the application of a thin film of functional material to any of a variety of substrates such as metal, concrete, masonry, wood, paper, fabric, leather or plastic.

Natural materials preceded petrochemicals by millennia in coating applications. The earliest cave paintings employed charcoal, iron oxides and/or chalk as pigments, and animal fats, blood, egg whites and yolks as binders.[1] Pitches and balsams comprised the protective coatings of ancient Egyptian ships.[2] Sap obtained from the *Rhus vernicifera* (varnish tree) was used as a vehicle in Japanese lacquers as early as 400 A.D.[3] Shellac (a resin secreted by the female lac bug, *Laccifer lacca*) finishes were one of the dominant wood finishes in the early 19th Century. Historically, resins such as copal, dammar, sandarac and rosin were key film-formers in varnishes. Milk-based paints

RSC Green Chemistry No. 29
Green Materials from Plant Oils
Edited by Zengshe Liu and George Kraus
© The Royal Society of Chemistry 2015
Published by the Royal Society of Chemistry, www.rsc.org

were in vogue before the advent of synthetic coatings and are still available commercially.[4] However, given their limited shelf-life, they are sold in powder form and are mixed with water just before application, particularly for the restoration of antiques and the reproduction of antique furniture. The use of a topcoat is recommended to seal the coated surface.

Of late, consumer and industrial interest in environmentally responsible coatings has grown tremendously. Developments in organic and polymer chemistry have enabled scientists to address the traditional limitations of natural derivatives. This chapter discusses the use of renewable resources (green materials) in coating technologies as a viable alternative to the petrochemical derivatives that form the backbone of the coatings industry today.

In 2004, the USA Department of Energy published a report identifying 12 sugar-based building block chemicals (1,4-diacids (succinic, fumaric, and malic), 2,5-furan dicarboxylic acid, 3-hydroxy propionic acid, aspartic acid, glucaric acid, glutamic acid, itaconic acid, levulinic acid, glycerol, sorbitol, xylitol/arabinitol, and 3-hydroxybutyrolactone) that can be produced from sugars *via* biological or chemical conversions and could be converted to high-value bio-based chemicals.[5] According to Lux Research, bio-based materials and chemical technologies have reached an inflection point with companies scaling to commercial production levels and growing revenues.[6] The report affirms that bio-based materials and chemical manufacturers continue to expand, with the industry growing and diversifying into new feedstock types, product types, and geographical areas. The leading growth category is predicted to be intermediate chemicals like adipic acid and lactic acid, with capacity growing from its current level of 2.0 million metric tons (MT) to 4.9 million MT in 2017, while the capacity of bio-derived polymers – currently at 1.1 million MT – is projected to grow 18 percent per year through 2017.

A number of companies offer coatings containing high levels of renewable resources. Cortec Corporation markets EcoLine® 3220, a 100% bio-based product that provides corrosion protection during storage and shipment, and EcoLine 3690, a bio-degradable, 76% bio-based temporary coating for severe marine and high humidity conditions.[7] BioSpan Technologies Inc. offers CR-3600® as a bio-degradable coating to protect metals from corrosion.[8] Eco Safety Products, LLC, markets Soycrete™ Architectural Concrete Stain and TimberSoy™ Natural Wood Stain (40% bio-based and recycled content), and Acri-Soy™ Penetrating Clear Sealer, a waterborne concrete and wood sealer (>20% bio-based content), and other bio-based sealers and coatings.[9] New Century Coatings offers Agristain™ wood and concrete stains based on byproducts from vitamin E production associated with various agricultural food stocks such as soybean, corn, palm, cashew nut and sunflower.[10] Benjamin Moore recently introduced Natura Renew™ which is claimed to be the first premium, bio-renewable coating with zero emissions, containing up to 40% bio-renewable materials.[11]

12.2   Renewable Resources Employed in Coatings

12.2.1   Vegetable Oils

Vegetable oils are an attractive resource for developing polymeric derivatives as they are widely grown across the world for edible and non-edible purposes. Technically, vegetable oils are triglyceride esters of fatty acids. Varying proportions of three unsaturated fatty acids (oleic, linoleic and linolenic) and two saturated fatty acids (palmitic and stearic) constitute the major vegetable oils. Oils such as castor oil, vernonia oil and lesquerella oil possess high levels of unique fatty acids such as ricinoleic acid, vernolic acid and lesquerolic acid, respectively. Vegetable oils are amenable to a wide variety of chemical modifications, *e.g.*, epoxidation, ozonolysis, sulfonation, maleinization, and hydroformylation. These processes enable the inherent auto-oxidative capability of unsaturated fatty acids to be supplemented with the reactivity of the functional groups introduced *via* chemical modification. Vegetable oils that inherently possess hydroxyl groups, *i.e.*, castor oil and lesquerella oil, offer an even simpler modification route, *e.g.*, reaction with diisocyanates to form polyurethanes.[12]

Alkyds (oil-modified polyesters) were one of the first modifications made to vegetable oils to render them useful for coatings. The first step involves conversion of the triglyceride to a mixture of hydroxyl-functional monoglycerides and diglycerides *via* glycerolysis. In the second step, these alcohols are combined with various diacids and/or anhydrides to yield the desired product. The diacid has a significant influence on the alkyd resin properties, *e.g.*, phthalic anhydride and maleic acid yield hard, brittle resins while adipic acid and sebacic acid result in softer resins. The oil content of an alkyd (oil length) is determined by dividing the amount of oil in the alkyd by the total weight of the alkyd solids, and expressing it as a percentage. Alkyds with oil lengths >60 are termed "long oil alkyds"; those with oil lengths between 40 and 60 are "medium oil alkyds", and those with oil lengths <40 are designated "short oil alkyds". While linseed oil, soybean oil, dehydrated castor oil, sunflower oil, safflower oil, tall oil and coconut oil are widely used in commercial alkyds, lesser known oils such as rubber seed oil, melon seed oil, tobacco seed oil, jatropha oil, karawila (*Momordica charantia*) seed oil, nahar seed (*Mesua ferrea*) oil, Africa locust bean (*Parkia biglobosa*) seed oil and yellow oleander (*Thevetia peruviana*) seed oil have also been used to synthesize alkyd coatings.[13–21] Another route for synthesizing vegetable oil-based polyesters is by reacting epoxidized oils with dicarboxylic acid anhydrides using catalysts such as tertiary amines, imidazoles or aluminum acetylacetonate.[22]

Epoxidation of fatty acids and vegetable oils involves the addition of an oxygen atom to the C=C bond thereby forming a three-membered epoxide (oxirane) group. Arkema's Vikoflex® line consists of epoxidized soybean fatty acid esters (7000 series) and epoxidized linseed fatty acid esters (9000 series) while epoxidized soybean oil and epoxidized linseed oil are sold under the trade name of Vikoflex™ 7170 and 7190, respectively. Boutevin *et al.* reacted epoxidized soybean oil with lactic, glycolic and acetic acids to yield bio-based polyols that contained an average hydroxyl functionality of 4–5 and an oligomer content close to 50%.[23]

12.2.2 Guayule Rubber

Guayule, *Parthenium argentatum gray*, is a shrub native to the southwestern USA and northern Mexico and is a source of natural rubber, *cis*-1,4-poly-isoprene. A number of guayule derivatives have been synthesized and incorporated into solvent-based, waterborne and powder coatings. For instance, hydroxylated guayule rubber and chlorinated maleinized guayule rubber were formulated into powder coatings, where they served as matting agents.[24,25] Improvements were also noted in toughness, flexibility and solvent resistance over the control formulations that did not contain the guayule rubber derivatives. Chlorinated maleinized guayule rubber has been validated to be a viable waterborne adhesion promoter for polypropylene, which is typically difficult to wet and adhere to because of its low surface energy.[26] Epoxidized guayule rubber has been formulated into an epoxy-phenolic solvent-based system and a polyester-epoxy powder coating system.[27]

12.2.3 Polyesters

Adipic acid, a common aliphatic diacid used in polyester synthesis, can be replaced by succinic acid, which has been available commercially from bio-based sources since the mid-90s. The use of bio-succinic acid in polyesters enables its renewable content to be as high as 66% and have a smaller environmental footprint than those made from petroleum-based raw materials.[28] Renewable, sustainable, 100% bio-based polyester polyols synthesized by combining bio-based raw materials – Susterra® propanediol (DuPont Tate & Lyle Bio Products) and bio-succinic acid (Myriant Corporation) are marketed by Piedmont Chemicals.[29] Similar 100% bio-based polyester polyols can be synthesized by combining bio-succinic acid with bio-based butanediol.[30] Croda markets dimer fatty-acid-based 100% renewable polyester polyols (Priplast®) and dimer fatty acids and diols (Pripol®) for the coatings and adhesive markets.[31]

12.2.4 Polyurethanes

Oil-modified urethane polymers are supplied commercially by a number of resin manufacturers, *e.g.*, Urotuf® from Reichhold, Desmophen® 1150, and Bayhydrol UH 2557, UH 2593/1 & UH XP 2592 from Bayer. Urethane alkyd-based coatings afford better durability and flexibility than alkyd finishes and are recommended for use in light industrial environments. BASF markets a dimer fatty acid-based diisocyanate (DDI 1410) that offers high flexibility, water insensitivity and lower toxicity than other aliphatic isocyanates.

Hyperbranched polyurethanes were synthesized from poly(ε-caprolactone) diol as a macroglycol, butanediol as a chain extender, and vegetable oil monoglyceride (*Mesua ferrea*, castor, and sunflower oils separately) as a bio-based chain extender.[32] *Mesua ferrea* oil-based polyurethane showed the highest thermal stability, whereas the castor-oil-based one exhibited the lowest thermal stability. However, the castor-oil-based polyurethane exhibited the highest tensile strength compared to the other vegetable-oil-based polyurethanes. Narine *et al.* synthesized a linear diisocyanate from oleic acid and reported that it provided polyurethanes with properties similar to those obtained with the widely used petroleum-based diisocyanate, 1,6-hexamethylene diisocyanate.[33] Polyester polyols produced from bio-technologically produced 1,3-propanediol and 1,18-octadecanedicarboxylic acid yielded polyurethane dispersions with bio-based content as high as 34%.[34] The use of 1,3-propanediol improved the strength-to-elasticity ratio, making it an attractive alternative to longer diols such as 1,6-hexanediol.

Shen *et al.* synthesized a diisocyanate from castor-oil-derived undecylenic acid by thiol-ene coupling (TEC) and Curtius rearrangement, and reacted it with castor oil and a castor-oil-based carboxylic acid functional chain extender prepared from castor oil and 3-mercaptopropionic acid *via* TEC to yield a fully bio-based polyurethane dispersion.[35] Gite *et al.* synthesized polyesteramide polyols by reacting dimer fatty acids with oleyl diethanolamide. The polyols were subsequently reacted with aromatic diisocyanates and could be formulated into wood coatings with good mechanical properties.[36] Koning *et al.* prepared waterborne polyurethane dispersions containing >92% renewable compounds by using dimer fatty-acid-based diisocyanate and 1,4:3,6-dianhydro-D-glucitol (isosorbide) as the bio-based building blocks. However, the films were too soft for many industrial applications.[37] Anti-bacterial soybean-oil-based cationic polyurethane coatings have been prepared using amino polyols.[38] With some strain-specific exceptions, these polyurethanes show good anti-bacterial properties towards a panel of bacterial pathogens comprised of *Listeria monocytogenes* NADC 2045, *Salmonella typhimurium* ATCC 13311 and *Salmonella minnesota* R613.

12.2.5 Emulsions

Emulsion polymers are traditionally designed with a glass transition temperature (T_g) near ambient temperature. In emulsions, particle coalescence, polymer chain entanglement, and film formation is inefficient below the minimum film-forming temperature (MFFT).[39] For efficient film formation to occur, emulsions often require coalescing solvents that are usually classified as volatile organic compounds (VOCs) as they evaporate subsequent to application. High T_g values promote good hardness and resistance characteristics, however, this is usually accompanied by high MFFT values. The MFFT-T_g difference is therefore a vital feature of all emulsions. Since the MFFT has to be set slightly above ambient temperature for air-drying applications, numerous efforts have been made to enhance or increase the ultimate T_g of the final film without affecting the latex's film-formation properties. Self-cross-linking latexes enable molecular weight advancement to occur subsequent to film formation and enhance the separation between MFFT and T_g. The auto-oxidative cross-linking capabilities of vegetable oils are an attractive option in this regard. While blending functionalized vegetable oil derivatives such as maleinized drying oil with latexes has shown promise,[40] they also exhibited issues such as phase separation and inhomogeneous cross-linking. Hybrid waterborne alkyd-acrylic dispersions (solid content 40%), free from surfactant and solvent, were synthesized by a melt co-condensation reaction between an acrylic pre-polymer bearing carboxylic groups and a long-oil alkyd resin.[41] Insertion of anhydride moieties within the acrylic pre-polymer ensured efficient coupling between the acrylic and alkyd resin and prevented phase separation. The authors reported that their coatings were stable for two months. In general, long-term stability is usually a challenge with alkyd emulsions due to the profusion of hydrolysis-sensitive ester groups.

Vegetable-oil-based macromonomers (VOMMs) constitute a series of vegetable oil acrylate and (meth)acrylate derivatives functionalized for efficient incorporation into emulsions. The synergistic combination of vegetable oil derivatives and an acrylic backbone provides storage-stable, auto-oxidatively cross-linking systems for architectural and/or industrial coatings with reduced or zero VOC emissions. Generically, VOMMs can be synthesized from any vegetable oil, independent of its composition. VOMMs have three distinct characteristics that are advantageous to environmentally responsible emulsions: (1) by virtue of their molecular length and large monomer size, they are excellent plasticizing monomers that facilitate coalescence without the necessity for solvent-based coalescing agents; (2) VOMMs readily co-polymerize with vinyl monomers through their acrylate functionality and are therefore retained during film formation reducing the T_g and MFFT; and (3) the allylic functionalities within the VOMM tail react auto-oxidatively during coalescence at ambient temperature, creating films with highly cross-linked networks that attain mechanical strength through room-temperature cross-linking of already high-molecular-weight entangled polymer chains.[42] VOMM-based latexes are amenable for formulating low-VOC coatings.[43] Kinetic investigation of VOMMs in bulk as well as emulsion polymerization has also been reported.[44–46]

The most well-studied VOMM, SoyAA-1, is based on soybean oil and offers a free radically co-polymerizable derivative from a renewable resource.[47] SoyAA-1 is synthesized by reacting soybean oil with N-methyl ethanolamine followed by (meth)acrylation with (meth)acrylic acid (Figure 12.1). The synthesis is not energy intensive and is accomplished at high yields (>95%).

SoyAA-1 is characterized by high vegetable oil content (>66% by weight), favorable hydrophobic–hydrophilic balance for facile emulsion synthesis under standard conditions, and is an excellent flexibilizing monomer (estimated T_g −67.5 °C). SoyAA-1 is co-polymerizable with a variety of common

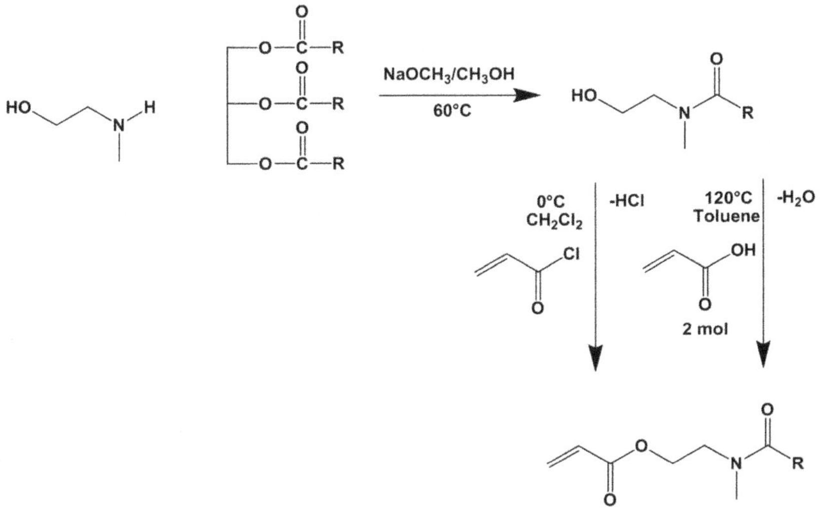

Figure 12.1   SoyAA-1 synthesis.

monomers employed in emulsion polymerization such as butyl acrylate, methyl methacrylate, styrene, and diacetone acrylamide/adipic dihydrazide.

Yabuuchi *et al.* synthesized poly(lactic acid) (PLA) macromonomers with a methacryloyl polymerizable group with different PLA chain lengths (average length $m = 4$, 6, 8, 12, 18, and 30) *via* ring-opening polymerization of L-lactide using hydroxyethyl methacrylate.[48] Radical co-polymerization of macromonomers with $m = 4$, 6, and 8 with vinyl monomers such as *n*-butyl methacrylate and *n*-butyl acrylate yielded stable mini-emulsions with physical properties appropriate for formulation into coatings. Solution co-polymerization of macromonomers ($m = 4$, 6, 8, 12, 18, and 30) with vinyl monomers resulted in a variety of co-polymers with PLA grafts while their solution homopolymerization yielded comb polymers.

A hydroxyl functional soybean oil amide was reacted with an isocyanate-terminated pre-polymer synthesized from polyethylene glycol and iso-phorone diisocyanate to yield an associative thickener.[49] Rheological data of different blends showed that low levels of soyamide-based hydrophobically modified ethoxylated urethane (HEUR) rheology modifiers imparted sig-nificant thickening to commercial latexes. Coatings formulated with the soyamide-based HEUR performed creditably with respect to gloss, viscosity, sag resistance, and flow and leveling when compared to commercial thickeners.

12.2.6   Coalescing Solvents

Bio-based coalescing solvents such as Myrifilm® (Myriant Chemicals) and Provichem® 2511 Eco (dimethyl succinate, Proviron) are characterized by their zero VOC content, low odor, bio-degradability, and non-hazardous air pollutant (HAP) characteristics, and are viable alternatives to petroleum-based coalescing solvents such as Texanol®. Dimethyl succinate is also used in the synthesis of dimethyl succinyloyl succinate, which is further processed to produce quinacridone pigments.[50] Methyl soyate (soybean oil methyl ester) offers very low flammability, a Kauri-butanol (Kb) value of 58, very high flash point (>360 °F), low VOC levels (<50 g L^{-1}), low toxicity, non-HAP status, and is useful as a coalescing solvent.[51] Ethyl lactate (ethyl α-hydroxy propionate) is derived from corn and is a non-HAP, 100% biodegradable, non-carcinogenic and non-ozone-depleting coalescing solvent (Kb value 500) used in specialty coatings, paint strippers and graffiti removers.[52–53] Technical grade D-limonene, the major component of the oil extracted from citrus rind, is a useful bio-based solvent with a Kb value of 67 and has better solvent properties than mineral spirits (Kb value 37).[54] Soy Technologies, LLC markets Soyanol™, a series of compositions containing soybean oil derivatives as viscosity and flow modifiers for solvent-based coatings and as coalescents for waterborne coatings.[55]

Reactive diluents function as solvents for resins and react into the final film rather than evaporating after coating application, thus they reduce viscosity without increasing VOC content. Reactive diluents prepared from tung oil *via* a Diels–Alder reaction with different dienophiles, methacryloxypropyl trimethoxysilane and triallyl ether acrylate improved the tensile strength and tensile modulus of a long oil soybean oil alkyd without significantly altering the elongation-at-break values.[56]

12.2.7   UV-curable Coatings

The use of epoxidized palm oil, epoxidized soybean oil, epoxidized linseed oil, epoxy norbornene linseed oil, and the naturally occurring epoxidized oil, vernonia oil, in cationic UV-cured coatings has been reported by various authors.[57–60] Radiation-curable acrylates have been synthesized by reactions of epoxidized vegetable oils and acrylic acid.[61,62] The saturated fatty acids provide flexibility, while the terminal methyl groups of the fatty acid chains play a significant role in the delocalization of electrons around the double bond during free radical production by UV radiation. Allnex markets a variety of vegetable oil derivatives for UV coatings, *e.g.*, epoxidized soybean oil acrylate (Ebecryl® 860) for improved flow and leveling, fatty-acid-modified polyester hexacrylate (Ebecryl® 450) for rapid cure, fatty-acid-modified epoxy diacrylate (Ebecryl® 3702) for promoting pigment wetting and flow and leveling, and fatty-acid-modified hexafunctional polyester acrylate oligomer (Ebecryl® 870) for fast cure response. Müller *et al.* described UV-cured wood coatings prepared with acrylated epoxidized linseed oil and characterized their performance properties such as gloss, scratch resistance, solvent resistance and adhesion.[63]

Yin *et al.* synthesized vegetable-oil-based photo-initiators by grafting acetic-acid-based thioxanthone and 4-(dimethylamino) benzoic acid to the backbone of epoxidized soybean oil.[64] The vegetable oil-based photo-initiators were more efficient at photo-polymerization than their low-molecular-weight analogs, especially in pigmented systems. Kim *et al.* described photo-cross-linked networks prepared from acrylated epoxidized soybean oil, acrylated castor oil and acrylated 7,10-dihydroxy-8(*E*)-octadecenoic acid with 2,5-furan diacrylate as a difunctional stiffener.[65] The addition of furan diacrylate increased the tensile strengths of the polymer films by a factor of 1.4–4.2.

Webster *et al.* studied epoxy homopolymerization of 100% bio-based epoxidized sucrose soyate esters and epoxidized tripentaerythritol soyate esters.[66] The petrochemical-based coatings exhibited lower modulus and T_g compared to coatings based on sucrose. The rigid ring structure found in the bio-based sucrose core provided rigidity on the macroscale *versus* the flexible ether bonds found in the tripentaerythritol core.

12.2.8 Powder Coatings

Bio-based terpolyesters based on isosorbide, 1,4:3,6-dianhydro-L-iditol (isoidide) and succinic acid were evaluated for their applicability in solvent-based and powder coatings.[67] Introduction of poly-functional monomers such as glycerol and citric acid led to coatings with enhanced performance, with respect to mechanical and chemical resistance, compared to formulations based on linear polymers. Citric-acid-modified resins exhibited rapid curing and produced dense networks with high storage moduli. Accelerated weathering experiments showed that isosorbide-based coating systems had weathering resistance similar to commercially available terephthalic acid-based formulations.

Haveren *et al.* synthesized co-polyesters and terpolyesters based on succinic acid and isosorbide with other renewable monomers such as 2,3-butanediol, 1,3-propanediol, and citric acid.[68] These bio-based polyesters provided functionalities and T_g values in the appropriate range for powder coatings. The branched polyester coatings exhibited significantly improved mechanical and chemical resistance compared to those formulated from linear polymers. Koning *et al.* observed that modifying linear OH-terminal bio-based polyesters with citric acid yielded acid-functional polyesters with significantly enhanced functionality that when cured with conventional epoxy curing agents and β-hydroxyalkylamides, afforded coatings with good chemical and mechanical stability.[69] Facile anhydride formation from citric acid around its melting temperature (153 °C) proved to be crucial in modifying sterically hindered secondary hydroxyl end-groups.

12.3 Conclusions

Of late, coatings companies and their suppliers have been increasingly keen to project an image of being environmentally responsible. This has compelled formulators to push the boundaries of technological advancement to incorporate increasing amounts of bio-based materials into resins and coatings without sacrificing performance. Yet, cost continues to be a significant factor in commercializing derivatives synthesized from natural materials. The examples discussed in this chapter offer only a brief glimpse into the potential of bio-based coatings. It is hoped that this field will grow significantly in the years to come and promote harmonious cooperation between nature and the coatings industry.

References

1. J. W. Gooch, in *Lead-Based Paint Handbook*, Springer, New York, 2002, p. 13.
2. W. H. Gardner, in *Protective and Decorative Coatings Paints, Varnishes, Lacquers and Inks*, ed. J. J. Mattiello, John Wiley & Sons, New York, 1947, ch. 1, p. 3.
3. E. Schulte, *Paint and Varnish Technology*, ed. William von Fischer, Reinhold Publishing, New York, 1948, ch. 1, p. 2.
4. www.milkpaint.com.
5. http://www1.eere.energy.gov/bioenergy/pdfs/35523.pdf.
6. http://www.pcimag.com/articles/98829-capacity-for-bio-based-materials-and-chemicals-to-nearly-double-by.
7. www.cortecvci.com.
8. www.biospantech.com.
9. www.ecosafetyproducts.com.
10. www.agristain.com.
11. http://www.biofuelsdigest.com/bio-based/2014/02/20/benjamin-moore-launches-natura-renew-the-industrys-first-premium-bio-renewable-paint.
12. M. Black and J. W. Rawlins, *Eur. Polym. J.*, 2009, **45**, 1433.
13. D. S. Ogunniyi and T. E. Odetoye, *Bioresour. Technol.*, 2008, **99**, 1300.
14. I. O. Igwe and O. Ogbobe, *J. Appl. Polym. Sci.*, 2000, **75**, 1441.
15. O. Saravari and S. Praditvatanakit, *Prog. Org. Coat.*, 2013, **76**, 698.
16. T. E. Odetoye, D. S. Ogunniyi and G. A. Olatunji, *Prog. Org. Coat.*, 2012, **73**, 374.
17. N. Dutta, N. Karak and S. K. Dolui, *Prog. Org. Coat.*, 2004, **49**, 146.
18. S. H. U. I. De Silva, A. D. U. S. Amarasinghe, B. A. J. K. Premachandr and M. A. B. Prashantha, *Prog. Org. Coat.*, 2012, **74**, 228.
19. E. T. Akintayo, *Bull. Chem. Soc. Ethiop.*, 2004, **18**, 167.
20. A. I. Aigbodiona and F. E. Okieimen, *Ind. Crop. Prod.*, 2001, **13**, 29.
21. M. M. Bora, P. Gogoi, D. C. Deka and D. K. Kakati, *Ind. Crop. Prod.*, 2014, **52**, 721.
22. S. Miao, P. Wang, Z. Su and S. Zhang, *Acta Biomater.*, 2013, **10**, 1692.
23. S. Caillol, M. Desroches, G. Boutevin, C. Loubat, R. Auvergne and B. Boutevin, *Eur. J. Lipid Sci. Technol.*, 2012, **114**, 1447.
24. A. Niroomand, T. P. Schuman and S. F. Thames, *J. Coat. Technol.*, 1996, **68**, 15.
25. W. A. Purvis and S. F. Thames, *J. Coat. Technol.*, 1996, **68**, 67.
26. M. D. Foster and S. F. Thames, *J. Coat. Technol.*, 1999, **71**, 91.

27. S. Gupta, S. K. Mendon and S. F. Thames, *J. Appl. Polym. Sci.*, 2001, **82**, 1718.
28. http://www.myriant.com/products/product-literature.cfm.
29. http://www.myriant.com/media/press-releases/piedmont-chemical-launches-renewablepolyols-leveraging-renewable-chemicals-from-dupont-and-myriant.cfm.
30. http://www.bio-amber.com/products/en/applications/polyurethanes/?id = 655.
31. http://www.crodacoatingsandpolymers.com/home.aspx?d = content& s = 139&r = 669&p = 4804.
32. H. Kalita and N. Karak, *J. Appl. Polym. Sci.*, 2014, **131**, 2014.
33. L. Hojabri, X. Kong and S. S. Narine, *Biomacromolecules*, 2009, **10**, 884.
34. http://www.wki.fraunhofer.de/en/services/ot/projects/polyurethane-dispersions.html.
35. C. Fu, Z. Zheng, Z. Yang, Y. Chen and L. Shen, *Prog. Org. Coat.*, 2014, **77**, 53.
36. S. D. Rajput, P. P. Mahulikar and V. V. Gite, *Prog. Org. Coat.*, 2014, **77**, 38.
37. Y. Li, B. A. J. Noordover, R. A. T. M. van Benthem and C. E. Koning, *Eur. Polym. J.*, 2014, **52**, 12.
38. Y. Xia, Z. Zhang, M. R. Kessler, B. Brehm-Stecher and R. C. Larock, *ChemSusChem*, 2012, **5**, 2221.
39. S. F. Thames, presented at the 25th International Waterborne, High-Solids and Powder Coatings Symposium, 1998.
40. B. G. Bufkin and J. R. Grawe, *J. Coat. Technol.*, 1978, **50**, 83.
41. M. Elrebii, A. Ben Mabrouk and S. Boufi, *Prog. Org. Coat.*, 2014, **77**, 757.
42. E. Kaya, S. K. Mendon, D. Delatte, J. W. Rawlins and S. F. Thames, *Macromol. Symp.*, 2013, **324**, 95.
43. K. L. Diamond, S. N. Shah, O. W. Smith and S. F. Thames, presented at the Annual meeting of the Partners in Environmental Technology and Pollution Prevention, 2002.
44. K. L. Diamond, S. N. Shah, S. K. Mendon, O. W. Smith and S. F. Thames, *Polym. Prepr.*, 2003, **44**, 109.
45. K. L. Diamond, S. N. Shah, O. W. Smith and S. F. Thames, *Int. J. Coat. Sci.*, 2003, **2**, 40.
46. C. C. Blackwell, S. N. Shah, O. W. Smith and S. F. Thames, *Polym. Prepr.*, 2003, **44**, 880.
47. D. Delatte, E. Kaya, L. G. Kolibal, S. K. Mendon, J. W. Rawlins and S. F. Thames, *J. Appl. Polym. Sci.*, 2014, **131**, DOI: 10.1002/app.40249.
48. K. Ishimoto, M. Arimoto, T. Okuda, S. Yamaguchi, Y. Aso, H. Ohara, S. Kobayashi, M. Ishii, K. Morita, H. Yamashita and N. Yabuuchi, *Biomacromolecules*, 2012, **13**, 3757.
49. M. Pramanik, S. K. Mendon and J. W. Rawlins, *J. Appl. Polym. Sci.*, 2013, **130**, 1530.
50. http://www.proviron.com/product/provichem-eco-dms.
51. http://www.cleanlink.com/casestudieswhitepapers/details/Biosolvents-Methyl-Soyate-is-the-Key-Ingredient.

52. http://www.pcimag.com/articles/green-solvents-agrochemicals-in-place-of-petrochemicals.

53. http://www.greenacetone.com/magento/index.php/chemistry.

54. http://www.floridachemical.com/whatisd-limonene.htm.

55. www.soytek.com.

56. K. Wutticharoenwong, J. Dziczkowski and M. D. Soucek, *Prog. Org. Coat.*, 2012, **73**, 283.

57. W. D. Wan Rosli, R. N. Kumar, S. Mek Zah and M. Mohd, *Hilmi, Eur. Polym. J.*, 2013, **39**, 593.

58. R. Raghavachar, G. Sarnecki, J. Baghdachi and J. Massingill, *J. Coat. Technol.*, 2000, **72**, 125.

59. J. Wu, Z. Zong, P. Chittavanich, J. Chen and M. D. Soucek, *UV-Curable, Seed Oil-Based Coatings by Cationic Photopolymerization: Part 2*, Radtech Report, xxx, 2013.

60. S. F. Thames and H. Yu, *Surf. Coat. Technol.*, 1999, **115**, 208.

61. S. F. Thames, H. Yu, T. P. Schuman and M. D. Wang, *Prog. Org. Coat.*, 1996, **28**, 299.

62. F. Habib and M. Bajpai, *Chem. Chem. Technol.*, 2011, **5**, 317.

63. A. R. Mahendran, G. Wuzella, N. Aust, A. Kandelbauer and U. Müller, *Prog. Org. Coat.*, 2012, **74**, 697.

64. A. Luo, X. Jiang and J. Yin, *Polymer*, 2012, **53**, 2183.

65. N. R. Jang, H.-R. Kim, C. T. Hou and B. S. Kim, *Polym. Adv. Technol.*, 2013, **24**, 814.

66. T. J. Nelson, T. P. Galhenage and D. C. Webster, *J. Coat. Technol. Res.*, 2013, **10**, 589.

67. B. A. J. Noordover, A. Heise, P. Malanowksi, D. Senatore, M. Mak, L. Molhoek, R. Duchateau, C. E. Koning and R. A. T. M. van Benthem, *Prog. Org. Coat.*, 2009, **65**, 187.

68. B. A. J. Noordover, V. G. van Staalduinen, R. Duchateau, C. E. Koning, R. A. T. M. van Benthem, M. Mak, A. Heise, A. E. Frissen and J. van Haveren, *Biomacromolecules*, 2006, **7**, 3406.

69. B. A. J. Noordover, R. Duchateau, R. A. T. M. van Benthem, W. Ming and C. E. Koning, *Biomacromolecules*, 2007, **8**, 3860.

Subject Index